TECHNOLOGY
1982 BK

PROCESS CONTROL INSTRUMENTATION TECHNOLOGY

SECOND EDITION

ELECTRONIC TECHNOLOGY SERIES

PROCESS CONTROL INSTRUMENTATION TECHNOLOGY

SECOND EDITION

Curtis D. Johnson
UNIVERSITY OF HOUSTON

JOHN WILEY & SONS
NEW YORK   CHICHESTER   BRISBANE   TORONTO   SINGAPORE

Library of Congress Cataloging in Publication Data:

Johnson, Curtis D., 1939–
    Process control instrumentation technology.

    (Electronic technology series, ISSN 0422-910X)
    Bibliography: p.
    Includes index.
    1. Process control. 2. Engineering instruments.
I. Title. II. Series.

TS156.8.J63   1982   660.2′83   81-10488
ISBN 0-471-05789-4   AACR2

Printed in the United States of America

10 9 8 7 6 5 4 3 2

PREFACE

The advancement of both knowledge and technique has resulted in the development of specialists in process control. Today, it is possible to delineate process control activities into three categories, each with specific requirements with regard to training. The process control system engineer is concerned with the design of an overall process control system that will provide the regulation specified by the manufacturing process experts. This activity requires a good theoretical understanding of stability, mode characteristics, and general process control loop dynamic characteristics. The process control technologist or applied engineer is responsible for the design of specific elements required to implement the overall design specified for the system. This involves a good, design-level understanding of the measurement, electronic, and pneumatic features inherent in process control loops. The process control technician or mechanic is responsible for the final assembly and testing of the elements of the process control loop and the system as a whole. This requires a good working knowledge of the electronics, pneumatics, and mechanical features of process control loop installations.

This book was developed primarily to fulfill the instructional needs of the last two of the above categories, that is, the applied process engineer, technologist, and advanced technician. Many fine textbooks exist that cover the stability and design criteria of process control systems, usually involving the use of Laplace transform methods. Often these books do not cover the system elements per se, such as measurement methods, signal conditioning, and final control elements. Indeed, there has been a serious lack of texts that cover process control elements with less than the mathematical rigor of Laplace transforms and yet enough math to allow the design of loop elements to satisfy specifications. This volume has been prepared to fill this need. Its overall objective is to provide instructional material for a general understanding of process control characteristics such as elements, modes, and stability along with detailed knowledge of measurement technique, control mode implementation, and final control element functions. In keeping with modern trends, the digital aspect of process control technology is stressed.

The student background is assumed to consist of that acquired by an advanced sophomore in a two-year engineering/technology program or a junior in a four-year program. Consequently, basic knowledge of physics, analog and digital electronics is assumed along with math through college algebra, although some knowledge of calculus would be beneficial.

Preparation of a second edition of a text in any technical discipline forces the author to make some very profound observations about the technical world. For one thing, it brings out very clearly aspects of the subject that are permanent—those areas that continue to be necessary to study. It also makes one conscious of the explosive developments characteristic of technology—those areas that require more detailed coverage. Certainly, the field of process control is characterized by both of these phenomena. Thus, the present edition of this text contains many features of the first edition, with improved explanations and descriptions. More importantly, however, the text has been expanded in those areas in which new developments in technology have changed the face of process control. Singular among these is the use of microcomputers to implement Direct Digital Control (DDC).

Users of the first edition will find the most obvious changes in Chapters 3, 5, 7, 8, and 10, reflecting both the expanded coverage of digital techniques and computers and the rearrangement of some subjects for more efficient presentation. Other additions to the text are in the areas of pressure and flow measurement, pneumatics, and final control operations.

I am indebted to those users of the first edition who communicated to me so many excellent suggestions on how to improve the text. A special tribute goes to Judy Green and the folks at John Wiley & Sons for making this task a not-so-painful, sometimes even enjoyable, experience.

Curtis D. Johnson

CONTENTS

CHAPTER

1

INTRODUCTION TO PROCESS CONTROL

INSTRUCTIONAL OBJECTIVES

This chapter will consider the overall process-control loop, its function and description. After you read it, you should be able to:

1. Draw a block diagram of a process-control loop with a description of each element.
2. List three dynamic variables.
3. Describe three criteria used to evaluate the response of a process-control loop.
4. Define analog signal processing.
5. Describe the two types of digital process control.
6. Define accuracy, hysteresis, and sensitivity.
7. List the SI units of measure for length, time, mass, and electric current.
8. Convert a physical quantity from SI to English units and vice versa.
9. Define the types of measurement time response.

1.1 INTRODUCTION

To study the elements of process control effectively, it is necessary to have an overall understanding of process-control principles. With such an understanding, the manner in which particular elements affect the overall control problem can be fully described. This chapter provides a general introduction to process control with emphasis on industrial applications. An equally important facet of the chapter is the definition of common terms and units of measure that are a necessary part of any study of the subject.

The elements of a complex system are much easier to understand if the operation of the whole system is considered first. Obviously, it would be quite frustrating to study all the elements of an automobile without having first noted that the result would be a transportation vehicle with certain characteristics. In keeping with this idea, this chapter discusses the overall process-control loop, its function, and description.

1.2 DEFINITION OF PROCESS CONTROL

The operations associated with process control have existed in nature since the first living creatures appeared on the earth. In this sense process control has probably been going on forever in some corner of the universe. We can recognize *natural process control* as an operation that *regulates* some internal physical characteristic important or critical to a living organism. Common examples of naturally regulated quantities are body temperature, fluid pressure in blood vessels, retinal light intensity, body fluid flow rate, and many others.

Humans found it necessary to regulate some of the external physical parameters in their environment to sustain life, thereby initiating *artificial process control*. This regulation was accomplished by observation of the parameter, comparison with some desired value, and action to bring the parameter to that desired value. Examples of such control are found in fires for light and heat, cooking of food, simple smelting, and so forth.

The actual term *process control* was developed when humans learned to adopt *automatic regulatory procedures* in the more efficient manufacture of products. Such procedures are automatic in the sense that a human is no longer either a necessary or integral part of the regulation. Our primary concern is an understanding of such automatic regulation applied to various industrial manufacturing processes. The initial objective is to provide a formal definition of process control and the terms used in its description. Consider first the terms:

DYNAMIC VARIABLE

Any physical parameter that can change either spontaneously or from external influences is a *dynamic variable*. The word *dynamic* conveys the idea of a time variable that can result from a number of unspecified or unknown influences; the word *variable* simply relates the capacity to vary from these influences. In process control we are interested in those dynamic variables that require regulation in some industrial application. Typical examples are temperature, pressure, flow rate, level, force, light intensity, and humidity. We call the regulated dynamic variables *controlled variables*.

REGULATION

The primary objective of process control is to cause some dynamic variable to remain fixed at or near some desired specific value. Since the variable itself is dynamic, we must constantly provide corrective action to keep the variable constant. The term *regulation* defines this operation of value maintenance. We say that process control regulates a dynamic variable.

To define process control let us refer to a simple example which illustrates the essential features of process control regulation. For this example (shown in Figure 1.1), a human is part of the control of liquid level in a tank. Let us work through the regulatory procedures of the system.

Here, the *process* is the ensemble of fluid, tank, fluid inlet, and outlet. The *dynamic variable* is the fluid level in the tank, and the *regulation* is the maintenance of the level at some prescribed height H from the bottom of the tank. Operationally, we see that the outlet valve A controls the rate of fluid *loss*. We may assume that the rate of flow into the tank is *not* under our control. If the system is left alone, the level will simply adopt any value of H commensurate with inward and outward flow. If fluid viscosity or some other parameter affecting inlet changes, a new level occurs. This is an *unregulated* or *open-loop* system. To provide regulation, we must *feed back* information on the fluid level to the controlling element, in this case the outlet valve A, to correct changes in level brought on by outside influences. With this *feedback* we have produced a *closed loop* or regulating process. In this example, we let a human operator provide the required feedback to close the loop. Thus, the sight tube T permits the operator to make a visual *measurement* of the fluid level in the tank. The operator now makes an *evaluation*, about whether the level is high, low, or just right. Whenever it is incorrect, the operator *adjusts* the control element valve A to effect a continuously correcting adjustment of fluid level. The three terms identified above—*measurement, evaluation,* and *control*— form the basis of a more formal description of a process-control loop.

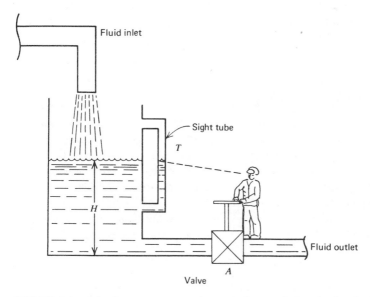

FIGURE 1.1   A basic process-control system for regulating the level of fluid in a tank.

1.3 ELEMENTS OF PROCESS CONTROL

To provide a practical, working description of process control, it is useful to describe the operations involved in more elementary terms. Such a description should be independent of a particular application (such as the example presented in the previous section) and thus be applicable to *all* control situations. A model may be constructed using blocks to represent each distinctive element of a loop. The characteristics of loop operation then may be developed from a consideration of the properties and interfacing of these elements. Numerous models have been employed in the history of process-control description; we will use one that seems most appropriate for a description of modern and developing technology of process control.

1.3.1 Identification of Elements

Perhaps the easiest method for identifying the elements of a process-control loop is to study the example presented in Figure 1.1. We can identify and describe the following elements.

PROCESS

The flow of liquid in and out of the tank, the tank itself, and the liquid all constitute a process to be placed under control with respect to the fluid level. In general, a process can consist of a complex assembly of phenomena that relate to some manufacturing sequence. Many dynamic variables may be involved in such a process, and it may be desirable to control all these variables at the same time. There are *single-variable* processes, in which only one variable is to be controlled, as well as *multivariable* processes, in which many variables, perhaps even interrelated, may require regulation.

MEASUREMENT

Clearly, to effect control of a dynamic variable in a process, we must have information on the variable itself. Such information is found from a *measurement* of the variable. In general, a measurement refers to the transduction of the variable into some corresponding *analog* of the variable, such as a pneumatic pressure, an electrical voltage, or current. A *transducer* is a device that performs the initial measurement and energy conversion of a dynamic variable into analogous electrical or pneumatic information. Further transformation or *signal conditioning* may be required to complete the measurement function. The result of the measurement is a transformation of the dynamic variable into some proportional information in a useful form required by the other elements in the process-control loop.

EVALUATION

The next step in the process-control sequence is to examine the measurement and determine what action, if any, should be taken. This part of the loop has many names; however, *controller* is the most common. We will refer to this operation as evaluation performed by a controller. The evaluation may be performed by an operator (as in the previous example), by electronic signal processing, by pneumatic signal processing, or by a computer. Computer use has been growing rapidly in the field of process control because it is easily adapted to the decisionmaking operations required in process control and because of its inherent capacity to handle control of multivariable systems. The controller requires an input of both a *measured represen-* *tation* of the dynamic variable and a representation of the *desired value* of the variable, expressed in the same terms as the measured value. The desired value of the dynamic variable is referred to as the *setpoint*. Thus, the evaluation consists of (a) a *comparison* of the controlled variable measurement and the setpoint and (b) a determination of action required to bring the controlled variable to the setpoint value.

CONTROL ELEMENT

The final element in the process-control loop is the device that exerts a direct influence on the process, that is, that provides those required changes in the dynamic variable to bring it to the setpoint condition. This element accepts an input from the controller, which is then transformed into some proportional operation performed on the process. In our previous example, the control element is the valve that adjusts the outflow of fluid from the tank. This element is also referred to as the *final control element*.

1.3.2 Block Diagram

The elements in a process-control loop are often shown in a diagram in which each element is represented as a block. Figure 1.2 is a block diagram constructed from the elements defined in the previous section. The controlled dynamic variable in the process is denoted by C in this diagram, whereas the measured representation of the controlled dynamic variable is labeled C_M. The controlled variable setpoint is

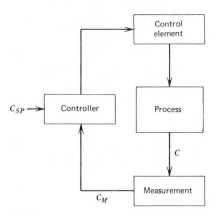

FIGURE 1.2   A process-control loop block diagram showing the four basic elements.

labeled C_{SP} and must be expressed in the same proportion as that provided by the measurement function.

In many cases, the controller block is presented in a fashion that carries over from electronic feedback or servo analysis. This presents the evaluation operation as a *summation point* that outputs an *error signal* $E = C_M - C_{SP}$ to the controller for comparison and action. Figure 1.3 shows the loop drawn using this very common representation of loop elements.

To further illustrate, the block diagram concept in Figure 1.4 shows a typical flow control system and its representation by a block diagram. In this example, we note that the dynamic variable is the flow rate which is converted to electric current as an analog.

The purpose of a block diagram approach is to allow the process-control system to be analyzed as the interaction of smaller and simpler subsystems. The whole is the sum of its parts. If the characteristics of each element of the loop can be determined, then the characteristics of the assembled loop can be established by analytical marriage of these subsystems. The historical development of the system approach in technology was dictated by this practical aspect: first, to specify the characteristics desired of a total system, and then, to delegate the development of subsystems that provide the overall criteria.

It becomes evident that the specification of a process-control system to regulate a variable C, within specified limits and with specified time responses, determines the characteristics the measurement system must possess. This same set of system specifications is reflected in the design of the controller and control element.

From this concept, we conclude that the analysis of a process-control system requires an understanding of the overall system behavior and the reflection of this behavior in the properties of the system elements. Most people find that an under-

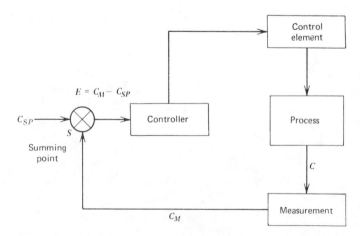

FIGURE 1.3 A process-control loop block diagram in terms of a summing point S and error signal E.

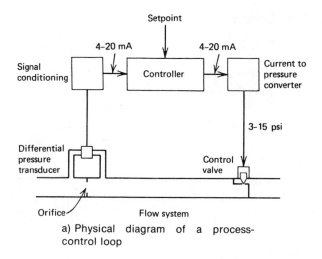

a) Physical diagram of a process-control loop

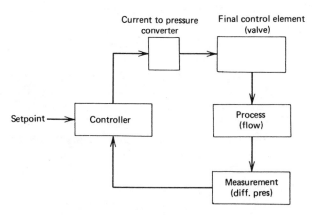

b) Block diagram of a process-control loop

FIGURE 1.4   A process-control system to regulate flow and the corresponding block diagram. Note the use of current and pressure transmission signals.

standing of the parts leads to a better understanding of the whole. We will proceed with this assumption as a guiding concept.

1.4 EVALUATION OF PROCESS CONTROL

Before considering the effects of element characteristics on the response of an overall process-control system, it is logical to consider how evaluations of both system and elements are defined. Such a base must provide the definition of various terms and conditions employed to perform these evaluations.

1.4.1 System Evaluation Criteria

The evaluation of system performance simply reflects the degree to which the process-control loop is performing the stated function of regulating some dynamic variable in the process. Such regulation is specified through two criteria. The first criterion is the *error* of the system in maintaining the controlled dynamic variable C at the specific value defined by the setpoint C_{SP}. The second criterion is the *dynamic response* of the system to any disturbance of the process or change of the process-control loop setpoint.

SYSTEM ERROR

The system error is a measure of the inherent error between the controlled variable setpoint value and the actual value of the dynamic variable *maintained* by the system. This error may be expressed by the percentage deviation of the actual C to the setpoint C_{SP}. Moreover, the error is not a function of adjustment or calibration of the system or system elements but, rather, represents the aggregate uncertainties of the entire system in regulating the dynamic variable. Only an improvement in the quality of system elements provides a reduction in this system error.

SETPOINT

The value of the dynamic variable desired in the process is referred to as the *setpoint*. Although the setpoint is expressed as some value of the dynamic variable, it must be translated into the same form as provided by the measurement of the dynamic variable. Thus, if a measurement converts a pressure into a proportional current, then the setpoint must be translated into a current expressed in the same proportionality. Generally, a setpoint C_{SP} is expressed with some allowable *deviation* $\pm \Delta C$ about the nominal value. Thus, when control has been achieved, the actual value of the controlled variable may be in the range from $(C_{SP} - \Delta C)$ to $(C_{SP} + \Delta C)$. Of course, the wider the allowable deviation, the easier the control is to achieve. Note that this deviation can never be less than the inherent system error.

DYNAMIC RESPONSE

The dynamic response of the system is the basic criteria on which system evaluation is performed. The purpose of regulation of a dynamic variable is to control change of the variable in time. Such changes can occur from spurious or transient influences or from intentional adjustments of the setpoint. Dynamic response is a measure of the systems reaction, as a function of time, in correcting transient inputs, or adjusting to a new setpoint.

CHANGE IN SETPOINT

In many cases, the setpoint of the dynamic variable may be adjusted to a new value. The process-control loop must respond by operating on the process to bring the dynamic variable to this new value. Depending on the nature of the process, the dynamic response of the loop may be adjusted either for *cyclic* or *damped* response.

In *cyclic* response, as shown in Figure 1.5a, the actual value of the dynamic variable may overshoot the new setpoint and perform a number of oscillations about this point before stabilizing. In the *damped* response mode, shown in Figure 1.5b, the dynamic variable value never overshoots the setpoint or performs oscillations, but approaches the operating point on an asymptoticlike curve. such a response might be required, for example, in a process where temperature must never rise above some critical value.

TRANSIENT RESPONSE

The basic purpose of a process-control loop is to maintain some dynamic variable at a prescribed operating point. The need for such a system implies that external influences would otherwise cause fluctuations of the variable. Transient response describes the ability of the dynamic response of the system to recover from a sudden influence on the process which causes an abrupt change in the controlled variable.

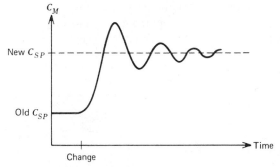

a) Cyclic response of a system to a set-point change

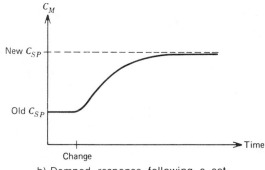

b) Damped response following a set-point change

FIGURE 1.5   A process-control system may be designed to exhibit either underdamped or damped response following a change in setpoint.

In Figure 1.6*a*, we see the underdamped or cyclic response pattern of a system in which the reaction of the control loop causes oscillations about the setpoint of the variable. Figure 1.6*b* shows the typical damped response for a transient process disturbance where the control system is able to bring the variable back to the operating point with no oscillations.

1.4.2 Dynamic Response Evaluation Criteria

A set of criteria for stability is used to evaluate the response of a process-control loop, whether the deviations are from changes in setpoint or transient inputs. These criteria are measures of loop characteristics and may be selected according to individual applications to fit the conditions of the process. In general, a *minimum value* of the criteria is the *optimum* response.

SETTLING TIME

A typical process-control loop will have a setpoint C_{SP} for the dynamic variable and an allowable deviation about this setpoint $\pm \Delta C$. In the event of a transient input or a change in setpoint, the settling time is the time required for the process-control loop to bring the dynamic variable back to within the allowable range, $C_{SP} \pm \Delta C$. This is illustrated in Figure 1.7*a*.

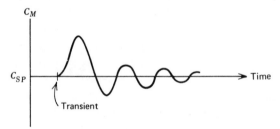

a) Cyclic recovery of a system to a transient input

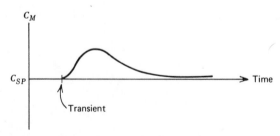

b) Damped recovery of a system to a transient input

FIGURE 1.6   A process-control system may recover from a transient input in either a cyclic or damped characteristic.

PEAK ERROR

In general, for a change in setpoint in a cyclic system or with transients, there will be a maximum deviation of the dynamic variable from the setpoint. In Figure 1.7*a* we see the designation of this maximum error as the peak error of the dynamic variable.

RESIDUAL ERROR

Any process-control loop has certain inevitable nonlinearities that prevent the system from returning the dynamic variable to the setpoint following a transient or from settling the system directly on the prescribed setpoint following a change. Such a condition gives rise to a *residual error* that represents a stabilized operating point of the dynamic variable. This error is acceptable within the allowable error range but offset from the optimum setpoint value, as shown in Figure 1.7*a*. This error can be reduced by further adjustment.

CYCLING

The cycling (oscillatory) behavior of a control system should be minimized and indeed in some cases cannot be allowed at all. For this reason a prescription of *minimum cycling* is often associated with the design of process-control systems. In this context, minimum cycling means that the system is adjusted so that the least number of oscillations occur following a transient or a change of setpoint.

MINIMUM AREA

The dynamic response of a system to transient inputs or setpoint changes brings about deviations from the desired setpoint in processes. In many cases, the extent and duration of these deviations may result in a bad end product or, at the very least, deviations of the end product from overall specifications. In other cases, the products of a manufacturing run produced during such deviations must be scrapped with the attendant loss of production and revenue. For this reason, process-control loops are often designed to minimize both the excursion of deviations from the operating setpoint and the *time span* during which such deviations occur. This is specified by minimizing the area of the deviation-time curve, as shown in Figure 1.7*b*. Thus, if $C(t)$ represents the dynamic variable and C_{SP} the setpoint, then the error at any time t is

$$E(t) = C(t) - C_{SP} \tag{1-1}$$

The area is indicated by an integral of the absolute magnitude of the error over the span of the disturbance

$$A = \int |E(t)| dt \tag{1-2}$$

The *minimum area criterion* specifies that this quantity A should be a minimum for the process-control loop.

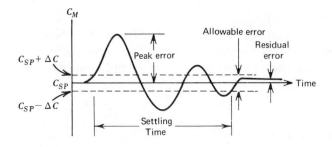

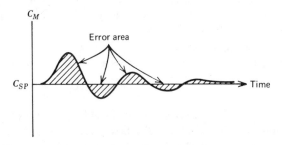

FIGURE 1.7 Numerous criteria are employed in the evaluation of a process-control loop response.

The evaluation criteria discussed above are overall measures of process-control loop behavior used to determine the adequacy of such a loop to perform some designed functions. Obviously, the loop characteristics are a product of the characteristics of the loop elements. The design of a process-control loop with a specified dynamic response is reflected in the design of loop elements through specification of characteristics, such as response time, accuracy, drift, and other factors. In later chapters we will be referring to a number of element characteristics that must be defined.

1.5 ANALOG VERSUS DIGITAL PROCESSING

The evolution of process control has seen the infusion of electronics technology into almost every facet because of low costs, reliability, miniaturization, and ease of interface. It is natural that the further development of digital electronics and associated computer technology has brought about the rapid introduction of digital techniques in the field of process control. Some aspects of process control, such as the initial transduction of a dynamic variable into electrical information, will probably always be of analog nature. It is inevitable, however, with the continued development of computers, miniaturized digital electronics, and associated technology, that the evaluation and controller phase of process control may be digitally performed. Let us contrast the two methods of processing.

1.5.1 Analog Processing

Consider the process-control loop shown in Figure 1.8, where a process temperature is to be controlled. Here a thermistor, whose resistance R is proportional to and
an analog of temperature, is used to measure the temperature. The resistance
change is converted to a voltage V via some unspecified signal conditioning. This
voltage is compared to some reference voltage (setpoint) by a differential amplifier
(evaluation), the output of which activates either a heater or cooler. The range of
allowed temperature is determined by the differential amplifier swing necessary to
trip either relay. Note that in this loop the temperature is represented by a proportional electrical signal. We then say that the electrical signal is an analog of the
temperature. In the case of the thermistor, the resistance is an analog of the temperature, and the signal conditioner output is a voltage that is an analog of the temperature.

In a similar fashion in a pneumatic system, we may have a fluid *pressure* that is an
analog of the dynamic variable. Note that the analog relationship between processing signals and dynamic variables need not be linear and in many cases is not. The

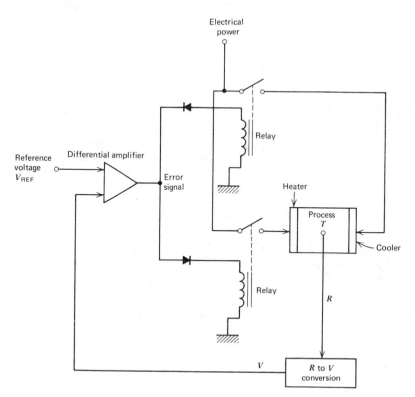

FIGURE 1.8   An analog process-control loop for regulation of temperature. Note that
all signals in the loop have some analog to temperature.

t factor here is that the processing signal is a *smooth* and unique *represen-tation* of the dynamic variable. In Figure 1.9 we see two examples of analog proportionalities between a dynamic variable C and an analog signal S. In case 1, we have a linear relationship and, in case 2, a nonlinear relationship but both are still analog representations of C.

1.5.2 Digital Processing

In digital processing, all information carried in the process-control loop is encoded into a signal that is binary in nature. Binary refers to a numbering system with a base of two, that is, *zero* and *one* are the only possible counting states.

The electrical signals on a wire are represented either by a binary "0," denoted also by low (L) and corresponds in transistor-transistor logic (TTL) to about 0 volts, or a binary "1," denoted by high (H) and corresponds in TTL to about +5 volts. Thus, in TTL, electrical voltages are only 0 or 5 volts and hence cannot be an analog representation of the dynamic variable. The value of the dynamic variable is represented as some *encoding* of the binary levels. The encoding itself is a corre-spondence between a set of binary numbers and the analog signal to be encoded. The set of binary numbers is commonly referred to as a word, which may contain many binary counts, called bits.

BINARY/DECIMAL ENCODING

Let us consider the encoding between a 4-bit binary word and a decimal voltage signal. Such a coding can be simply the representation of numbers between the two systems, as shown in Table 1.1. Thus, if we wished to encode a 5-volt signal, where each bit corresponded to 1 volt, we would represent this in a 4-bit binary word as "0101."

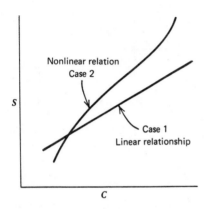

FIGURE 1.9 The relationship between a dynamic variable C and its transduced equivalent S may be linear as in case 1, or nonlinear as in case 2.

TABLE 1.1   DECIMAL-BINARY ENCODING

Voltage	Binary Word
0	0000
1	0001
2	0010
3	0011
4	0100
5	0101
6	0110
7	0111
8	1000
9	1001
10	1010

SIGNAL TRANSMISSION

The encoding of a signal into a binary word implies that a sequence of 0 and 1 levels corresponds to the value of the signal. There are two methods by which a digitally encoded signal can be transmitted through the process-control loop. One method, referred to as the *parallel transmission mode*, provides a separate wire for each binary number in the word. Thus, a 4-bit word requires four wires, an 8-bit word, eight wires, and so on. The alternate method is *serial transmission mode* of binary numbers over a single wire where the binary levels are provided in a time sequence over the wire. These two methods are illustrated in Figure 1.10 for the digital encoding of the 5-volt level. In the parallel mode, the levels can remain set for any time on the lines, whereas in the serial mode, the levels appear only as a pulse train and must be interpreted in turn as they appear, introducing timing requirements.

ANALOG AND DIGITAL CONVERTERS

The transducer is most often a device that converts a dynamic variable to an electrical or pneumatic analog. Furthermore, the final control element is typically a device that converts a controller analog signal to some effect on the dynamic variable in the process. If digital processing is to be employed in the process-control loop, we must have a means of converting between the analog and digital representations of the dynamic variable and controller outputs. The analog to digital converter (ADC) performs the function of conversion of an analog input to a digitally encoded signal. These devices are designed to output a digital word coded to a specified range of signal inputs. Thus, an ADC might accept a 0–10 volt input encoded into a 4-bit word. A 4-bit word can represent 16 different states; hence, each bit represents

$$\frac{10 \text{ volts}}{16 \text{ states}} = 0.625 \text{ volts/state or volts/bit}$$

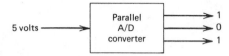

a) Parallel transmission mode

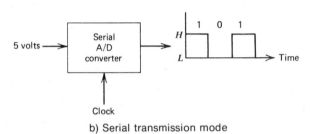

b) Serial transmission mode

FIGURE 1.10 Digital information may be transmitted on parallel lines, one for each bit, or serially where the information appears on a single line in a time sequence.

If more voltage resolution is necessary, we may use a word with more bits. A 6-bit word would have 64 states and

$$\frac{10 \text{ volts}}{64 \text{ states}} = 0.15625 \text{ volts/bit}$$

The digital-to-analog converter (DAC) provides the reverse action of converting the digitally encoded word into an appropriate analog output; that is, it decodes. Here again, each bit, by design, will correspond to a certain level of output, and thus the resolution, or the smallest increment, is the level of one state change. Note that the example of ADC from voltage is only one example of the type of analog signal that can be used. Other converters may use analog signals of frequency variation, current, or even resistance as the primary analog signal.

There are two approaches to digital processing in industrial-control situations. One involves using digital logic circuits or computers to supervise analog process-control loops and the other direct digital control of a dynamic variable.

SUPERVISORY DIGITAL CONTROL

The expression, supervisory digital control, is used to describe a situation where digital techniques are employed to supervise analog control loops. Generally, in these cases, a computer is employed to adjust the setpoint of an analog controller. This approach is employed to take advantage of the well-known and trusted performance of analog control, while at the same time using the various advantages of computer supervision. In particular, the computer can examine many variables and solve complicated control equations to determine and then set optimum setpoints of several analog loops associated with some process. This is particularly important when interaction exists between variables, such as changing the temperature setpoint causing the pressure to vary. The block diagram of a supervisory system is shown in Figure 1.11. In general, the logic or computer supervisor will input the measured quantity and programming and output a calculated setpoint. In the analog loop described in Section 1.5.1, a hybrid conversion could be developed by using ADCs and DACs to provide temperature input and reference voltage output to the analog loop as illustrated in Figure 1.12.

DIRECT DIGITAL CONTROL

The method of process control described by the term *direct digital control* (DDC) applies to those cases in which digital logic circuits or a computer are an integral part of the loop. In Figure 1.13 we see a block diagram of this approach to processing. Essentially, the evaluation and controller function is taken over by digital logic circuits or programming of a computer. We distinguish these two approaches to DDC by calling logic circuit methods *hardware program* controlling and computer methods *software program* controlling. Figure 1.14 shows how the temperature-control problem defined in Section 1.5.1 is implemented by DDC. Here, the control function, setpoint, and deviation about the nominal are all defined by the program. Direct digital control has the capacity to enable control of multivariable processes with interaction between elements. This approach is gaining in accep-

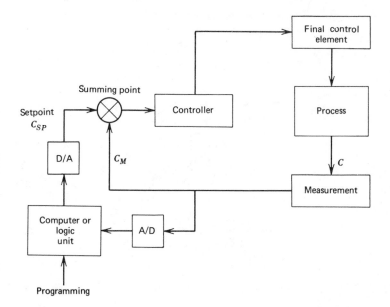

FIGURE 1.11   Implementation of the process-control loop using a computer or digital logic circuits in a supervisory capacity.

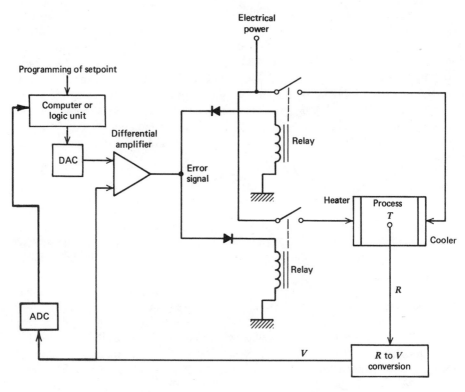

FIGURE 1.12   An example of supervisory digital control in the case of the temperature control problem of Section 1.5.1.

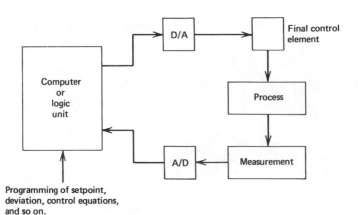

FIGURE 1.13   Block diagram of a process-control loop which is under direct digital control (DDC).

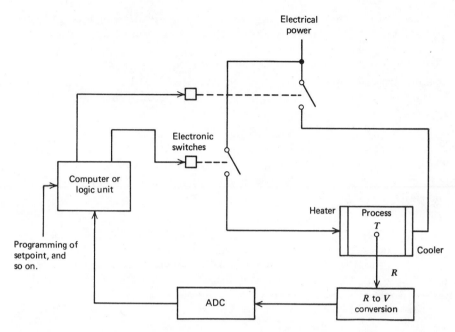

FIGURE 1.14 Implementation of the problem of Section 1.5.1 using direct digital control.

tance as the reliability of computers improves and backup methods are being developed to avoid process shutdown due to a computer failure. The development of the so-called "computer on a chip" in the form of a single integrated circuit (IC) microprocessor has given considerable impetus to the use of DDC.

1.6 UNITS, STANDARDS, AND DEFINITIONS

As in any other technological discipline, the field of process control has associated with it many sets of units, standards, and definitions to describe its characteristics. Some of these are a result of historical use, some are for convenience, and some are very confusing. As the discipline grew, there have been efforts to standardize terms so that professional workers in process control could effectively communicate among themselves and with specialists in other disciplines. In this section, we summarize the present state of affairs relative to the common units, standards, and definitions.

1.6.1 Units

To assure precise technical communication among individuals employed in technological disciplines, it is essential to use a well-defined set of units of measurement. The metric system of units provides for such communication and has been adopted by most technical disciplines and will probably be employed by the world public in general in the near future. In process control a particular set of metric units, which was developed by an international conference, has been adopted; it is called the International System (SI, Système International D'Unités) of units. Because much technical work in the United States is still done in the English system of units, it is necessary to perform transformations between these systems.

SI

The SI units are based on *defined* units for eight physical quantities. All other units of measure in the SI can be derived from this basic set. In Appendix A-1 a comprehensive list of the SI units is presented. The basic set of eight fundamental units is:

1. *Length* is defined by the *meter* (m) measured as 1,650,763.73 wavelengths in vacuum of radiation from a $2p_{10}$ to $5d_5$ transition in Krypton-86.
2. *Mass* is defined by the *kilogram* (kg) measured by a standard kilogram maintained in the Bureau International des Poids et Mesures in Seures, France. Note that kilogram is the whole name of this unit and does *not* imply 1000 from the prefix kilo.
3. *Time* is defined as the *second* (s) measured as 9,192,631,770 cycles of cesium-133 radiation from a $(F = 4, M_F = 0)$ to $(F = 3, M_F = 0)$, transition.
4. *Electric current* is defined by the *ampere* (A).
5. *Temperature* is defined by the *kelvin* (K).
6. *Luminance* is defined by the *candela* (cd).
7. *Plane angle* is defined by the *radian* (rad).
8. *Solid angle* is defined by the *steradian* (sr).

All other SI units can be derived from these eight, although in some cases a special name is assigned to such a derived quantity. Thus, a force is measured by the newton (N), where $1 \text{ N} = 1 \text{ kg m/s}^2$, energy is measured by the joule (J) or watt-second (W-s), given by $1 \text{ J} = 1 \text{ kg m}^2/\text{s}^2$, and so on as for other units, as shown in Appendix A-1.

OTHER UNITS

Although the SI system will be used in this text, other units remain in common use in some technical areas. The reader should therefore be able to identify and translate among the SI system and other systems. The centimeter-gram-second system (CGS) and English system are given in Appendix A-1 in relationship to the SI system. The following examples illustrate some typical translations of units:

EXAMPLE 1.1

Express a pressure of $p = 2.1 \times 10^3$ dyn/cm^2 in pascals. Note that $1 \text{ Pa} = 1 \text{ N/m}^2$.

SOLUTION

From Appendix A-1, we find 10^2 cm $= 1$ m and 10^5 dyne $= 1$ newton; thus,

$$p = (2.1 \times 10^3 \text{ dyn/cm}^2) \left(10^2 \frac{\text{cm}}{\text{m}}\right)^2 \left(\frac{1 \text{ N}}{10^5 \text{ dyn}}\right)$$

$p = 210$ pascals

EXAMPLE 1.2

Find the number of feet in 5.7 m.

SOLUTION

Reference to the table of conversions in Appendix A-1 shows that 1 m = 39.37 in; therefore

$$\ell = (5.7 \text{ m}) \left(39.37 \ \frac{\text{in}}{\text{m}}\right) \left(\frac{1 \text{ ft}}{12 \text{ in}}\right) = 18.7 \text{ ft}$$

EXAMPLE 1.3

Express 6 ft in meters.

SOLUTION

Using 39.37 in/m

$$(6 \text{ ft})(12 \text{ in/ft}) \left(\frac{1 \text{ m}}{39.37 \text{ in}}\right) = 1.829 \text{ m}$$

EXAMPLE 1.4

Find the mass in kg of a 2-lb object.

SOLUTION

We first find the mass in slugs

$$m = \frac{2 \text{ lb}}{32.17 \text{ ft/s}^2}$$

$m = 0.062$ slugs

Where

1 slug = 1 lb ft/s^2

Then, from Appendix A-1, we have

$$m = (0.062 \text{ slugs}) \left(14.59 \ \frac{\text{kg}}{\text{slug}}\right)$$

$m = 0.905$ kg

METRIC PREFIXES

With the wide variation of variable magnitudes that occurs in industry, there is a need to abbreviate large and small numbers. Scientific notation allows the expression of such numbers through powers of 10. A set of standard metric prefixes has been adopted by the IEEE (Std. No. 268A) to express these powers of 10, which are employed to simplify the expression of large or small numbers. These prefixes are given in Appendix A-1.

EXAMPLE 1.5

Express 0.0000215 s and 3,781,000,000 W using decimal prefixes.

SOLUTION

We first express quantities in scientific notation and then find the appropriate decimal prefix from Appendix A-1.

$$0.0000215 \text{ s} = 21.5 \times 10^{-6} \text{ s} = \textbf{21.5 } \boldsymbol{\mu}\textbf{s}$$

and

$$3,781,000,000 \text{ W} = 3.781 \times 10^{9} \text{ W} = \textbf{3.781 GW}$$

1.6.2 Standard Signals

Certain standard signal levels exist in process control that have been adopted to provide an easier interface of control loop elements, among other reasons. The most common standard concerns the manner of signal transmission, such as from the measurement element to the process controller.

CURRENT SIGNAL TRANSMISSION

When a control loop has been implemented using analog electrical signals, it is most common to transmit the analog signal as a current level. Whenever a process-control loop is designed, some operating range is specified for the dynamic variable to be regulated. It may be necessary to define a setpoint anywhere within this range. One standard specifies that the signal conditioning be such that a *4–20 mA* current range on the signal transmission wires represents the specified range of dynamic variable. Another common standard provides that a *10–50 mA* range provide this same information. There are three significant points regarding the use of current transmission to represent the dynamic variable.

1. *Load impedance.* By using a current to carry analog information about the dynamic variable, we avoid errors introduced by attaching different loads to the transmitting circuit. Thus, changing lead resistance or inserting a series resistance in the leads will not change the current delivery. Generally, the transmitting circuits are designed to work into any load from 0 Ω to about 1000 Ω.

2. *Interchangeability*. By using a specified current range to represent the dynamic variable range, we provide for interchangeability of the controller in the process-control loop. Once the dynamic variable range has been translated to a fixed current range, then all control can be based on a setpoint and deviation as some percentage of this range. Thus, a controller only sees a 4–20 mA signal, for example, and it does not matter what specific dynamic variable this represents.

3. *Measurement/power supply*. Generally speaking, the current signal lines are also the power delivery lines to energize the transducer and local signal conditioning. Thus, only two wires are necessary to connect a transducer and signal conditioning measurement system to the rest of the loop. The signal conditioning is designed so that the circuit draws more or less current from the power source in proportion to the value of the dynamic variable. This is shown in Figure 1.15, where the controller signal is taken as some voltage drop across a series resistor in the line.

PNEUMATIC SIGNAL TRANSMISSION

The final control element in a process-control loop is a pneumatic device. In some cases, an *entire* process-control loop may be pneumatic. In either instance, the standard transmission signal is a pressure level of a range 3–15 *pounds per square inch (psi)*. Often, a current to pressure converter is employed to scale the 4–20 *mA* signal to a 3–15 psi signal. Note that the 3–15 psi standard remains very common in the process-control industry despite the desired conversion to SI. We point out, however, that this would be expressed in SI as 2.07×10^4 to $10.34 \times 10^4 \, N/m^2$ or perhaps more conveniently as 20.7–103.4 kPa.

1.6.3 Definitions

In this section, several terms commonly applied to process control are defined. Effective communication demands a common understanding of specialized words and expressions. This has led to the establishment of standard definitions of commonly used terms.

ERROR

The algebraic difference between the *indicated* value and the *actual* value of a measured variable is called the *error*. Thus, error can be either negative or positive. This deviation represents an *uncertainty* in knowledge of the actual value of a variable on the basis of a measurement.

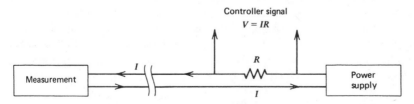

FIGURE 1.15   A two-wire system can be used for both the power leads to a measurement system and also carry information on the measurement through a current.

ACCURACY

This term is used to specify the maximum overall error to be expected from a device, such as measurement of a variable. Accuracy usually is expressed as the *inaccuracy* and can appear in several forms:

1.   Measured variable, as the accuracy is ± 2°C in some temperature measurement. Thus, there would be an uncertainty of ±2°C in any value of temperature measured.
2.   Percentage of the instrument full-scale (FS) reading. Thus, an accuracy of ±0.5% FS in a 5-volt full-scale range meter would mean the inaccuracy or uncertainty in any measurement is ± 0.025 volts.
3.   Percentage of instrument span, that is, percentage of the range of instrument measurement capability. Thus, for a device measuring ± 3% of span for a 20–50 psi range of pressure, the accuracy would be (±0.03)(50 − 20) = ±0.9 psi.
4.   Percentage of the actual reading. Thus, for a ±2% of reading voltmeter, we would have an inaccuracy of ±0.04 volts for a reading of 2 volts.

EXAMPLE 1.6

A temperature transducer has a span of 20°–250°C. A measurement results in a value of 55°C for the temperature. Specify the error if the accuracy is (a) ±0.5% FS, (b) ±0.75% of span, and (c) ±0.8% of reading. What is the possible temperature in each case?

SOLUTION

Using the definitions given above we find
   (a) Error = (±0.005)(250°C) = ±**1.25°C**. Thus, the actual temperature is in the range 53.75–56.25°C.
   (b) Error = (±0.0075)(250 − 20)°C = ±1.725°C. Thus, the actual temperature is in the range 53.275–56.725°C.
   (c) Error = (±0.008)(55°C) = ±0.44°C. Thus, the temperature is in the range 54.56°C–55.44°C.

TRANSFER FUNCTION

This term describes the relationship between the input and output of any element in the process-control loop. Often, a static (time-independent) transfer function is specified separately from the dynamic (time-dependent) transfer function. Thus, if a pressure transducer produces an output of 2 volts for every 1 psi pressure input, the transfer function would be 2 volts/psi. If this were only the static transfer function, we might also need to specify the time response of the transducer. It also is possible that this transfer function is only valid over some range of the input pressure. In most cases, the transfer function itself has an uncertainty associated with it.

EXAMPLE 1.7

A temperature transducer has a transfer function of 5 mV/°C with an accuracy of ±1%. Find the possible range of the transfer function.

SOLUTION

The transfer function range will be $(\pm 0.01)(5 \text{ mV/°C}) = \pm 0.05 \text{ mV/°C}$. Thus, the range is 4.95–5.05 mV/°C.

EXAMPLE 1.8

Suppose a reading of 27.5 mV results from the transducer used in Example 1.7. Find the temperature that could provide this reading.

SOLUTION

Since the range of transfer function is 4.95–5.05 mV/°C, the possible temperature range that could be inferred from a reading of 27.5 mV is

$$(27.5 \text{ mV}) \left(\frac{1}{4.95 \text{ mV/°C}} \right) = \mathbf{5.56°C}$$

$$(27.5 \text{ mV}) \left(\frac{1}{5.05 \text{ mV/°C}} \right) = \mathbf{5.45°C}$$

Thus, we can be certain only that the temperature is between 5.45°C and 5.56°C.

The application of digital processing has necessitated an accuracy definition compatible with digital signals. In this regard we are most concerned with the error involved in the digital representation of analog information. Thus, the accuracy is quoted as the percentage deviation of the analog variable per bit of the digital signal. As an example, an A/D converter may be specified as 0.635 volts per bit ±1%. This means that a bit will be set for an input voltage change from 0.635 ± 0.006 volts or 0.629–0.641 V.

SYSTEM ACCURACY

Often, one must consider the overall accuracy of *many* elements in a process-control loop to represent a dynamic variable. Generally, the best way to accomplish this is to express the accuracy of each element in terms of the transfer functions. For example, suppose we have a process with two transfer functions which act on the dynamic variable to produce an output voltage as shown in Figure 1.16. We can describe the output as

$$V \pm \Delta V = (K \pm \Delta K)(G \pm \Delta G) C \qquad (1\text{-}3)$$

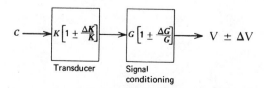

FIGURE 1.16   Each element of a process-control loop contributes uncertainty to a measurement through uncertainties in their transfer functions.

Where

$$V = \text{output voltage}$$
$$\pm \Delta V = \text{uncertainty in output voltage}$$
$$K, G = \text{nominal transfer functions}$$
$$\Delta K, \Delta G = \text{uncertainties in transfer functions}$$
$$C = \text{dynamic variable}$$

From Equation (1-3) we can find the output uncertainty to be

$$\Delta V = \pm GC \; \Delta K \pm KC \; \Delta G \pm \Delta K \; \Delta GC \qquad (1\text{-}4)$$

Equation (1-4) can be further simplified by noting that the nominal output is $V = KGC$ and by ignoring second-order errors

$$\frac{\Delta V}{V} = \pm \left(\frac{\Delta K}{K} + \frac{\Delta G}{G} \right) \qquad (1\text{-}5)$$

Where

$$\frac{\Delta V}{V} = \text{fractional uncertainty in } V$$
$$\frac{\Delta K}{K}, \frac{\Delta G}{G} = \text{fractional uncertainties in transfer functions}$$

We can best interpret Equation (1-5) as stating that the worst case accuracy would be the sum of the uncertainties of each transfer function.

EXAMPLE 1.9

Find the system accuracy of a flow process if the transducer transfer function is 10 mV/m³/s ± 1.5% and the signal conditioning system transfer function is 2 mA/mV ± 0.5%.

SOLUTION

Here we have a direct application of

$$\frac{\Delta V}{V} = \pm \left[\frac{\Delta K}{K} + \frac{\Delta G}{G} \right]$$

$$\frac{\Delta V}{V} = \pm [0.015 + 0.005] \qquad (1\text{-}5)$$

$$\frac{\Delta V}{V} = \pm 0.02 = \pm 2\%$$

So that the net transfer function is 20 mA/m³/s ± 2%.

SENSITIVITY

Sensitivity is a measure of the change in output of an instrument for a change in input. Generally speaking, *high* sensitivity is desirable in an instrument because a large change in output for a small change in input implies that a measurement may be taken easily. Sensitivity must be evaluated together with other parameters, such as *linearity* of output to input, *range*, and *accuracy*. The value of the sensitivity is generally indicated by the transfer function. Thus, when a temperature transducer outputs 5 mV per degree celsius, the sensitivity is 5 mV/°C.

HYSTERESIS AND REPRODUCIBILITY

Frequently, an instrument will not have the same output value for a given input in repeated trials. Such variation can be due to inherent uncertainties which imply a limit on *reproducibility* of the device. This variation is random from measurement to measurement and is not predictable. A similar effect is related to the history of a particular measurement taken with an instrument. In this case, a different reading results for a specific input, depending on whether the input value is approached from higher or lower values. This effect, called *hysteresis*, is shown in Figure 1.17, where the output of an instrument has been plotted against input. We see that if the input parameter is varied from low to high, curve *A* gives values of the output. If the input parameter is decreasing, then curve *B* relates input to output. Hysteresis usually is specified as a percentage of full-scale maximum deviation between the two curves. This effect is predictable if measurement values are always approached from one direction, since hysteresis will not cause measurement errors.

RESOLUTION

Inherent in many measurement devices is a minimum measurable value of the input variable. Such a specification is called the *resolution* of the device. This represents a characteristic of the instrument that can be changed only by redesign. A good example is a wire-wound potentiometer where the slider moves across wind-

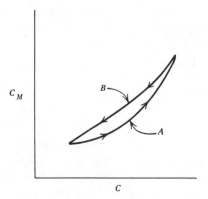

FIGURE 1.17   Hysteresis is a predictable error resulting from differences in transfer functions when a reading is taken from above or below the value to be measured.

ings to vary resistance. If one turn of the winding represents a change of ΔR ohms, then the potentiometer cannot provide a resistance change *less* than ΔR. We say that the potentiometer *resolution* is ΔR. This is often expressed as a percentage of the full-scale range.

EXAMPLE 1.10

A force transducer measures a range of 0–150 N with a resolution of 0.1% FS. Find the smallest change in force that can be measured.

SOLUTION

Because the resolution is 0.1% FS, we have a resolution of $(0.001)(150\text{ N}) = 0.15\text{ N}$, which is the smallest measurable change in force.

In some cases, the resolution of a measurement system is limited by the sensitivity of associated signal conditioning. When this occurs, the resolution can be improved by employing better conditioning.

EXAMPLE 1.11

A transducer has a transfer function of 5 mV/°C. Find the required voltage resolution of the signal conditioning if a temperature resolution of 0.2°C is required.

SOLUTION

A temperature change of 0.2°C will result in a voltage change of

$$\left(5\ \frac{\text{mV}}{\text{°C}}\right)(0.2\text{°C}) = 1.0\text{ mV}$$

Thus, the voltage system must be able to resolve 1.0 mV.

In analog systems the resolution of the system is usually determined by the smallest measurable change in the analog output signal of the measurement system. In digital systems, the resolution is a well-defined quantity which is simply the change in dynamic variable represented by a *one-bit change* in the binary word output. In these cases, resolution can be improved only by a different coding of the analog information or adding more bits to the word. (This will be discussed further in Section 3.4.)

LINEARITY

In both transducer and signal conditioning, output is represented in some functional relationship to the input. The only stipulation is that this relationship be unique, that is, that for each value of the input variable there exists one unique value of the output variable. For simplicity of design, a linear relationship between input and output is highly desirable. Indeed, many nonlinear devices are employed for transduction simply because no other linear devices can be found! Figure 1.9

illustrates cases of both linear and nonlinear relationships. We see that linear relationships can be represented by the equation of a straight line:

$$C_M = mC + C_0 \qquad (1\text{-}6)$$

Where

C = variable to be measured
m = slope of straight line
C_0 = offset or intercept of straight line
C_M = output of measure.

No simple relationship such as Equation (1-6) can usually be found for the non-linear cases, although in some cases *approximations* of a linear or quadratic nature are fitted to portions of these curves as will be shown in Chapter 4.

1.6.4 Process-Control Drawings

A standard form and set of symbols are used to prepare drawings of process-control systems. This is much the same as a *schematic* in electricity used to make drawings of some circuit. In process control, the drawing is often referred to as a piping and instrumentation drawing (P&ID), although the symbols and diagram may be used in processes for which there is no piping. It is important for anyone working in the process industries to recognize the nature of the P&ID and the symbols employed. Appendix A-5 presents a detailed account of the set of symbols and abbreviations. Here, we merely note the basic structure of such diagrams. Thus, Figure 1.18 shows some part of a typical P&ID. First, note that the P&ID contains a diagram of the *entire* process, that is, the reactants and products, as well as the instrumentation and signals that form the various process-control loops. Thus, someone with a good understanding of the P&ID will be able to see how the overall manufacturing sequence works as well as to recognize the individual process-control loops. The actual process flow lines, such as reactant flow and crackers on a conveyor, are shown as heavy solid lines in the diagram.

INSTRUMENT SIGNAL LINES

As shown in Figure 1.18, the instrumentation signal lines are presented in a form that shows whether they are pneumatic or electric. A cross-hatched line is used to show all lines which are pneumatic, as a 3–15 psi signal, for example. The common electric current line, as 4–20 mA, is presented as a dashed line.

INSTRUMENTATION

A balloon symbol with an enclosed two- or three-letter code is used to represent the instrumentation associated with the process-control loops. Thus, the balloon in Figure 1.18 with TT enclosed is a temperature transducer while that with TC enclosed is the temperature controller associated with that loop. Special items such as control valves and in-line instruments (for example, flow meters) have special symbols as shown in Figure 1.18 by the control valves and flow transducers.

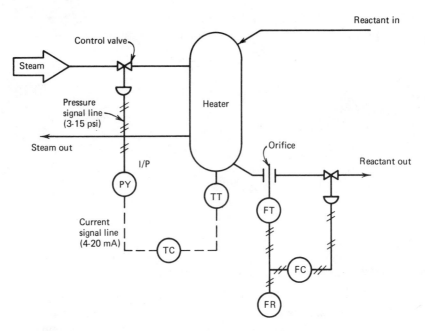

FIGURE 1.18 A P & ID uses special symbols and lines to show the devices and interconnections in a process-control system.

1.7 TIME RESPONSE

In Section 1.4, the effect of sudden changes of the dynamic variable on the process-control loop performance was briefly noted. Clearly, the nature of this system response depends on the nature of the process being controlled and on the behavior of each element to a sudden change. The words *time response* are used to describe the behavior of a system element output whenever the input changes in time. In many cases, the choice of signal conditioning is based in part on the transducer time response; in other cases, the time response of the signal conditioning itself may be important to the overall measurement process. In general, however, the signal conditioning system consists of electronic circuitry designed to have the desired time behavior, depending on the application. The most common way of representing the time response of a transducer is to specify the transducer output following a sudden, discontinuous change in transducer input. This specification usually is sufficient to define the effect that time response has on the process-control system. We differentiate between two responses: *nonoscillating* and *oscillating*.

1.7.1 Nonoscillating Response

In many transducers, output lags behind the input so that a delay exists between the input and indicated output. Furthermore, the transducer output changes smoothly in time from the predisturbance value to the postdisturbance value, even though the disturbance was discontinuous. It is interesting to note that the nature of the transducer *time-response* curve is the same for virtually *all* transducers, even though they may measure different physical dynamic variables.

In Figure 1.19, a typical response curve for a transducer output is shown when the input has been changed suddenly at $t = 0$. It is found that this curve can be described quite well by an exponential equation for the output response $R(t)$ as a function of time

$$R(t) = R_I + (R_F - R_I) [1 - e^{-t/\tau}] \tag{1-7}$$

Where

R_I = initial transducer output response
R_F = final transducer output response
τ = transducer time constant

Note that the transducer output is *in error* during the transition time of the output from R_I to R_F. The actual dynamic variable value was changed instantaneously to a new value at $t = 0$. Equation (1-7) describes transducer output very well except during the initial time period, that is, at the *start* of the response near $t = 0$. In particular, the actual transducer response generally starts the change with a *zero* slope, whereas Equation (1-7) predicts a finite starting slope.

TIME CONSTANT

Equation (1-7) relates initial transducer output, final transducer output, and a constant that is a specification of the transducer. The time constant τ conveys that the significance of this quantity can be found by consideration of Equation (1-7) for the case where the initial transducer output is zero. In this special case, the transducer output response is

$$R(t) = R_F(1 - e^{-t/\tau}) \tag{1-8}$$

Now suppose we wish to find the value of the output exactly τ seconds after the change occurs, then

$$\begin{aligned} R(\tau) &= R_F(1 - e^{-1}) \\ R(\tau) &= 0.6321 \, R_F \end{aligned} \tag{1-9}$$

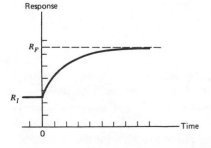

FIGURE 1.19   Characteristic exponential time response of a transducer.

Thus, we see that one time constant (1τ) represents the time at which the output value has changed by approximately 63% of the total change.

The time constant τ is sometimes referred to as the 63% time, the *response time*, or the e-folding time. For a step change, the output response is defined as reaching its final value after five time constants, 5τ or from Equation (1-8)

$$R(5\tau) = 0.993\,R_F$$

EXAMPLE 1.12

A transducer measures temperature linearly with a transfer function of 33 mV/°C and has a 1.5 s time constant. Find the transducer output 0.75 s after the input changes from 20°C to 41°C. Find the temperature this represents.

SOLUTION

We first find the initial and final values of the transducer output:

$$R_I = (33 \text{ mV/°C})\,(20°C)$$
$$R_I = 660 \text{ mV}$$
$$R_F = (33 \text{ mV/°C})(41°C)$$
$$R_F = 1353 \text{ mV}$$

Now,

$$R(t) = R_I + (R_F - R_I)\,[1 - e^{-t/\tau}]$$
$$R(0.75) = 660 + (1353 - 660)[1 - e^{-0.75/1.5}] \quad (1\text{-}7)$$
$$R(0.75) = \mathbf{932.7 \text{ mV}}$$

This corresponds to a temperature of

$$T = \frac{947.4 \text{ mV}}{33 \text{ mV/°C}}$$
$$T = \mathbf{28.3°C}$$

In many cases, the transducer output may be inversely related to the input. Equation (1-7) still describes the time response of the element where now the final output is less than the initial output.

REAL-TIME EFFECTS

The concept of the exponential time response and associated time constant is based on a sudden discontinuous change of the input value. In the real world such instantaneous changes occur rarely, if ever, and thus, we have presented a *worst-case* situation in the time response. In general, a transducer should be able to track any changes in the physical dynamic variable in a time less than one time constant.

1.7.2 Oscillating Response

In some transducers, a discontinuous change in the input will cause the output response to oscillate for a short period of time before settling down to an output that corresponds to the new input. Such oscillation (and the decay of the oscillation itself) is a function of the transducer itself. This output transient generated by the transducer is an error and must be accounted for in any measurement involving a transducer with this behavior.

Figure 1.20 shows a typical output curve that might be expected from a transducer having an oscillation characteristic for a discontinuous change in the input. It is not possible to describe this behavior by an analytic expression with the same degree of accuracy as for the nonoscillating response. However, the behavior can be seen to exhibit a proportionality so that the response with time is

$$R(t) \propto R_0 e^{-at} \sin (2\pi f_n t) \tag{1-10}$$

Where

$R(t)$ = the transducer output
 a = output damping constant
 f_n = natural frequency of the oscillation
 R_0 = amplitude

This equation shows the basic damped oscillation output of the device. The damping constant a and natural frequency f_n are characteristics of the transducer itself and must be considered in many applications.

In general, such a transducer can be said to track the input when the input changes in a time that is *greater* than the period represented by the natural frequency. The damping constant defines the time one must wait after a disturbance at $t = 0$ for the transducer output to be a true indication of the transducer input. Thus, we see that in a time of $(1/a)$ the amplitude of the oscillations would be down to e^{-1} or approximately 37%. More will be said of the effects of natural frequency and damping in the treatment of specific transducers that exhibit this behavior.

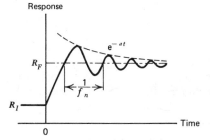

FIGURE 1.20   Characteristic oscillatory time response of a transducer.

1.8 SIGNIFICANCE AND STATISTICS

Process control is vitally concerned with the value of variables, since the stated objective is to regulate the value of selected variables. It is therefore very important that the true significance of some measured value be understood. We have already seen that inherent errors may lend uncertainty to the value indicated by a measurement. In this section we need to consider another feature of measurement that may be misleading about the actual value of a variable, as well as a method to help interpret the significance of measurements.

1.8.1 Significant Figures

In any measurement we must be careful not to attach more significance to a variable value than the instrument can support. This is particularly true with the growing use of digital reading instruments and calculators with 8–12 digit readouts. Suppose, for example, that a digital VOM measures a resistance to be 125 kΩ. Even if we ignore the instrument accuracy, this does not mean that the resistance is 125,000 Ω. Rather, it means that the resistance is closer to 125,000 than it is to 124,000 or 126,000 Ω. We can use the 125 kΩ number in subsequent calculations, but we cannot draw conclusions about results having more than three numbers, that is, three significant figures. The significant figures are the digits (places) actually read or known from a measurement or calculation.

SIGNIFICANCE IN MEASUREMENT

When using a measuring instrument, the number of significant figures is indicated either by readability in the case of analog instruments or by the number of digits in a digital instrument. This is not to be confused with accuracy, which supplies an uncertainty to the reading itself. The following example illustrates how significant figures in measurement and accuracy are treated in the same problem.

EXAMPLE 1.13

A digital Multimeter measures the current through a 12.5-kΩ resistor at 2.21 mA, using the 10-mA scale. The instrument accuracy is $\pm 0.2\%$ FS. Find the voltage across the resistor and the uncertainty in the value obtained.

SOLUTION

First, we note that the significant figures of the numbers given is three, so no result we find can be significant to more than three digits. Then we see that the given accuracy becomes an uncertainty in the current of ± 0.02 mA. From Ohm's law we find the voltage as

$$V = IR = (2.21 \text{ mA})(12.5 \text{ k}\Omega) = 27.625 \text{ volts}$$

But in terms of significant figures we give this as V = 27.6 volts. The accuracy means the current could vary from 2.19 to 2.23 mA, which introduces an uncertainty of ±0.25 volts. Thus, the complete answer is 27.6 ± 0.3 volts, because we must express the uncertainty so that our significance is not changed.

SIGNIFICANCE IN CALCULATIONS

In calculations one must be careful to not obtain a result that has more significance than the numbers employed in the calculation. The answer can have no more significance than the least of the numbers used in the calculation.

EXAMPLE 1.14

A transducer has a specified transfer function of 22.4 mV/°C for temperature measurement. The measured voltage is 412 mV. What is the temperature?

SOLUTION

Using the values given we find

$$T = (412 \text{ mV})/(22.4 \text{ mV/°C}) = 18.392857°C$$

This was found using an 8-digit calculator, but the two given values are significant to only three places. This result can be significant to only three places; the answer is 18.4°C.

SIGNIFICANCE IN DESIGN

The reader should be aware of the difference in significant figures associated with measurement and conclusions drawn from measurement and significant figures associated with *design*. A design is a hypothetical development that makes implicit assumptions about selected values in the design. If the designer specifies a 1.1-kΩ resistor, the assumption is that it is exactly 1100 Ω. If the designer specifies that there are 4.7 volts across the resistor, then there is exactly 4.7 volts and the current can be calculated as 4.2727272 mA. Now, suppose we *measure* the resistor when the design is built and find it to be 1.1 kΩ (two significant figures) and measure the voltage and find it to be 4.7 volts (two significant figures). In this case, we report the calculated current of 4.3 mA, since we are dealing with two significant figures.

1.8.2 Statistics

Often, knowledge of the value of a variable can be improved by the use of elementary statistical analysis of measurements. This is particularly true where random errors in measurement cause a distribution of readings of the value of some variable.

ARITHMETIC MEAN

If many measurements of some variable are taken, the arithmetic mean is calculated to obtain an average value for the variable. There are many instances in process control when such an average value is of interest. For example, one may wish to control the average temperature in a process. The temperature might be measured in 10 locations and averaged to give a controlled variable value for use in the control loop.

Another common application is the calibration of transducers and other process instruments. In such cases, the average gives information about the transfer function. In digital or computer process control it is often easier to use the average value of process variables. The arithmetic mean of a set of n values, given by $x_1, x_2, x_3, \cdots, x_n$ is defined by the equation

$$\bar{x} = \frac{x_1 + x_2 + x_3 + \cdots + x_n}{n} \tag{1-11}$$

Where

$$\bar{x} = \text{arithmetic mean}$$
$$n = \text{number of values to be averaged}$$
$$x_1, x_2 \ldots x_n = \text{individual values}$$

We often use the symbol Σ to represent a sum of numbers such as that used in Equation (1-11). Here we would write the equation as

$$\bar{x} = \frac{\Sigma x_i}{n} \tag{1-12}$$

Where

$$x_i = \text{symbolic for all the values } x_1 - x_n$$

STANDARD DEVIATION

It often is not sufficient to know the value of the arithmetic mean of a set of measurements. To properly interpret the measurements, it may be necessary to know something about how the individual values are spread out about the mean. Thus, although the mean of the set $(50,40,30,70)$ is 47.5 and the mean of the set $(5,150,21,14)$ is also 47.5, the second group of numbers is obviously far more spread out. The standard deviation is a measure of this spread. Given a set of n values $x_1, x_2, \cdots, x_n$, we first define a set of deviations by the difference between the individual values and the arithmetic mean of the values, $\bar{x}$. The deviations are

$$d_1 = x_1 - \bar{x}$$
$$d_2 = x_2 - \bar{x}$$
$$\text{and so on}$$
$$d_n = x_n - \bar{x}$$

The set of these n deviations is now used to define the standard deviation according to the equation

$$\sigma = \sqrt{\frac{d_1^2 + d_2^2 + d_3^2 + \cdots + d_n^2}{n - 1}} \qquad (1\text{-}13)$$

or, using the summation symbol

$$\sigma = \sqrt{\frac{\Sigma d_i^2}{n - 1}} \qquad (1\text{-}14)$$

Now, of course, the larger the standard deviation, the more spread out the numbers from which it is calculated.

EXAMPLE 1.15

Temperature was measured in eight locations in a room, and the values obtained were 21.2°, 25.0°, 18.5°, 22.1°, 19.7°, 27.1°, 19.0°, and 20.0°C. Find the arithmetic mean of the temperature and the standard deviation.

SOLUTION

Using Equation (1-11) we have

$$\bar{T} = \frac{21.2 + 25 + 18.5 + 22.1 + 19.7 + 27.1 + 19 + 20}{8}$$

$$\bar{T} = \textbf{21.6°C} \text{ (remember significant figures)}$$

The standard deviation is found from Equation (1-13):

$$\sigma = \sqrt{\frac{(21.2 - 21.6)^2 + (25 - 21.6)^2 + \cdots + (20 - 21.6)^2}{(8 - 1)}}$$

$$\sigma = 3.04°C$$

INTERPRETATION OF STANDARD DEVIATION

A more quantitative evaluation of spreading can be made if we make certain assumptions about the set of data values used. In particular, we assume that the errors are truly random, and that we have taken a large sample of readings. We then can claim that the standard deviation and data are related to a special curve called the *normal probability curve*, or *bell curve*. If this is true, then,

1. 68% of all readings lie within $\pm 1\sigma$ of the mean.
2. 95.5% of all readings lie within $\pm 2\sigma$ of the mean.
3. 99.7% of all readings lie within $\pm 3\sigma$ of the mean.

This gives us the added ability to make quantitative statements about how the data is spread about the mean. Thus, if one set of pressure readings has a mean of 44 psi with a standard deviation of 14 psi and another a mean of 44 psi with a standard deviation of 3 psi, we know the latter is much more peaked about the mean. In fact, 68% of all the readings in the second case lie from 41 to 47 psi, while in the first case 68% of readings lie from 30 to 58 psi.

EXAMPLE 1.16

A control system was installed to regulate the weight of potato chips dumped into bags in a packaging operation. Given samples of 15 bags drawn from the operation before and after the control system was installed, evaluate the success of the system. Do this by comparing the arithmetic mean and standard deviations before and after. The bags should be 200 g.

Samples before: 201, 205, 197, 185, 202, 207, 215, 220, 179, 201, 197, 221, 202, 200, 195

Samples after: 197, 202, 193, 210, 207, 195, 199, 202, 193, 195, 201, 201, 200, 189, 197

SOLUTION

In the before case we use Equations (1-12) and (1-14) to find the mean and standard deviation and get

$$\overline{W}_b = 202 \text{ g}$$
$$\sigma_b = 11 \text{ g}$$

Now the mean and standard deviation are found for the after case:

$$W_a = 199 \text{ g}$$
$$\sigma_a = 5 \text{ g}$$

Thus, we see that the control system has brought the average bag weight closer to the ideal of 200 g and that it has also cut the spread by a factor of two. In the before case, 99% of the bags weighed 202 ± 33 g, whereas in the case with the control system, 99% of the bags weighed in the range of 199 ± 15 g.

SUMMARY

This chapter presented an overview of the process-control loop and its elements. Subsequent chapters will examine the topics discussed in more detail and provide a more quantitative understanding.

The following list is provided to help the reader view the chapter from the key points.

1.   Process control itself has been described as suitable for application to any situation where some dynamic variable is regulated to some desired value or range of values. Figure 1.2 shows a block diagram in which the elements of measurement, controller, and controlled element are connected to provide the required regulation.

2.   Numerous criteria have been discussed that allow the evaluation of process-control

loop performance of which the settling time, residual error, peak error, and minimum area are the most indicative of loop characteristics.

3. Both analog and digital processing are used in process-control applications. The current trend is to make analog measurements of the dynamic variable, digitize these, and use a digital controller for evaluation. The basic technique of digital encoding allows each bit of a binary word to correspond to a certain quantity of the measured variable. The arrangement of "0" and "1" states in the word then serves as the encoding.

4. The SI system of units forms the basis of computations in this text as well as in the process-control industry in general. It also is still necessary to understand conversions to other systems, notably the English system (see Appendix A-1).

5. A standard, adopted for analog process-control signals, is the 4–20 mA current range to represent the span of measurements of the dynamic variable.

6. The definitions of accuracy, resolution, and other terms used in process control are necessary and are similar to those in related fields.

7. The concept of transducer time response was introduced. The *time constant* becomes part of the dynamic properties of a transducer.

8. The use of significant figures is important to properly interpret measurements and conclusions drawn from measurements.

9. Statistics can help interpret the validity of measurements through the use of the arithmetic mean and the standard deviation.

10. The P&ID drawing and symbols were introduced as the typical representation used to display process-control systems.

PROBLEMS

1.1 Describe a common room air-conditioning system in terms of a process-control loop.

1.2 Identify the elements of a process-control loop by which an operator steers an automobile.

1.3 A process-control loop has an allowable band of ±5°C about a nominal of 175°C. Given the response in Figure 1.21 to a transient, specify the *maximum error, settling time,* and *residual error.*

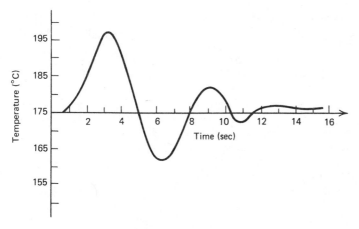

FIGURE 1.21 Figure for Problem 1.3.

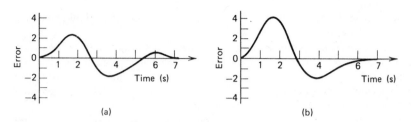

FIGURE 1.22   Figure for Problem 1.4.

1.4   Given the two dynamic response curves of Figure 1.22 show which response would be preferred for satisfaction of a minimum area specification.

1.5   An analog process-control loop linearly converts a 0–300 m³/s flow rate into a 0–50 mA analog current signal. Calculate the analog current for 225 m³/s.

1.6   Find the 4-bit binary word that would represent a decimal 14 if Table 1.1 were to be continued.

1.7   Calculate the volts per bit of an 8-bit ADC for 0–10 volt encoding.

1.8   For the process loop shown in Figure 1.8 and described in Section 1.5.1, assume the relays close at 1.5 V and open at 1.1 V. The amplifier gain is 10, the reference voltage is 3 V, and the measurement system outputs 150 mV/°C. Calculate temperatures at which the heater (a) turns on, (b) turns off, and the temperatures at which the cooler (c) turns on and (d) turns off.

1.9   An instrument has an accuracy of ±0.5% FS for a measurement range of 0–1500 Ω. Calculate the uncertainty in an indicated measurement of 397 Ω.

1.10   A transducer has a transfer function of 0.5 mV/°C with an accuracy of ±1% in this function. If the temperature is 60°C, calculate the nominal transducer output and uncertainty.

1.11   An accelerometer is used to measure the constant acceleration of a vehicle which covers a ¼ mile in 7.2 s.

   (a)   Using $x = \frac{1}{2}at^2$ to relate distance x; acceleration a; and time t, find the acceleration in ft/s².

   (b)   Express this in m/s².

   (c)   Find the velocity in m/s after the ¼ mile using $v^2 = 2ax$.

   (d)   Find the vehicle energy in joules at the end of the ¼ mile if the vehicle weighs 2000 lb, where energy $W = \frac{1}{2}mv^2$ and m is the mass.

1.12   A temperature transducer has a transfer function of 0.15 mV/°C and a time response of $\tau = 3.3$ s. If a step input changing from 22°C to 50°C is applied, specify the *indicated* temperature after 0.5 s, 2 s, 3.3 s, and 9 s, respectively.

1.13   A pressure transducer measures 44 psi just before a sudden change to 70 psi input. The transducer measures 52 psi 4.5 s after the change. Find the transducer time constant.

1.14   Draw Figure 1.4a in the standard form of P&ID symbols.

1.15   Flow rate was monitored for a week and the following values were recorded as gal/min: 10.1, 12.2, 9.7, 8.8, 11.4, 12.9, 10.2, 10.5, 9.8, 8.9, 11.5, 10.3, 9.3, 7.7, 10.2, 10.0, 11.3. Find the mean and standard deviation of these data.

1.16   The transfer function of a temperature transducer is given as 44.5 mV/°C. The output voltage was measured to be 8.86 volts on a three-digit voltmeter. What would you report the temperature to be?

1.17   A man lists his height as 6 ft, 2 in, his weight as 208 lb. What is his height in meters, his weight in newtons, and his weight in kilograms?

1.18   In a circuit, the design calls for a 1.5 kΩ resistor to have a voltage of 4.7 volts across its terminals. What current would be expected through the resistor? A measurement of the resistance shows a value of 1500 Ω, and the voltage is measured to be 4.7 volts. What is the current through the resistor?

1.19   A level range of 4.50–10.6 ft is converted to the standard 3–15 psi signal in a pneumatic process installation. Find an equation that relates level to pressure. What is the pressure for a level of 9.2 ft?

1.20   A light-intensity transducer has a time constant of 35 ms. How long after a sudden change in intensity will the indicated intensity be 80% of the correct value? The initial intensity is zero.

2

ANALOG SIGNAL CONDITIONING

INSTRUCTIONAL OBJECTIVES

The purpose of this chapter is to familiarize the reader with the essential background for basic signal conditioning applications in process control rather than to produce experts in the subject. In view of this, attention has been given to only the most basic techniques. After you have read the chapter, you should be able to:

1. Define the common types of analog signal conditioning.
2. Design a Wheatstone bridge for resistance measurement.
3. Draw a diagram of a current balance bridge and describe its operation.
4. Define the principles of operation of a potentiometer circuit.
5. Define the operation of a silicon controlled rectifier.
6. Design a high input impedance op amp d-c amplifier for specific gain.
7. Analyze a simple op amp circuit for its transfer characteristics.
8. Explain the purpose of compensation leads in a bridge circuit.
9. Design a voltage to current converter for specified voltage input and current output.
10. Define the basic linearization procedure.

2.1 INTRODUCTION

The wide variety of transducers needed to transform the wide variety of dynamic variables in process-control systems into electrical analogs produces an equally wide variety of resultant signal characteristics. *Signal conditioning* refers to operations performed on such signals to convert them to a form suitable for *interface* with other elements in the process-control loop. In this chapter, we are concerned only with *analog* conversions, where the conditioned output is still an analog representation of the dynamic variable. A majority of transducer outputs are of this form. Even in applications involving digital processing, some type of analog conditioning usually is required before analog to digital conversion is made. Specifics of digital signal conditioning are considered in Chapter 3.

2.2 PRINCIPLES OF ANALOG SIGNAL CONDITIONING

A transducer measures a dynamic variable by converting information (about that variable) into a dependent signal of either electrical or pneumatic nature. To develop such transducers, we take advantage of fortuitous circumstances in nature where a dynamic variable influences some characteristic of a material. Consequently, there is little choice of the type or extent of such proportionality. For example, once we have researched nature and found that cadmium sulfide resistance varies inversely and nonlinearly with light intensity, we must then learn to employ this device for light measurement within the confines of that dependence. Analog signal conditioning accomplishes this by providing the operations necessary to transform a transducer output into a form which is required to interface with other elements of the process-control loop. We will confine our attention to electrical transformations.

We often describe the effect of the signal conditioning by the term *transfer function*. By this term we relate the effect of the signal conditioning on the input signal. Thus, a simple voltage amplifier has a transfer function of some constant which, when multiplied by the input voltage, gives the output voltage.

It is possible to categorize signal conditioning in several general types.

2.2.1 Signal Level Changes

The simplest method of signal conditioning is to change the level of a signal. The most common example is the necessity to either amplify or attenuate a voltage level. Generally, process-control applications result in slowly varying low frequency signals where d-c or low-frequency response amplifiers can be employed. An important factor in the selection of an amplifier is the input impedance which the amplifier offers to the transducer (or any other element that serves as an input). In process control, the signals are always representative of a dynamic variable, and any loading effects obscure the correspondence between the measured signal and the dynamic variable value. In some cases, such as accelerometers and optical detectors, the frequency response of the amplifier is very important.

2.2.2 Linearization

As pointed out earlier, the process-control designer has little choice of the characteristics of a transducer output versus dynamic variable. Often the dependence that exists between input and output is nonlinear. Even those devices that are approximately linear may present problems when precise measurements of the dynamic variable are required. One of the functions of analog signal conditioning is to maximize linearization of a transducer's response.

Linearization may be provided by an amplifier whose gain is a function of voltage level so as to linearize the overall variation of input voltage to output voltage. An example of this occurs quite frequently in a transducer where the output is exponential with respect to the dynamic variable. In Figure 2.1 we see such a case (contrived) where the voltage of a transducer is assumed to be exponential with respect to light intensity I. We may write this variation as

$$V_I = V_0\, e^{-\alpha I} \tag{2-1}$$

Where

V_I = output voltage at intensity I
V_0 = zero intensity voltage
α = exponential constant
I = light intensity

To linearize this signal we employ an amplifier whose output varies as the natural logarithm or inverse of the input

$$V_A = K\, \ell n\, (V_{IN}) \tag{2-2}$$

Where

V_A = amplifier output voltage
K = calibration constant
V_{IN} = amplifier input voltage = V_I [in Eq. (2-1)]

By substituting Equation (2-1) into Equation (2-2) where $V_{IN} = V_I$, we find

$$V_A = K\, \ell n\, (V_0) - \alpha K I \tag{2-3}$$

where all terms have been defined above.

The amplifier output *does* vary linearly with intensity but with an offset $K\, \ell n\, V_0$ and a scale factor of αK as shown in Figure 2.2. Further signal conditioning may be employed, if required, to eliminate the offset and provide any desired calibration of voltage versus intensity.

Other types of linearization are possible including eliminating small variations (in response) away from linearity.

2.2.3 Conversions

Often, signal conditioning is used to convert one type of electrical variation into another. Thus, a large class of transducers provides change of resistance with changes in a dynamic variable. In these cases, it is necessary to provide a circuit to convert this resistance change either to a voltage or current signal. This is generally

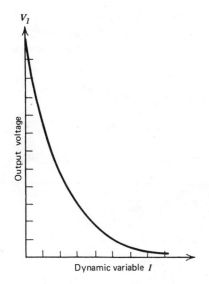

FIGURE 2.1 An example of a nonlinear transducer output. Here, light intensity is assumed to produce an output voltage.

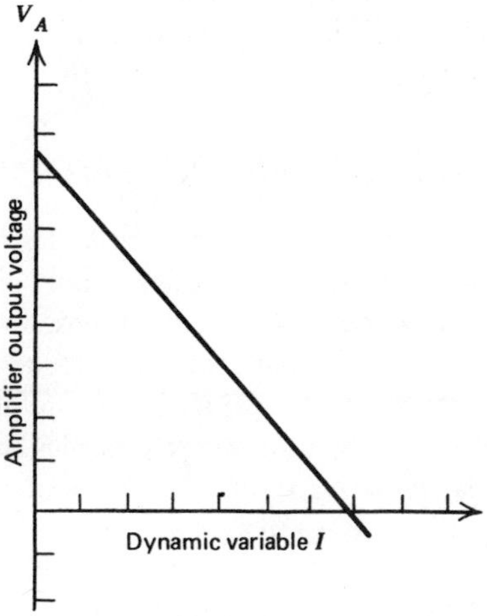

FIGURE 2.2 Proper signal conditioning has produced an output voltage which now varies linearly with light intensity.

accomplished by bridges when the fractional resistance change is small and/or by amplifiers whose gain varies with resistance.

SIGNAL TRANSMISSION

An important type of conversion is associated with the process-control standard of transmitting signals as 4 mA to 20 mA current levels in a wire. This gives rise to the need for converting resistance and voltage levels to an appropriate current level at the transmitting end and for converting the current back to voltage at the receiving end. Of course, current transmission is used because such a signal is independent of load variations other than accidental shunt conditions which may draw off some current. Thus, voltage to current and current to voltage converters are often required.

2.2.4 Filtering and Impedance Matching

Two other common signal conditioning requirements are filtering and matching impedance.

Often, spurious signals of considerable strength are present in the industrial environment, such as the 60 Hz and 400 Hz standard-line frequency signals. Motor start transients also may cause pulses and other unwanted signals in the process-control loop. In many cases, it is necessary to use high pass, low pass, or notch *filters* to eliminate unwanted signals from the loop. Such filtering can be accomplished by *passive* filters using only resistors, capacitors, inductors, or *active* filters, using gain and feedback.

Impedance matching is an important element of signal conditioning when transducer internal impedance or line impedance may cause errors in measurement of a dynamic variable. Both active and passive networks also are employed to provide such matching.

2.3 BRIDGE AND POTENTIOMETER CIRCUITS

Bridge and potentiometer circuits are two passive measurement techniques that have been extensively used for signal conditioning for many years. Although modern active circuits often replace these techniques, there are still many applications where their particular advantages make them useful.

Bridge circuits are primarily used as an accurate means of measuring changes in impedance. Such circuits are particularly useful when the fractional *changes* in impedance are *very small*. Potentiometeric circuits are used to measure voltages both with great accuracy and at very high impedance.

2.3.1 Bridge Circuits

Bridge circuits are passive networks used to measure impedances by a technique of potential matching. In these circuits, a set of accurately known impedances is adjusted in value in relation to an unknown until a condition exists where the potential difference between two points in the network is zero, that is, a *null*. This condition defines an equation used to find the unknown impedance in terms of the known values. A distinct advantage of such measurement techniques is that they are based on reaching the null condition, that is, zero voltage or current, as opposed to an absolute measurement. It is usually much easier to refine and improve techniques for detection of a null than for measurement of some other specific value. This leaves the accuracy of the measurement predominantly dependent on the accuracy with which the known impedances have been determined.

WHEATSTONE BRIDGE

The simplest and most common bridge circuit is the d-c Wheatstone bridge as shown in Figure 2.3. This network is used in signal conditioning applications where a transducer changes resistance with dynamic variable changes. Many modifications of this basic bridge are also employed for other specific applications. In Figure 2.3 the object labeled D is a *null detector* used to compare the potentials of points a and b of the network. In most modern applications the null detector is a very high input impedance differential amplifier. In some cases, a highly sensitive galvanometer with a relatively low impedance may be used, especially for calibration or single measurement instruments.

For our initial analysis, assume the null detector impedance is infinite, that is, an open circuit.

In this case the potential difference, ΔV between points a and b, is simply

$$\Delta V = V_a - V_b \tag{2-4}$$

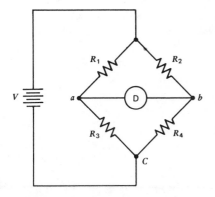

FIGURE 2.3   The basic d-c Wheatstone bridge.

Where

V_a = potential of point a with respect to c
V_b = potential of point b with respect to c

The values of V_a and V_b can now be found by noting that V_a is just the supply voltage, V, divided between R_1 and R_3

$$V_a = \frac{VR_3}{R_1 + R_3} \qquad (2\text{-}5)$$

In a similar fashion V_b is a divided voltage given by

$$V_b = \frac{VR_4}{R_2 + R_4} \qquad (2\text{-}6)$$

Where

V = bridge supply voltage
R_1, R_2, R_3, R_4 = bridge resistors as given in Figure 2.3.

If we now combine Equations (2-4), (2-5), and (2-6), the voltage difference or voltage offset, can be written

$$\Delta V = \frac{VR_3}{R_1 + R_3} - \frac{VR_4}{R_2 + R_4} \qquad (2\text{-}7)$$

After some algebra, the reader can show that this equation reduces to

$$\Delta V = V \frac{R_3 R_2 - R_1 R_4}{(R_1 + R_3) \cdot (R_2 + R_4)} \qquad (2\text{-}8)$$

Equation (2-8) shows how the difference in potential across the detector is a function of the supply voltage and the values of the resistors. Because a difference appears in the *numerator* of Equation (2-8), it is clear that a particular combination of resistors can be found which will result in zero difference and zero voltage across the detector, that is, a null. Obviously, this combination, from examination of Equation (2-8), is

$$R_3R_2 = R_1R_4 \qquad (2\text{-}9)$$

Equation (2-9) indicates that whenever a Wheatstone bridge is assembled and resistors are adjusted for a detector null, the resistor values must satisfy the indicated equality. It does *not* matter if the supply voltage drifts or changes, the null is maintained. Equations (2-8) and (2-9) underlie the application of Wheatstone bridges to process-control applications using high input impedance detectors.

EXAMPLE 2.1

If a Wheatstone bridge shown in Figure 2.3 nulls with $R_1 = 1000\ \Omega$, $R_2 = 842\ \Omega$, and $R_3 = 500\ \Omega$, find the value R_4.

SOLUTION

Since the bridge is nulled, find R_4 using

$$R_1R_4 = R_3R_2$$
$$R_4 = \frac{R_3R_2}{R_1} = \frac{(500\ \Omega)\,(842\ \Omega)}{1000\ \Omega} \qquad (2\text{-}9)$$
$$R_4 = 421\ \Omega$$

EXAMPLE 2.2

The resistors in a bridge are given by $R_1 = R_2 = R_3 = 120\ \Omega$ and $R_4 = 121\ \Omega$. If the supply is 10.0 volts, find the voltage offset.

SOLUTION

Assuming the detector impedance to be very high, we find the offset from

$$\Delta V = V\,\frac{R_3R_2 - R_1R_4}{(R_1 + R_3)\cdot(R_2 + R_4)}$$
$$\Delta V = 10V\,\frac{(120\ \Omega)(120\ \Omega) - (120\ \Omega)(121\ \Omega)}{(120\ \Omega + 120\ \Omega)\cdot(120\ \Omega + 121\ \Omega)} \qquad (2\text{-}8)$$
$$\Delta V = -\,20.8\ \text{mV}$$

GALVANOMETER DETECTOR

The use of a galvanometer as a null detector in the bridge circuit introduces some differences in our calculations because the detector resistance may be low and because we must determine the bridge offset as current offset. When the bridge is nulled, Equation (2-9) still defines the relationship between the resistors in the bridge arms. Equation (2-8) must be modified to allow determination of current drawn by the galvanometer when a null condition is *not* present. Perhaps the easiest way to determine this offset current is first to find the Thévenin equivalent circuit between points *a* and *b* of the bridge (as drawn in Figure 2.3 circuit with the galvanometer removed). The Thévenin voltage is simply the open circuit voltage difference between points *a* and *b* of the circuit. But wait! Equation (2-5) *is* the open circuit voltage, so,

$$V_{Th} = V \frac{R_3 R_2 - R_1 R_4}{(R_1 + R_3)(R_2 + R_4)} \tag{2-10}$$

The Thévenin resistance is found by replacing the supply voltage by its internal resistance and calculating the resistance between terminals *a* and *b* of the network. We may assume that the internal resistance of the supply is negligible compared to the bridge arm resistances. It is left as an exercise for the reader to show that the Thévenin resistance seen at points *a* and *b* of the bridge is

$$R_{Th} = \frac{R_1 R_3}{R_1 + R_3} + \frac{R_2 R_4}{R_2 + R_4} \tag{2-11}$$

The Thévenin equivalent circuit for the bridge enables us to easily determine the current through any galvanometer with internal resistance R_G as shown in Figure 2.4. In particular, then, the offset current is

$$I_G = \frac{V_{Th}}{R_{Th} + R_G} \tag{2-12}$$

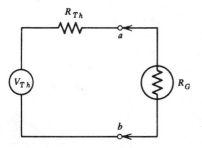

FIGURE 2.4   When a galvanometer is used for a null detector it is convenient to use the Thévenin equivalent circuit of the Wheatstone bridge.

Using this equation in conjunction with Equation (2-9) defines the Wheatstone bridge response whenever a galvanometer null detector is used.

EXAMPLE 2.3

A bridge circuit has resistances of $R_1 = R_2 = R_3 = 2.00 \, k\Omega$ and $R_4 = 2.05 \, k\Omega$ and a 5.00 V supply. If a galvanometer with a 50.0 Ω internal resistance is used for a null detector, find the offset current.

SOLUTION

From Equation (2-10), the offset voltage is V_{Th}.

$$V_{Th} = 5 \, V \, \frac{(2 \, k\Omega) \, (2 \, k\Omega) - (2 \, k\Omega) \, (2.05 \, k\Omega)}{(2 \, k\Omega + 2 \, k\Omega) \, (2 \, k\Omega + 2.05 \, k\Omega)} \tag{2-10}$$

$$V_{Th} = -30.9 \, mV$$

We next find the bridge Thévenin resistance.

$$R_{Th} = \frac{(2 \, k\Omega) \, (2 \, k\Omega)}{(2 \, k\Omega + 2 \, k\Omega)} + \frac{(2 \, k\Omega) \, (2.05 \, k\Omega)}{(2 \, k\Omega + 2.05 \, k\Omega)} \tag{2-11}$$

$$R_{Th} = 2.01 \, k\Omega$$

Finally, the current is

$$I_G = \frac{-30.9 \, mV}{2.01 \, k\Omega + 0.05 \, k\Omega} \tag{2-12}$$

$$I_G = -15.0 \, \mu A$$

BRIDGE RESOLUTION

The resolution of the bridge circuit is a function of the resolution of the null detector used to determine the bridge offset. Thus, referring primarily to the case where a voltage offset occurs, we define the resolution in resistance as that resistance change in one arm of the bridge which causes an offset voltage that is equal to the resolution of the null detector. If a null detector can measure a null to 100 μV, this sets a limit on the minimum measurable resistance change in a bridge using this detector. In general, once given the detector resolution, we may use Equation (2-8) to find the change in resistances from null which causes this offset.

EXAMPLE 2.4

A bridge circuit has $R_1 = R_2 = R_3 = R_4 = 120 \, \Omega$ resistances and a 10-volt supply. Clearly, the bridge is nulled as Equation (2-9) shows. If a detector of 10 mV resolution is employed to detect the null, find the resolution in resistance change in R_4.

SOLUTION

We can simply use Equation (2-7) with R_4 unspecified and find the change in R_4 which will produce a 10 mV offset voltage; as

$$10 \text{ mV} = \frac{(120 \text{ }\Omega) (10 \text{ V})}{120 \text{ }\Omega + 120 \text{ }\Omega} - \frac{R_4 (10 \text{ V})}{120 \text{ }\Omega + R_4}$$

solving for R_4, we get

$$R_4 = 119.52 \text{ }\Omega$$

so the bridge resolution is

$$\Delta R = 0.48 \text{ }\Omega$$

In this example, we see that a minimum resistance change of 0.48 Ω must occur before the detector indicates a change in offset voltage.

One may also view this as an overall *accuracy* of the instrument, since it can also be said that ΔR represents the uncertainty in any determination of resistance using the given bridge and detector.

The same arguments can be applied to a galvanometer measurement where the resolution is now limited by the minimum measurable current.

LEAD COMPENSATION

In many process-control applications, a bridge circuit may be located at considerable distance from the transducer whose resistance changes are to be measured. In such cases, the remaining fixed bridge resistors can be chosen to account for the resistance of leads required to connect the bridge to the transducer in providing a null. Furthermore, any measurement of resistance can be adjusted for lead resistance to determine the actual transducer resistance. Another problem exists which is not so easily handled, however. There are many effects that can change the resistance of the long lead wires on a transient basis, such as frequency, temperature, stress, and chemical vapors. Such changes normally are picked up by the bridge response and interpreted as changes in the transducer output. This problem is reduced using *lead compensation* wherein any changes in lead resistance are introduced equally into *two* (both) arms of the bridge circuit thus causing no effective change in bridge offset. Lead compensation is shown in Figure 2.5. Here we see that R_4, which is assumed to be the transducer, has been removed to a remote location with lead wires (1), (2), and (3). Wire (3) is the power lead and has no influence on the bridge balance condition. Notice that if wire (2) changes in

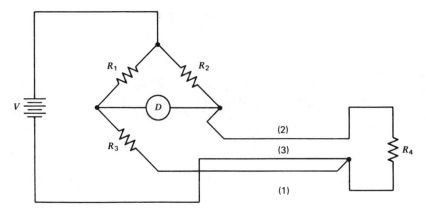

FIGURE 2.5 For remote transducer applications a compensation system is used to avoid errors from lead resistance.

resistance due to spurious influences, it introduces this change into the R_4 leg of the bridge. Wire (1) is exposed to the same environment and changes by the same amount but is in the R_3 leg of the bridge. Effectively, both R_3 and R_4 are identically changed, and thus Equation (2-9) shows that no change in the bridge null occurs. This type of compensation is often employed where bridge circuits must be used with long leads to the active element of the bridge.

CURRENT BALANCE BRIDGE

One disadvantage of the simple Wheatstone bridge is the need to obtain a null by variation of resistors in bridge arms. In the past, many process-control applications used a feedback system in which the bridge offset voltage was amplified and used to drive a motor whose shaft altered a variable resistor to renull the bridge. Such a system does not suit modern technology of electronic processing because it is not very fast, is subject to wear, and generates electronic noise. A technique that provides for an electronic nulling of the bridge and which uses only fixed resistors (except as may be required for calibration) can be used with the bridge. This method uses a *current* to null the bridge. A closed-loop system can even be constructed which provides the bridge with a *self-nulling* ability.

The basic principle of the current balance bridge is shown in Figure 2.6. Here, the standard Wheatstone bridge is modified by splitting one arm resistor into two, R_4 and R_5. A current I is fed into the bridge through the junction of R_4 and R_5 as shown. We now stipulate that the size of the bridge resistors are such that the current flows predominantly through R_5. This can be provided for by any of several requirements. The least restrictive is to require

$$R_4 \gg R_5 \qquad (2\text{-}13)$$

Often, if a high impedance null detector is used, then the restriction of Equation (2-13) becomes

$$(R_2 + R_4) >> R_5 \qquad (2\text{-}14)$$

Assuming that either conditions of Equations (2-13) or (2-14) are satisfied, the voltage at point b is the sum of the divided supply voltage plus the voltage dropped across R_5 from the current I.

$$V_b = \frac{V(R_4 + R_5)}{R_2 + R_4 + R_5} + IR_5 \qquad (2\text{-}15)$$

The voltage of point a is still given by Equation (2-5). Thus, the bridge offset voltage is given by $\Delta V = V_a - V_b$, or

$$\Delta V = \frac{VR_3}{R_1 + R_3} - \frac{V(R_4 + R_5)}{R_2 + R_4 + R_5} - IR_5 \qquad (2\text{-}16)$$

This equation shows that a null is reached by adjusting the magnitude and polarity of the current I until IR_5 equals the voltage difference of the first two terms. If one of the bridge resistors now changes, the bridge can be renulled by change of the current I. In this manner, the bridge is electronically nulled from any convenient current source. In most applications the bridge is nulled at some nominal set of resistances with zero current. Changes of a bridge resistor are detected as a bridge

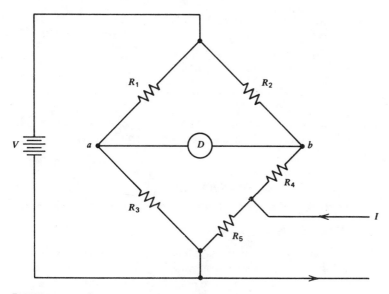

FIGURE 2.6   The current balance bridge.

offset signal which is used to provide the renulling current. The action is explained in Example 2.5:

EXAMPLE 2.5

A current balance bridge, as shown in Figure 2.6, has resistors $R_1 = R_2 = 10\ k\Omega$, $R_4 = 950\ \Omega$, $R_3 = 1\ k\Omega$, $R_5 = 50\ \Omega$ and a high impedance null detector. Find the current required to null the bridge if R_3 changes by 1Ω. The supply voltage is 10 volts.

SOLUTION

First, we note that for the nominal resistance values given, the bridge is at a null with $I = 0$. This can be seen since

$$V_a = \frac{(10\ V)\ (1\ k\Omega)}{10\ k\Omega\ +\ 1\ k\Omega} \qquad (2\text{-}5)$$

$$V_a = 0.9091\ volts$$

and, with $I = 0$, Equation (2-15), gives

$$V_b = \frac{(10\ V)\ (950\ \Omega\ +\ 50\ \Omega)}{10\ k\Omega\ +\ 950\ \Omega\ +\ 50\ \Omega} \qquad (2\text{-}15)$$

$$V_b = 0.9091\ volts$$

Now, when R_3 increases by $1\ \Omega$ to $1001\ \Omega$, V_a becomes

$$V_a = \frac{(10\ V)\ (1001)}{10\ k\Omega\ +\ 1001}$$

$$V_a = 0.9099\ volts$$

which shows that the voltage at b must increase by 0.0008 volts or 0.8 mV to renull the bridge. This can be provided by a current, from Equation (2-16) with $\Delta V = 0$, of $50\ I = 0.8$ mV.

$$I = 16.0\ \mu A$$

POTENTIAL MEASUREMENTS USING BRIDGES

A bridge circuit is also useful to measure small potentials at a very high impedance. This can be accomplished using either a conventional Wheatstone bridge or a current balance bridge. This type of measurement is performed by placing the potential to be measured in series with the null detector, as shown in Figure 2.7.

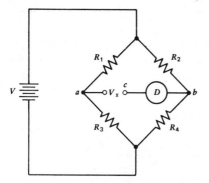

FIGURE 2.7   Using the basic Wheatstone bridge for potential measurement.

The null detector responds to the potential between points c and b. In this case, V_b is given by Equation (2-6) and V_c by Equation (2-17)

$$V_c = V_x + V_a \qquad (2\text{-}17)$$

where V_a is given by Equation (2-5), and V_x is the potential to be measured. The voltage appearing across the null detector is

$$\Delta V = V_c - V_b = V_x + V_a - V_b$$

A null condition is established when $\Delta V = 0$; furthermore, no current flows through the unknown potential when such a null has been found. Thus, a measurement of V_x can be made by varying bridge resistors to provide a null with V_x in the circuit and solving for V_x using the null condition

$$V_x + \frac{R_3 V}{R_1 + R_3} - \frac{VR_4}{R_2 + R_4} = 0 \qquad (2\text{-}18)$$

A similar analysis using a current balance bridge and fixed bridge resistors provides a null condition which can be solved for V_x in terms of the nulling current I,

$$V_x + \frac{R_3 V}{R_1 + R_3} - \frac{V(R_4 + R_5)}{R_2 + R_4 + R_5} - IR_5 = 0 \qquad (2\text{-}19)$$

Notice that if the fixed resistors are chosen to null the bridge with $I = 0$ when $V_x = 0$, then the two middle terms in Equation (2-13) cancel leaving a very simple relationship between V_x and the nulling current

$$V_x - IR_5 = 0 \qquad (2\text{-}20)$$

EXAMPLE 2.6

A bridge circuit for potential measurement nulls when $R_1 = R_2 = 1$ kΩ, $R_3 = 605$ Ω, and $R_4 = 500$ Ω with a 10-volt supply. Find the unknown potential.

SOLUTION

Here we simply use Equation (2-18) to solve for V_x.

$$V_x + \frac{(605)(10)}{605 + 1000} - \frac{(10)(500)}{1000 + 500} = 0 \tag{2-18}$$

$$V_x + 3.769 - 3.333 = 0$$
$$V_x = -0.436 \text{ volts}$$

EXAMPLE 2.7

A current balance bridge is used for potential measurement. The fixed resistors are $R_1 = R_2 = 5$ kΩ, $R_3 = 1$ kΩ, $R_4 = 990$ Ω, and $R_5 = 10$ Ω with a 10-volt supply. Find the current necessary to null the bridge if the potential is 12 mV.

SOLUTION

First, an examination of the resistances shows that the bridge is nulled when $I = 0$ and $V_x = 0$ since, from Equation (2-19),

$$\frac{VR_3}{R_1 + R_3} = \frac{10(1 \text{ k})}{1 \text{ k} + 5 \text{ k}} = 1.667 \text{ volts} \tag{2-19}$$

and

$$\frac{V(R_4 + R_5)}{R_2 + R_4 + R_5} = \frac{10(990 + 10)}{5 \text{ k} + 990 + 10} = 1.667 \text{ volts} \tag{2-19}$$

Thus, we can use Equation (2-20)

$$12 \text{ } mV + 10 \text{ } I = 0$$
$$I = 1.2 \text{ mA}$$

A-C BRIDGES

The bridge concept described in this section can be applied to the matching of impedances in general as well as to resistances. In this case, the bridge is represented as in Figure 2.8 and employs an a-c excitation, usually a sine wave voltage signal. The analysis of bridge behavior is basically the same as in the previous treatment but impedances replace resistances. The bridge offset voltage is then represented as

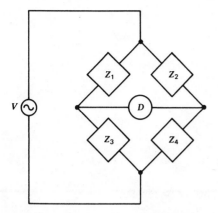

FIGURE 2.8   A general a-c bridge circuit.

$$\Delta E = E \frac{Z_3 Z_2 - Z_1 Z_4}{(Z_1 + Z_2)(Z_3 + Z_4)} \tag{2-21}$$

ΔE = a-c offset voltage

Where

$$E = \text{sine wave excitation voltage}$$
$$Z_1, Z_2, Z_3, Z_4 = \text{bridge impedances}$$

A null condition is defined as before by a zero offset voltage $\Delta E = 0$. From equation (2-21) this condition is met if the impedances satisfy the relation

$$Z_3 Z_2 = Z_1 Z_4 \tag{2-22}$$

Note that this condition is analogous to Equation (2-9) for resistive bridges.

A special note is necessary concerning the achievement of a null in a-c bridges. In some cases, the null detection system is phase sensitive with respect to the bridge excitation signal. In these instances, it is necessary to provide a null of both the inphase and quadrature (90° out-of-phase) signals before Equation (2-22) applies.

EXAMPLE 2.8

An a-c bridge employs impedances as shown in Figure 2.9. Find the value of R_x and C_x when the bridge is nulled.

SOLUTION

Because the bridge is at null, we have

$$Z_2 Z_3 = Z_1 Z_x \tag{2-22}$$

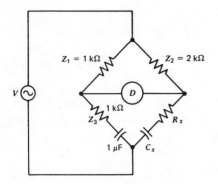

FIGURE 2.9   A-C bridge circuit and components for Example 2.8.

or

$$R_2 \left(R_3 - \frac{j}{\omega C} \right) = R_1 \left(R_x - \frac{j}{\omega C_x} \right)$$

$$R_2 R_3 - j\frac{R_2}{\omega C} = R_1 R_x - \frac{jR_1}{\omega C_x}$$

The real and imaginary parts must be independently satisfied so that

$$R_x - \frac{R_2 R_3}{R_1} = 0$$

$$R_x = \frac{(2\ \text{k}\Omega)(1\ \text{k}\Omega)}{1\ \text{k}\Omega}$$

$$R_x = 2\ \text{k}\Omega$$

and

$$C_x = C\,\frac{R_1}{R_2}$$

$$C_x = (1\ \mu\text{F})\,\frac{1\ \text{k}\Omega}{2\ \text{k}\Omega}$$

$$C_x = 0.5\ \mu\text{F}$$

2.3.2 Potentiometer Circuit

Voltage measurements in process control often must be made at very high impedance and with a high degree of accuracy. Many modern circuits using active devices
have been developed in recent years to perform such measurements. For many
years the only reliable method for such measurements, which was both accurate
and high impedance, was the potentiometer. Basically, this circuit is a calibrated

voltage divider which measures an unknown potential by adjusting a known, divided voltage until it matches the unknown. This technique can be understood from an examination of Figure 2.10. The divider is constructed by the series R_1, R_2, and R connected to the working supply voltage, V_W. Resistor R_2 is a precision, fixed unit and R_1 is a precision and linear variable resistor. The calibration resistor R is a variable (whose actual value is never used in any calculation), and V_W is any source that has adequate voltage (as will be defined later) and is stable. The supply V_{REF} is a calibration standard having an accurately known voltage. Units D_1 and D_2 are both null detectors and may be either galvanometers or high impedance voltage detectors. V_x is the unknown voltage to be measured.

Calibration of the voltage divider is accomplished by closing switch S_1 and adjusting R until detector D_1 indicates a null. Under this condition we will have established that $V_a = V_{REF}$ to within the accuracy of the null detector. This effectively calibrates the divider circuit since V_a is divided between the precision resistors R_1 and R_2. The wiper of R_1 sweeps out voltages between zero at the bottom and V_b at the top of the variable resistor. Voltage V_b is found from

$$V_b = \frac{R_1 V_a}{R_1 + R_2} \qquad (2\text{-}23)$$

Since $V_a = V_{REF}$, we have identification of V_b directly in terms of V_{REF}. Now if the wiper of R_1 is a fraction α from the ground side, the resistance above the wiper is $(1 - \alpha)R$. If an unknown voltage is attached as shown in Figure 2.10 and the wiper is adjusted until detector D_2 indicates a zero, the wiper voltage and unknown potential are the same. Thus, the unknown voltage is given by

$$V_x = \alpha V_b$$

Where

α = fraction of R for which a null occurs
V_b = point b voltage given by Equation (2-23)

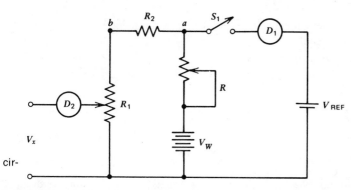

FIGURE 2.10   A basic potentiometer circuit.

In many cases the variable resistor R_1 is scaled by divisions, such as 1000 readable divisions. In this case, α is simply the number of divisions that result in a null of detector D_2. Note that once the divider has been calibrated, the reference voltage V_{REF} and detector D_1 are no longer needed.

EXAMPLE 2.9

A potentiometer circuit has $R_1 = 1$ kΩ with 1000 divisions, $R_2 = 2500$ Ω, and a reference of $V_{REF} = 1.00329$ V with a null for $\alpha = 225$ divisions. Find the unknown potential.

SOLUTION

From Figure 2.10 we find V_b, where calibration has set $V_a = 1.00329$, thus

$$V_b = \frac{R_1 V_a}{R_1 + R_2}$$

$$V_b = \frac{(1000)\,(1.00329)}{1000 + 2500} \tag{2-23}$$

$$V_b = 0.28665 \text{ volts}$$

with $\alpha = 225$ divisions, we see that the V_x is

$$V_x = \frac{225}{1000}(0.28665)$$

$$V_x = 0.0645 \text{ volts}$$

EXAMPLE 2.10

Design a potentiometer that will measure 0–100 mV with a 1000 division variable resistor of 1 kΩ. Use a 6-volt working battery and a 1.35629-volt reference cell.

SOLUTION

Our first goal is to determine a value for R_2 which will provide $V_b = 100$ mV. This can be found from Equation (2-23).

$$0.1 \text{ volts} = \frac{(1.35629)(1 \text{ k}\Omega)}{R_2 + 1 \text{ k}\Omega}$$

solving for R_2 we get

$$R_2 = 12.5629 \text{ K}\Omega$$

Now R can be found by knowing that $6 - 1.35629 \simeq 4.64$ V must be dropped across R at the current of the divider. This divider current is

$$I_D = \frac{V_{REF}}{R_1 + R_2}$$

$$I_D = \frac{1.35629}{1 \text{ k}\Omega + 12.563 \text{ k}\Omega}$$

$$I_D = 0.1 \text{ mA}$$

Then we find

$$R \simeq 46.4 \text{ k}\Omega$$

We select a **50 kΩ** variable resistor to provide this resistance.

2.4 OPERATIONAL AMPLIFIERS

As discussed in Section 2.2., there are many diverse requirements for signal conditioning in process control. In Section 2.3 we considered two common, passive circuits which can provide some of the required signal operations, the bridge and potentiometer. Historically, the detectors used in bridge circuits and potentiometers used in process-control systems consisted of tube and transistor circuits. In many other cases where impedance transformations, amplification, and other operations were required, a circuit was designed which depended on discrete electronic components. With the remarkable advances in electronics and integrated circuits (ICs), the requirement to implement designs from discrete components has given way to easier and more reliable methods of signal conditioning. Many special circuits and general purpose amplifiers are now contained in integrated circuit (IC) packages producing a quick solution to signal conditioning problems together with small size, low-power consumption, and low cost.

In general, the application of ICs requires familiarity with an available line of such devices, their specifications and limitations, before they can be applied to a specific problem. Apart from these specialized ICs there also is a type of amplifier that finds wide application as the building block of signal conditioning applications. This device, called an operational amplifier (op amp), has been in existence for many years, first as constructed from tubes, then from discrete transistors, and now as integrated circuits. Although many lines of op amps with diverse specifications exist from many manufacturers, they all have common characteristics of operation that can be employed in basic designs relating to any general op amp.

2.4.1 Op Amp Characteristics

Taken alone, an op amp is an exceedingly simple and apparently useless electronic amplifier. In Figure 2.11a we see the standard symbol of an op amp with the designations of (+) input, (−) input, and the output. The (+) input is also called the *noninverting* input and the (−) the *inverting* input. The relation of op amp input to output is very simple indeed, as will be seen by consideration of an idealization of its description.

IDEAL OP AMP

To describe the response of an ideal op amp, we label V_1 the voltage on the (+) input, V_2 the voltage on the (−) input terminal, and V_0 the output voltage. Ideally, if $V_1 - V_2$ is positive ($V_1 > V_2$), then V_0 saturates positively. If $V_1 - V_2$ is negative ($V_2 > V_1$), then V_0 saturates negatively as shown in Figure 2.11b. The (−) input is called the inverting input. If the voltage on this input is more positive than that on the (+) input, the output saturates negatively. This ideal amplifier has infinite gain since an infinitesimal difference between V_1 and V_2 results in a saturated output.

Other characteristics of ideal op amps are (1) an infinite impedance between inputs and (2) a zero output impedance. Basically, then, the op amp is a device which has only two output states, $+V_{sat}$ and $-V_{sat}$. In practice, the device is always used with feedback of output to input. Such feedback permits implementation of many special relationships between input and output voltage.

IDEAL INVERTING AMPLIFIER

To see how the op amp is used, let us consider the circuit of Figure 2.12. Here resistor R_2 is used to feed back the output to the inverting input of the op amp, and R_1 connects the input voltage V_{in} to this same point. The common connection is called the summing point. We can see that with no feedback and the (+) grounded, $V_{in} > 0$ saturates the output negative, whereas $V_{in} < 0$ saturates the output positive. With feedback, the output adjusts to a voltage such that:

1. The summing point voltage is equal to the (+) op amp input level, zero in this case.
2. No current flows through the op amp input terminals because of the assumed infinite impedance.

In this case, the sum of currents at the summing point must be zero.

$$I_1 + I_2 = 0 \qquad (2\text{-}24)$$

Where

I_1 = current through R_1
I_2 = current through R_2

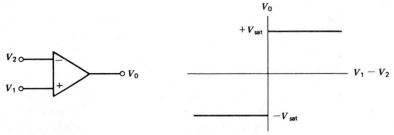

a) Operational amplifier
(op amp) symbol

b) Relation between input and output

FIGURE 2.11   Symbol and ideal characteristics of an op amp.

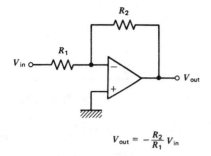

$$V_{out} = -\frac{R_2}{R_1} V_{in}$$

FIGURE 2.12   Inverting amplifier.

Since the summing point potential is assumed to be zero, we have

$$\frac{V_{in}}{R_1} + \frac{V_{out}}{R_2} = 0 \tag{2-25}$$

From Equation (2-25), we can write the circuit response as

$$V_{out} = -\frac{R_2}{R_1}V_{in} \tag{2-26}$$

Thus, the circuit of Figure 2.12 is an inverting *amplifier* with gain R_2/R_1 which is shifted 180° in phase (inverted) from the input. This device is also an *attenuator* by making $R_2 < R_1$.

A similar approach may be applied to the ideal analysis of many other op amp circuits where steps (1) and (2), given above, lead to equations such as Equations (2-24) and (2-25). We must note, however, that the inverting amplifier of Figure 2.12 has an input impedance of R_1 which, in general, may not be high. Thus, although blessed with the virtue of variable gain or attenuation, the circuit does not have inherently high input impedance.

NONIDEAL EFFECTS

Analysis of op amp circuits with nonideal response is performed by considering the following parameters:

1.   *Finite Open Loop Gain.* A real op amp has finite voltage gain as shown by the amplifier response in Figure 2.13a. The voltage gain is defined as the change in output voltage, ΔV_0, produced by a change in differential input voltage $\Delta[V_1 - V_2]$.
2.   *Finite Input Impedance.* A real op amp has an input impedance and, as a consequence, a finite voltage across and current through input terminals.
3.   *Nonzero Output Impedance.* A real op amp has a nonzero output impedance, although this low output impedance is typically only a few ohms.

In most modern applications these nonideal effects can be ignored in designing op amp circuits. For example, consider the circuit of Figure 2.13b where the finite impedances and gain of the op amp have been included. We can employ standard

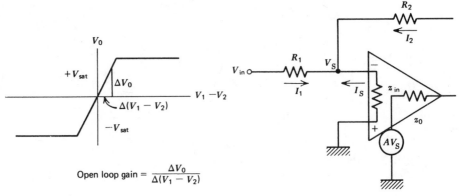

a) Nonideal characteristics of an op amp

b) Nonideal effects

FIGURE 2.13   Types of nonideal effects in op amp and circuit analysis.

circuit analysis to find the relationship between input and output voltage for this circuit. Summing the currents at the summing point gives

$$I_1 + I_2 + I_s = 0$$

Then, each current can be identified in terms of the circuit parameters to give

$$\frac{V_{in} - V_s}{R_1} + \frac{V_0 - V_s}{R_2} - \frac{V_s}{z_{in}} = 0$$

Finally, V_0 can be related to the op amp gain as

$$V_0 = AV_s - \left(\frac{V_0 - V_s}{R_2}\right) z_0$$

Now, combining the equations above, we find

$$V_0 = -\frac{R_2}{R_1}\left(\frac{1}{1 - \mu}\right) V_{in} \tag{2-27}$$

Where

$$\mu = \frac{\left(1 + \dfrac{z_0}{R_2}\right)\left(1 + \dfrac{R_2}{R_1} + \dfrac{R_2}{z_{in}}\right)}{\left(A + \dfrac{z_0}{R_2}\right)} \tag{2-28}$$

If we assume that μ is very small compared with unity, then Equation (2-27) reduces to the ideal case given by Equation (2-26). Indeed, if typical values for an

IC op amp are chosen for a case where $R_2/R_1 = 100$, we can show that $\mu \ll 1$. For example, a common, general purpose IC op amp shows

$$A = 200,000$$
$$z_0 = 75 \ \Omega$$
$$z_{in} = 2 \ M\Omega$$

If we use a feedback resistance R_2 of 100 kΩ and substitute the above values into Equation (2-28), we find $\mu \simeq 0.0005$ which shows that the gain from Equation (2-27) differs from the ideal by only 0.05%. This was, of course, only one example of the many op amp circuits which are employed, but in virtually all cases a similar analysis shows that the ideal characteristics may be assumed.

2.4.2 Op Amp Specifications

There are characteristics of op amps other than those given in the previous section which enter into design applications. These characteristics are given in the specifications for particular op amps together with the open loop gain and input and output impedance previously defined. Several of these characteristics are:

- **Input offset voltage.** In many cases, the op amp output voltage may not be zero when the voltage across the input is zero. The voltage that must be applied across the input terminals to drive the output to zero is the *input offset voltage.*
- **Input offset current.** Just as a voltage offset may be required across the input to zero the output voltage, so a net current may be required between the inputs to zero the output voltage. Such a current is referred to as an input offset current. This is taken as the difference of the two input currents.
- **Input bias current.** This is the average of the two input currents required to drive the output voltage to zero.
- **Slew rate.** If a voltage is suddenly applied to the input of an op amp, the output will saturate to the maximum. For a step input the *slew rate* is the rate at which the output changes to the saturation value. This typically is expressed as volts per microsecond. (V/μs).
- **Unity gain frequency bandwidth.** The frequency response of an op amp is typically defined by a Bode plot of open-loop voltage gain versus frequency. Such a plot is very important for the design of circuits which deal with a-c signals. It is beyond the scope of this text to consider the details of such design employing Bode plots. Instead, we note that the gross frequency behavior can be seen by determination of the frequency at which the open-loop gain of the op amp has become unity, thus defining the *unit gain frequency bandwidth.*

2.5 OP AMP CIRCUITS IN INSTRUMENTATION

As the op amp has become familiar to the individuals working in process control and instrumentation technology, a large variety of circuits has been developed with direct application in this field. In general, it is much easier to develop a circuit for a specific service using op amps than discrete components; with the development of low cost, IC op amps, it is also a practical design. Perhaps one of the greatest disadvantages is the requirement of a bipolar power supply for the op amp. This section presents a number of typical circuits and their basic characteristics together with a derivation of the circuit response assuming an ideal op amp.

2.5.1 Voltage Follower

In Figure 2.14 we see an op amp circuit having unity gain and very high input impedance. The input impedance is essentially the input impedance of the op amp itself which can be greater than 100 MΩ. The voltage output tracks the input over a range defined by the plus and minus saturation voltage outputs. Current output is limited to the short circuit current of the op amp, and output impedance is typically less than 100 Ω. In many cases a manufacturer will market an op amp voltage follower whose feedback is provided internally. Such a unit is usually specifically designed for very high input impedance. The unity gain voltage follower is essentially an impedance transformer in the sense of converting a voltage at high impedance to the same voltage at low impedance.

2.5.2 Inverting Amplifier

The inverting amplifier has already been discussed in connection with our treatment of op amp characteristics. Equation (2-26) shows that this circuit inverts the input signal and may have either attenuation or gain depending on the ratio of input resistance R_1 and feedback resistance R_2. The circuit for this amplifier is shown in Figure 2.12. It is important to note that input impedance of this circuit is essentially equal to R_1, the input resistance. In general, this resistance is not large, and hence the input impedance is not large.

SUMMING AMPLIFIER

A common modification of the inverting amplifier is an amplifier which sums or adds two or more applied voltages. This circuit is shown in Figure 2.15 for the case of summing two input voltages. The transfer function of this amplifier is given by

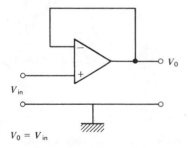

$V_0 = V_{in}$

FIGURE 2.14 An op amp voltage follower. This circuit has very high input impedance; depending on the op amp, it may be 10^6–10^{11} Ω. This circuit serves as an impedance transformer.

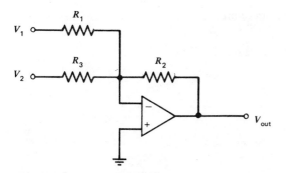

FIGURE 2.15   Summing amplifier.

$$V_{out} = -\left[\frac{R_2}{R_1}V_1 + \frac{R_2}{R_3}V_2\right]$$ (2-29)

The sum can be scaled by proper selection of resistors. For example, if we make $R_1 = R_2 = R_3$, then the output is simply the (inverted) sum of V_1 and V_2. The average can be found by making $R_1 = R_3$ and $R_2 = R_1/2$.

2.5.3 Noninverting Amplifier

A noninverting amplifier may be constructed from an op amp as shown in Figure 2.16. The gain of this circuit is found by summing the currents at the summing point S, and using the fact that the summing point voltage is V_{in} so that no voltage difference appears across the input terminals.

$$I_1 + I_2 = 0$$

Where

  I_1 = current through R_1
  I_2 = current through R_2

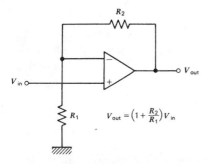

FIGURE 2.16   Noninverting amplifier.

But these currents can be found from Ohm's law such that this equation becomes

$$\frac{V_{in}}{R_1} + \frac{V_{in} - V_{out}}{R_2} = 0$$

Solving this equation for V_{out}, we find

$$V_{out} = \left(1 + \frac{R_2}{R_1}\right) V_{in} \tag{2-30}$$

Equation (2-30) shows that the noninverting amplifier has a gain which depends on the ratio of feedback resistor R_2 and the ground resistor R_1, but this gain can never be used for voltage attenuation. We note also that because the input is taken directly into the noninverting input of the op amp, the input impedance is very high since it is effectively equal to the op amp input impedance.

EXAMPLE 2.11

Design a high impedance amplifier with a voltage gain of 42.

SOLUTION

We use the noninverting circuit of Figure 2.16 with resistors selected from

$$V_{out} = \left[1 + \frac{R_2}{R_1}\right] V_{in}$$
$$\tag{2-30}$$
$$42 = 1 + \frac{R_2}{R_1}$$
$$R_2 = 41\,R_1$$

so we could choose $R_1 = 1$ kΩ, which requires $R_2 = 41$ kΩ.

2.5.4 Differential Amplifier

Frequently, in the instrumentation associated with process control, differential voltage amplification is required, as for a bridge circuit, for example. A differential amplifier is constructed using an op amp as shown in Figure 2.17a. Analysis of this circuit shows that the output voltage is given by

$$V_{out} = \frac{R_2}{R_1}(V_2 - V_1) \tag{2-31}$$

This circuit has a variable gain or attenuation given by the ratio of R_2 and R_1 and responds to the difference in voltage inputs as required. It is very important that the resistors in Figure 2.17a indicated to have the same value be carefully matched to assure rejection of voltage common to both inputs. A significant disadvantage of this circuit is that the input impedance at each input terminal is not large, being $R_1 + R_2$ at the V_2 input and R_1 at the V_1 input. In order to employ this circuit when a high input impedance differential amplification is desired, voltage followers may be employed before each input as shown in Figure 2.17b. This circuit makes a very versatile gain, high input impedance differential amplifier for use in instrumentation systems.

2.5.5 Voltage to Current Converter

Because signals in process control are most often transmitted as a current, specifically 4–20 mA, it is often necessary to employ a linear voltage to current converter. Such a circuit must be capable of sinking a current into a number of different loads without changing the voltage to current transfer characteristics. An op amp circuit to provide this function is shown in Figure 2.18. An analysis of this circuit shows that the relationship between current and voltage is given by

$$I = -\frac{R_2}{R_1 R_3} V_{in} \tag{2-32}$$

provided that the resistances are selected so that

$$R_1(R_3 + R_5) = R_2 R_4 \tag{2-33}$$

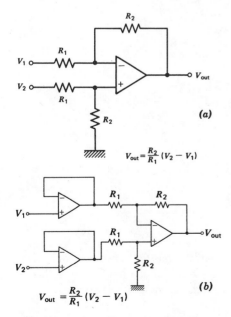

FIGURE 2.17   Differential amplifiers. (a) Differential Amplifier (b) Instrumentation Amplifier.

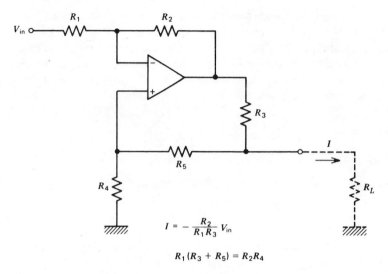

$$I = -\frac{R_2}{R_1 R_3} V_{in}$$

$$R_1(R_3 + R_5) = R_2 R_4$$

FIGURE 2.18   Voltage to current converter.

the circuit can deliver current in either direction, as required by a particular application.

The maximum load resistance and maximum current are related and are determined by the condition that the amplifier output saturates in voltage. Analysis of the circuit shows that when the op amp output voltage saturates the maximum load resistance and maximum current are related by

$$R_{ML} = \frac{(R_4 + R_5)\left[\dfrac{V_{SAT}}{I_M} - R_3\right]}{R_3 + R_4 + R_5} \tag{2-34}$$

R_{ML} = maximum load resistance

V_{SAT} = op amp saturated on voltage

I_M = maximum current

Note that a study of Equation (2-34) shows that the *maximum* load resistance is always less than V_{SAT}/I_M. The *minimum* load resistance is zero.

2.5.6 Current to Voltage Converter

At the receiving end of the process-control signal transmission system we often need to convert the current back into a voltage. This can be done most easily with the circuit shown in Figure 2.19. This circuit provides an output voltage given by

$$V_{out} = IR \tag{2-35}$$

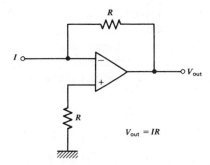

FIGURE 2.19   Current to voltage converter.

provided the op amp saturation voltage has not been reached. The resistor R in the noninverting terminal is employed to provide temperature stability to the configuration.

2.5.7 Sample and Hold

When the measurement must interface with a digital process in a control or measurement situation, it is often necessary to provide a fixed value to an analog-to-digital converter (ADC). Thus, if a measurement is to be made at some time, it may be that during the procedure of A/D conversion the measured value changes. Such a variation can cause error in the conversion process. To alleviate this, an op amp is employed in a sample-and-hold configuration. This circuit, shown in Figure 2.20, can take a very fast sample of an input voltage signal and then hold this value, even though the input signal may change, until another sample is required. The method exploits the charge-storing ability of a capacitor and the inherently high impedance of an op amp. As shown in the simple circuit example of Figure 2.20, when switch

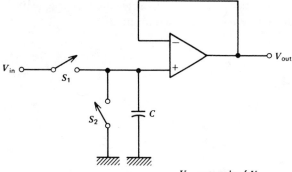

FIGURE 2.20   Sample and hold circuit. Close S_1 to take a sample and open to hold the sample. Close S_2 to reset.

1 is closed, the capacitor quickly charges to the input voltage level. If switch 1 is now opened, the voltage follower op amp allows a measure of the capacitor voltage to be taken at the output without changing the capacitor charge. When a *new* sample must be taken, switch 2 is first closed to discharge the capacitor and hence reset the circuit. The switches used are usually electronic switches activated by digital logic levels.

2.5.8 Integrator

The last regular op amp circuit to be considered is the *integrator*. This configuration, shown in Figure 2.21, consists of an input resistor and feedback capacitor. Using the ideal analysis we can sum the currents at the summing point as

$$\frac{V_{in}}{R} + C\,\frac{dV_{out}}{dt} = 0 \tag{2-36}$$

which can be solved by integrating both terms so that the circuit response is

$$V_{out} = -\frac{1}{RC}\int V_{in}\, dt \tag{2-37}$$

This shows that the output voltage varies as an integral of the input voltage with a scale factor of $-1/RC$. This circuit is employed in many cases where an integration of a transducer output is desired.

Other functions also can be implemented, such as a highly linear ramp voltage. If the input voltage is constant, $V_{in} = K$, then equation (2-37) reduces to

$$V_{out} = -\frac{K}{RC}\,t \tag{2-38}$$

which is a linear ramp, a negative slope of K/RC. Some mechanism of reset through discharge of the capacitor must be provided because otherwise V_{out} will rise to the output saturation value and remain fixed there in time.

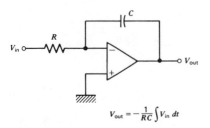

$$V_{out} = -\frac{1}{RC}\int V_{in}\, dt$$

FIGURE 2.21 Integrator circuit. A switch is placed across the capacitor to reset the integrator.

EXAMPLE 2.12

Use an integrator to produce a linear ramp voltage rising at 10 volts per ms as in Figure 2.21.

SOLUTION

An integrator circuit produces a ramp of

$$V_{out} = -\frac{V_{in}}{RC}t \tag{2-38}$$

when the input voltage is constant. If we make $RC = 1$ ms and $V_{in} = -10$ V, then we have

$$V_{out} = (10 \cdot 10^{+3})t$$

which is a ramp rising at 10 volts/ms. A choice of $R = 1$ kΩ and $C = 1$ μF will provide the required RC product.

2.5.9 Linearization

The op amp represents a very effective device to implement linearization. Generally, this is achieved by placing a *nonlinear* element in the feedback loop of the op amp as shown in Figure 2.22. The summation of currents provides that

$$\frac{V_{in}}{R} + F(V_{out}) = 0 \tag{2-39}$$

Where

V_{in} = input voltage
R = input resistance
$F(V_{out})$ = nonlinear variation of current with voltage

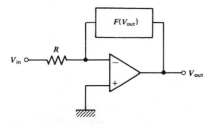

FIGURE 2.22   A nonlinear amplifier is constructed by placing any nonlinear element in the feedback of the op amp.

Now if Equation (2-39) is solved (in principle) for V_{out}, we get

$$V_{out} = G\left(\frac{V_{in}}{R}\right)$$ (2-40)

Where

V_{out} = output voltage

$G\left(\frac{V_{in}}{R}\right)$ = a nonlinear function of the input voltage, actually the inverse function of $F(V_{out})$.

Thus, as an example, if a diode is placed in the feedback as shown in Figure 2.23, then the function $F(V_{out})$ is an exponential

$$F(V_{out}) = F_0 \exp(\alpha V_{out})$$ (2-41)

Where

F_0 = amplitude constant
α = exponential constant

The inverse of this is a logarithm and thus Equation (2-40) becomes

$$V_{out} = \frac{1}{\alpha}\ell n(V_{in}) - \frac{1}{\alpha}\ell n(F_0 R)$$ (2-42)

which thus constitutes a (linear) logarithmic amplifier.

Different feedback devices can produce amplifiers which only smooth out non-linear variations or provide specified operations such as the logarithmic amplifier.

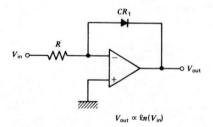

$V_{out} \propto \ell n(V_{in})$

FIGURE 2.23 When a diode is placed in the feedback leg of an op amp, a nonlinear amplifier is formed whose output is proportional to the natural logarithm of the input.

2.5.10 Special Integrated Circuits (ICs)

A vast line of special integrated circuits (ICs) is available from many manufacturers and is useful to the process-control instrumentation designer. Such special purpose devices include:

1. High gain differential instrumentation amplifiers.
2. Current to voltage converters.
3. Modulator/demodulators.
4. Bridge and null detectors.
5. Phase sensitive detectors.

In the following chapters we often require signal conditioning which can be implemented through the use of these special ICs. In general, we shall indicate the details of a signal conditioning design, but the reader should always be aware that use of special purpose ICs may render such a detailed design unnecessary.

EXAMPLE 2.13

Design a current to voltage converter to provide a 0–10 mA current for a 0–1 volt input. Specify the maximum load resistance. The op amp saturates at ±10 volt output.

SOLUTION

If we make $R_1 = R_2$, then Equation (2-32) reduces to

$$I = \frac{1}{R_3} V_{in} \tag{2-32}$$

where now Equation (2-33) specifies $R_3 + R_5 = R_4$. So we choose $R_1 = R_2 = 1\ k\Omega$ and then, from Equation (2-32),

$$R_3 = \frac{1\ V}{10\ mA} = 10\ \Omega \tag{2-32}$$

If we now make $R_5 = 0$, which is allowed, then $R_4 = 100\ \Omega$ also. The maximum load resistance is now found from Equation (2-34)

$$R_{ML} = \frac{100}{200}[1000 - 100]\Omega \tag{2-34}$$

which gives

$$R_{ML} = 450\ \Omega$$

2.6 INDUSTRIAL ELECTRONICS

The signal conditioning discussed thus far in this chapter has referred mainly to measurement signal modifications. It also is often necessary to perform a type of signal conditioning on the controller output to activate the final control element. For example, the 4 to 20 mA controller output may be required to adjust heat input to a large, heavy-duty oven for baking crackers. Such heat may be provided by a 2-kW electrical heater. Clearly, some sort of conditioning is required to allow such a high power system to be controlled by a low-energy current signal. In this section we present two devices which are commonly used in process control to provide a mechanism by which such energy conversion can occur. The intent here is not to give you all the information needed to construct practical circuits to use these devices, but to make you familiar with them and their specifications.

2.6.1 Silicon Controlled Rectifier (SCR)

The SCR has become a very important part of high-power electrical signal conditioning and control. In some regards, it is a solid state replacement for the relay, although there are some problems if that analogy is taken too far. The standard diode is, in the ideal sense, a device which will conduct current in only one direction. The SCR, again in the ideal sense, is like a diode which will not conduct in either direction until it is turned on or "fired." In Figure 2.24 we see the schematic symbol of an SCR. Note the similarity to a diode but with the added terminal, called the *gate*. If the SCR is forward biased, that is, positive voltage on the anode with respect to the cathode, it will *not* be conducting. Now suppose a voltage is placed on the gate with respect to the cathode. There will be some positive value of this voltage—the trigger voltage—at which the SCR will start conducting and behave then like a normal diode. Even if the gate voltage is taken away, it will continue to conduct like a diode; that is, once turned on it will stay on, regardless of the gate. The only way to turn the SCR back "off" is to have the forward bias condition taken away. This means the voltage must drop below the forward voltage drop of the SCR so that the current drops below a minimum value, called the holding current, or the polarity from anode to cathode must actually reverse. The fact that the SCR cannot be turned off easily limits its use in dc applications to those cases when some method of reducing the forward current to below the holding value can be provided. In ac circuits, the SCR will automatically turn off in every half cycle when the ac voltage applied to the SCR reverses polarity.

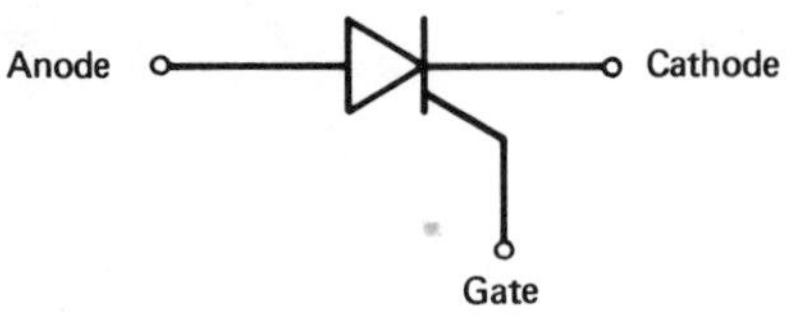

FIGURE 2.24 Symbol for an SCR.

Characteristics and specifications of SCRs are given below:

1. *Maximum Forward Current.* There is a maximum current which the SCR can carry in the forward direction without damage. This value varies from a few hundred milliamps to more than a thousand amps for large industrial types.

2. *Peak Reverse Voltage.* Like a diode there is a peak reverse bias voltage which can be applied to the SCR without damage. The value varies from a few volts to several thousand volts.

3. *Trigger Voltage.* The minimum gate voltage to drive the SCR into conduction varies between types and sizes from a few volts to 40 volts.

4. *Trigger Current.* There is a minimum current which the source of trigger voltage must be able to provide before the SCR can be fired. This varies from a few milliamps to several hundred milliamps.

5. *Holding Current.* This refers to the minimum anode to cathode current necessary to keep the SCR conducting in the forward conducting state. The value varies from 20 to 100 mA.

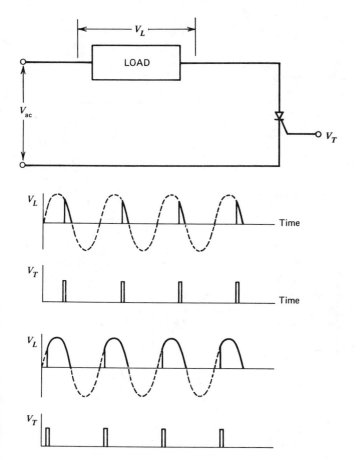

FIGURE 2.25   Half-wave SCR operation. Changing time of V_T application changes the dc rms voltage applied to the load, V_L.

AC OPERATION

Figure 2.25 illustrates the operation of an SCR in varying the rms dc voltage in half-wave operation. The trigger voltage is developed by some circuit which produces a pulse at a certain selected phase of the applied ac signal. Thus, the SCR turns on in a repetitive fashion as shown. The SCR is turned back off, of course, in each half cycle when the ac polarity reverses. Note that by changing the part of the positive half cycle when the trigger is applied, the effective (rms) value of dc voltage applied to the load can be increased. Of course, with this circuit the maximum possible rms dc voltage is that which would be developed by a half-wave rectifier. If more power is required, the SCR can be used in a full-wave bridge type of circuit. Figure 2.26 shows this type of circuit and the voltages versus time which result. The trigger voltages must now be generated in each half cycle and applied to the corresponding SCR trigger (gate) terminal. In a process-control application, the control-

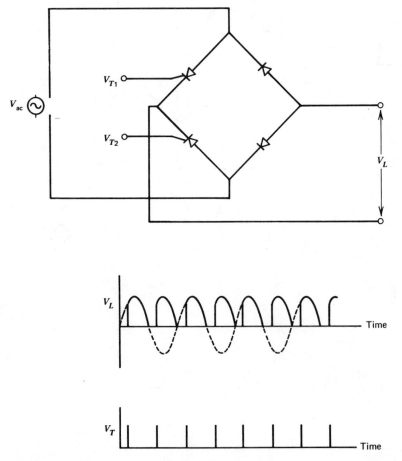

FIGURE 2.26   Full-wave SCR circuit. The effective rms dc voltage applied to the load is increased because both cycles of the ac are used.

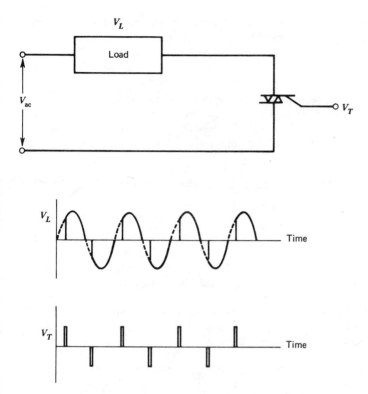

FIGURE 2.27   The TRIAC can conduct in both directions so that the load voltage remains ac, but the rms value is determined by the time at which the trigger voltages are applied.

ler output signal would be used to drive a circuit which changed the time at which the pulses were applied to the gates and thus changed the power applied to the load. Note that the voltage applied to the load is pulsating dc. This configuration could not be used with a load which required ac voltage for operation.

2.6.2 TRIAC

An extension of the SCR discussed in the previous section is a device which can be triggered to conduct in either direction. The TRIAC can be thought of as two SCRs connected in parallel and reversed but with the gates connected. A positive trigger will cause it to conduct in one direction, and a negative trigger will cause it to conduct in the other direction. The TRIAC thus can be used in pure ac applications. Figure 2.27 shows the symbol of a TRIAC and a circuit for a typical application. Note that the voltage across the load, as shown, remains ac. The effective ac rms value of voltage applied to the load can be changed by changing the time in the phase of the cycles when the TRIAC gate is pulsed. The trigger voltage generated must be bipolar, one pulse in one polarity and the next of the opposite polarity.

Specifications of TRIACs are similar to those of SCRs; maximum rms current, peak reverse voltage, trigger voltage, and trigger current.

SUMMARY

The signal conditioning discussed in this chapter relates to the standard techniques employed for providing signal compatibility and measurement in analog systems. The reader was introduced to the basic concepts that form the foundation of such analog conditioning.

To present a complete picture of analog signal conditioning, the following points were considered:

1.   The need for analog signal conditioning was reviewed and resolved into the requirements of signal-level changes, linearization, signal conversions, and filtering and impedance matching.

2.   Bridge circuits are a common example of a conversion process where a changing resistance is measured either by a current or by a voltage signal. Many modifications of the bridge are used, including *electronic balancing* and techniques of *lead compensation*.

3.   The potentiometer circuit has been a standard of accurate high-impedance voltage measurement for many years. The basic principle is *potential matching,* using a calibrated resistance voltage divider.

4.   Operational amplifiers (op amps) are a very special signal conditioning building block around which many special function circuits can be developed. The device was demonstrated in applications involving amplifiers, converters, linearization circuits, integrators, and several other functions.

5.   Silicon controlled rectifiers (SCRs) and TRIACs are semiconductor devices, similar to diodes, which can control large energy ac or dc signals using low-level inputs.

PROBLEMS

2.1   A Wheatstone bridge, as shown in Figure 2.3, nulls with $R_1 = 227\ \Omega$, $R_2 = 448\ \Omega$, and $R_3 = 1414\ \Omega$. Find R_4.

2.2   A bridge circuit used for a nominally 50 Ω transducer must have a resolution of 0.1 Ω in resistance measurement. If we use $R_1 = R_2 = 100\ \Omega$, $V = 10$ V, and let $R_3 = 100\ \Omega$ precision pot, then what voltage resolution must the null detector have?

2.3   An a-c Wheatstone bridge with the impedance arms as capacitors nulls when C_1 is 0.4 μF, C_2 is 0.31 μF, and $C_3 = 0.27\ \mu$F. Find C_4.

2.4   Design a potentiometer which uses a 9-V working cell, a 1.35002-V reference cell, and measures 0–100 mV, using a 1 kΩ linear pot.

2.5   A transducer must be loaded by a 1.5 kΩ resistance and a gain of $+100$ is required. Design an op amp configuration to provide this specification.

2.6   Specify the components of a differential op amp circuit with a gain of 22.

2.7   Using an integrator with $RC = 10$ s and a following amplifier, specify an op amp voltage-ramp generator with an output of 0.5 V/s.

2.8   A control system needs the average temperature from three locations. Temperature is available as voltages V_1, V_2, and V_3 for the three sources. Design a 2-op amp circuit which outputs the average voltage.

2.9   For a voltage to current converter, such as Figure 2.18, find resistors so that $I = (2.1 \times 10^{-3})V_{in}$. If the op amp saturates at ± 10 V and the circuit should deliver 5 mA maximum, find the maximum load resistance.

(Many solutions are possible; one is: $R_1 = R_3 = 1$ kΩ, $R_2 = 2.1$ kΩ, $R_5 = 500$ Ω, $R_4 = 714.3$ Ω, $R_{ML} = 548.4$ Ω.)

2.10 A current balance bridge, such as shown in Figure 2.6, has resistances of $R_1 = R_2 = 1$ kΩ, $R_4 = 590$ Ω, $R_5 = 10$ Ω with a 10-V supply. (a) Find the value of R_3 which balances the bridge with no current. (b) Find the value of R_3 which balances the bridge with 0.25 mA of current.

2.11 Prove that Equation (2-8) results from Equation (2-7).

2.12 Derive Equation (2-11).

2.13 Assume some load resistance R_L and then show that Equation (2-32) results from an analysis of the voltage to current converter, assuming Equation (2-33) is satisfied.

2.14 Use an inverting amplifier, an integrator, and a summing amplifier to construct an output voltage given by

$$V_{out} = 10V_{in} + 4\int V_{in}dt$$

2.15 A bridge circuit is to be used with a transducer located 100 meters away. The lead wire connected to the transducer has a resistance of 0.45 Ω/ft and a lead compensated circuit is *not* used. If the bridge nulls with $R_1 = 3400$ Ω, $R_2 = 3445$ Ω, and $R_3 = 1560$ Ω, what is the value of R_4? (See Figure 2.3 for designations.)

2.16 A bridge such as that in Figure 2.3 has $R_1 = R_2 = R_3 = 120.0$ Ω. Design a circuit using op amps that provides approximately -5.0 to $+5.0$ volt output as R_4 varies from 100 to 140 Ω. Is the output voltage linear with respect to R_4? Plot voltage versus R_4. Use a voltage source of 10 volts.

2.17 Figure 2.28 shows an SCR used in a circuit to provide variable electrical power to a resistive load. The SCR triggers when the voltage on the gate is $+2.5$ volts. Plot the voltage across the capacitor as a function of time and the voltage across the load on the same graph. (Note: The SCR gate goes to ground when the SCR is triggered.)

2.18 A process signal is in the 4 to 20-mA representation. The set point is 9.5 mA. Use a current to voltage generator and a summing amplifier to get an error signal with a scale factor of 0.5 volts per 1.0 mA error.

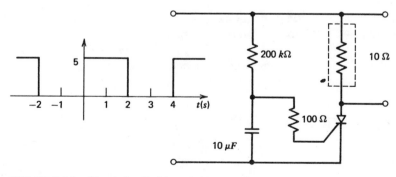

FIGURE 2.28 Circuit for Problem 2.18.

CHAPTER

3

DIGITAL
SIGNAL
CONDITIONING

INSTRUCTIONAL OBJECTIVES

In this chapter, the basic principles of digital electronics and digital signal processing are considered with particular emphasis on digital-to-analog (D/A) and analog-to-digital (A/D) conversion techniques, and data acquisition systems. After you have read it, you should be able to:

1. Develop a Boolean equation for a simple process control alarm problem.
2. Implement a process-control design of an alarm by digital circuits and comparators.
3. Define the representation of fractional binary and decimal numbers.
4. Diagram a basic DAC and describe its operation.
5. Diagram a successive approximation ADC and describe its operation.
6. Define the conversion resolution of ADCs and DACs.
7. Design a data acquisition system.

3.1 INTRODUCTION

Perhaps the best way to start this chapter is to briefly consider why we are interested in *digital* signal conditioning. An overall survey of electronics applications in industry shows that conversions to digital techniques are occurring rapidly. There are many reasons for this conversion, but two in particular are important. One is the reduction in *uncertainty* when dealing with digitally encoded information over analog information. We did not say *accuracy*, we said *uncertainty*. If a system presents analog information, great care must be taken to account for electrical noise influence, drift of amplifier gains, loading effects, and a host of other problems familiar to the analog electronics designer. In a digitally encoded signal, however, a wire carries either a high or low level and is not particularly susceptible to the above problems associated with analog processing. Thus, there is an inherent certainty in representing information by digitally encoding due to the *isolation* of digital representation from spurious influences. The *accuracy* of this signal in representing the information is a separate matter discussed later in this chapter (Section 3.3).

A second reason for conversion to digital electronics is the growing desire to use digital computers in the industrial process. The digital computer by nature requires information encoded in digital format before it can be used. The question of the need for digital signal conditioning becomes a question of why computers are so widely used in industry. This is indeed a very complex story and volumes can be written in reply. To mention a few reasons we can note, as discussed in Chapter 10: (1) the ease with which a computer controls a multivariable process-control system, (2) through computer programming, nonlinearities in a transducer output can be linearized, (3) complicated control equations can be solved to determine required control functions, and (4) the ability to microminiaturize rather complex digital processing circuits as integrated circuits (ICs). Indeed, with the development of the microprocessor chip, to be discussed in Chapter 10, an entire computer can now be implemented on one printed circuit (PC) board. This technology not only reduces physical size but also both power consumption and failure rate.

With growing use of computers in process-control technology, it is now clear that any individual trained to work in this field must also be versed in the technology of digital electronics. The basic question is how far such preparation should extend into this related complex field of study. The answer is that a process-control technologist must understand the elements and characteristics of process-control loops. In this context, digital electronics is used as a tool to implement necessary features of process control and therefore should be understood to the extent of knowing how such devices affect the characteristics of the loop. Consider that one does not need to know detailed physics of stretched wires to understand the application of strain gages in order to use these devices successfully in process control. Similarly, one need not know the internal design of logic gates and microcomputers to use such devices in process control. In this light, the objectives of this chapter have been carefully chosen to provide the reader with sufficient background in digital technology to understand its application to process control.

3.2 DIGITAL FUNDAMENTALS

A working understanding of the application of digital techniques to process control requires a foundation of basic digital electronics. The design and implementation of control logic systems and microcomputer control systems require a depth of understanding which can only be obtained after several courses devoted to the subject. In this text, we wish to provide a sufficient background so that the reader can appreciate the essential features of digital electronic design and its application to process control.

3.2.1 Digital Information

The use of digital techniques in process control requires that dynamic variable measurements and control information be encoded into a digital form. Digital signals themselves are simply two-state (binary) levels of voltage on a wire as discussed in Section 1.5.2. We speak then of the digital information as a high state (H or 1) or a low state (L or 0) on a wire that carries the digital signal.

DIGITAL WORD

Given the simple binary information which is carried by a digital signal, it is clear that a more complicated arrangement must be used to describe analog information. Generally, this is done by using an assemblage of digital levels to construct a *word*. The individual digital levels are referred to as *bits* of the word. Thus, for example, a 6-bit word consists of 6 independent digital levels as **101011** which can be thought of as a 6-digit base two number. An important consideration then is how the process-control information is encoded into this digital word.

DECIMAL WHOLE NUMBERS

One of the most common schemes for encoding analog data into a digital word is to use the straight counting of decimal (or base 10) and binary representations. The principles of this are reviewed in Appendix A-2 together with octal and hexadecimal representations.

EXAMPLE 3.1

Find the base 10 equivalent of the binary number **00100111**.

SOLUTION

Note that as in the base 10 system, zeros preceding the first significant digit do not contribute. Thus, the binary number is actually **100111** and so $n = 5$. To find the decimal equivalent we use Appendix A-2 and

$$
\begin{aligned}
N_{10} &= a_5 2^5 + a_4 2^4 + \cdots a_1 2^1 + a_0 2^0 \\
N_{10} &= (1)2^5 + (0)2^4 + (0)2^3 + (1)2^2 + (1)2^1 + (1)2^0 \\
N_{10} &= 32 + 4 + 2 + 1 \\
N_{10} &= 39
\end{aligned}
\tag{A2-1}
$$

EXAMPLE 3.2

Find the binary equivalent of the base 10 number 47.

SOLUTION

Starting the successive division, we get:

$$\tfrac{47}{2} = 23 \text{ with a remainder of } \tfrac{1}{2} \text{ so that } a_0 = 1,$$

then

$$\tfrac{23}{2} = 11 \text{ with a remainder of } \tfrac{1}{2} \text{ so that } a_1 = 1$$

then

$$\tfrac{11}{2} = 5 + \tfrac{1}{2} \therefore a_2 = 1$$
$$\tfrac{5}{2} = 2 + \tfrac{1}{2} \therefore a_3 = 1$$
$$\tfrac{2}{2} = 1 + 0 \therefore a_4 = 0$$
$$\tfrac{1}{2} = 0 + \tfrac{1}{2} \therefore a_5 = 1$$

We find then that base 10 number 47 becomes a binary number 101111_2.

The representation of negative numbers in binary format takes on several forms as discussed in Appendix A-2.

OCTAL AND HEX NUMBERS

It is quite cumbersome for humans to work with digital words expressed as numbers in the binary representation. For this reason, it has become common to use either the octal (base 8) or hexadecimal (base 16, called hex) representations which are reviewed in Appendix A-2. Octal numbers are conveniently formed from groupings of three binary digits, since 000_2 is 0_8 and 111_2 is 7_8. Thus, a binary number like 101011_2 is equivalent to 53_8. Hex numbers are formed easily from groupings of four binary digits, since 0000_2 is 0h and 1111_2 is FFh. Note that the small h is used to designate a hex number instead of a subscript of 16. Also recall the hex counting sequence is 0, 1, 2, 3, 4, 5, 6, 7, 8, 9, A, B, C, D, E, and F to cover the possible states. Since microcomputers most frequently use either 4-bit, 8-bit, or 16-bit words, the hex notation is very commonly used with these machines. In hex, a binary number like 10110110_2 would be written B6h.

3.2.2 Fractional Binary Numbers

Although not as commonly used, it is possible to define a fractional binary number in the same manner as whole numbers using only the 1 and 0 of this counting system. Such numbers, just as in the decimal framework, represent divisions of the counting system to values less than unity. A correlation can be made to decimal numbers in a similar fashion to Equation (A2-1), as

$$N_{10} = b_1 2^{-1} + b_2 2^{-2} + \cdots b_m 2^{-m} \tag{3-1}$$

Where

$$N_{10} = \text{base 10 number less than one}$$
$$b_1 b_2 \cdots b_{m-1} b_m = \text{base 2 number less than one}$$
$$m = \text{number of digits in base 2 number}$$

EXAMPLE 3.3

Find the base 10 equivalent of the binary number 0.11010_2.

SOLUTION

This can be found most easily by using

$$N_{10} = b_1 2^{-1} + b_2 2^{-2} + \cdots b_m 2^m \qquad (3\text{-}1)$$

with

$m = 5$
$N_{10} = (1)2^{-1} + (1)2^{-2} + (0)2^{-3} + (1)2^{-4} + (0)2^{-5}$
$N_{10} = \frac{1}{2} + \frac{1}{4} + \frac{1}{16}$
$N_{10} = 0.8125_{10}$

Conversion of a base 10 number which is less than 1 to a binary equivalent employs a procedure where repeated multiplication by 2 is performed. The result of each multiplication will be a fractional part and either a 0 or 1 whole number part, which determines whether that digit is a 0 or 1. The first multiplication gives the most significant bit b_1, and the last gives either a 0 or 1 for the least significant bit b_m.

EXAMPLE 3.4

Find the binary and octal equivalent of 0.3125_{10}.

SOLUTION

Using successive multiplication, we find:

$$
\begin{array}{ll}
2(0.3125) = 0.6250 & \text{so } b_1 = 0 \\
2(0.625) = 1.250 & \text{so } b_2 = 1 \\
2(0.025) = 0.5 & \text{so } b_3 = 0 \\
2(0.5) = 1.0 & \text{so } b_4 = 1
\end{array}
$$

Thus, we find that 0.3125_{10} is equivalent to 0.0101_2. Note that it can be represented as 0.010100_2 because trailing zeros are not significant in a number less than one and thus as 0.24 octal, since $010_2 = 2_8$ and $100_2 = 4_8$.

3.2.3   Boolean Algebra

In process control, as well as in many other technical disciplines, some action can be taken on the basis of an evaluation of observations made in the environment. In driving an automobile, for example, we are constantly observing such external factors as traffic, lights, speed limits, pedestrians, street conditions, low flying aircraft, and such internal factors as how fast we wish to go, where we are going, and many others. We evaluate these factors and take actions predicated on the evaluations. We may see that the light is green, streets good, speed low, no pedestrians or aircraft, we are late and thus conclude that an action of pressing on the accelerator is required. Then we may observe a parked police radar unit with all other factors the same, negate the above conclusion, and apply the brake. Note that many of these parameters can be represented by a *true* or *not true* observation and, in fact, with enough definition, all the observations could be reduced to simple *true* or *false* conditions. When we learn to drive, we are actually setting up internal response to a set of such true/false observations in the environment. In the industrial world, an analogous condition exists relative to the external and internal influences on a manufacturing process, and when we control a process we are in effect teaching a control system response to a set of true/false observations. This teaching may consist of designing electronic circuits which can *logically* evaluate the set of true/false conditions and initiate some appropriate action. To design such an electronic system, we must first be able to mathematically express the inputs, the logical evaluation, and the corresponding outputs. *Boolean algebra* is a mathematical procedure which allows the combinations of true/false conditions in various logical operations so that conclusions can be drawn. For purposes of this text, we do not require expertise in Boolean technique but rather an operational familiarity with it to apply it to a process-control environment.

Before a particular problem in industry can be implemented using digital electronics, it must be analyzed in terms that are amendable to the binary nature of digital techniques. Generally, this is accomplished by stating the problem in the form of a set of true/false-type conditions which must be applied to derive some desired result. These sets of conditions then are stated in the form of one or more Boolean equations. We will see in the next section that a Boolean equation is in a form that is readily implemented with existing digital circuits. The mathematical approach of Boolean algebra allows us to write an analytical expression to represent these stipulations. The fundamentals of Boolean algebra are summarized in Appendix A-2.

Let us consider a simple example of how a Boolean equation may result from a practical problem. Consider a mixing tank for which there are three variables of interest: liquid level, pressure, and temperature. The problem is that we must signal an alarm when certain combinations of conditions occur between these variables. Referring to Figure 3.1, we denote level by A, pressure by B, and temperature by C, and assume that breakpoint values have been assigned for each so that the Boolean variables are either 1 or 0 as the physical quantities are above or below the breakpoint values. The alarm will be triggered when the Boolean variable D goes to the logic true state. The alarm conditions are:

1. Low level with high pressure.
2. High level with high temperature.
3. High level with low temperature and high pressure.

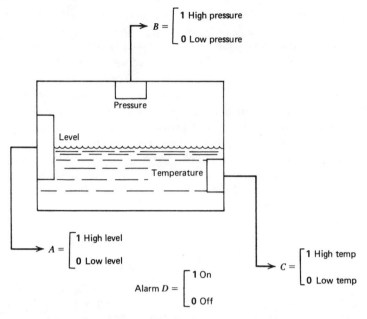

FIGURE 3.1   Conditions for the application example of Section 3.2.3.

We now define a Boolean expression that will give a $D = 1$ for each condition

1. $D = \bar{A} \cdot B$ will give $D = 1$ for condition (1).
2. $D = A \cdot C$ will give $D = 1$ for condition (2).
3. $D = A \cdot \bar{C} \cdot B$ will give $D = 1$ for condition (3).

The final logic equation results by combining all three conditions so that if any are true, the alarm will sound ($D = 1$). This is accomplished with the OR operation

$$D = \bar{A} \cdot B + A \cdot C + A \cdot \bar{C} \cdot B \tag{3-2}$$

This equation would now form the starting point for design of electronic digital circuitry which would perform the indicated operations.

3.2.4 Digital Electronics

The electronic building blocks of digital electronics are designed to operate on the binary levels present on digital signal lines. These building blocks are based on families of types of electronic circuits, as discussed in Appendix A-2, which have their specific stipulations of power supplies and voltage levels of the 1 and 0 states. The basic structure involves the use of AND/OR logic and NAND/NOR logic to implement Boolean equations.

EXAMPLE 3.5

Develop a digital circuit using AND/OR gates which implements the equation developed in Section 3.2.3.

SOLUTION

The problem posed in Section 3.2.3 (with Figure 3.1) has a Boolean equation solution of

$$D = \bar{A} \cdot B + A \cdot C + A \cdot \bar{C} \cdot B \tag{3-2}$$

The implementation of this equation using AND/OR gates is shown in Figure 3.2. Note that the AND, OR, and inverter are used in a straightforward implementation of the equation.

EXAMPLE 3.6

Repeat Example 3.5 using NAND/NOR gates.

SOLUTION

One way to implement the equation in NAND/NOR would be to provide inverters after every gate to, in effect, convert the devices back to AND/OR gates. In this

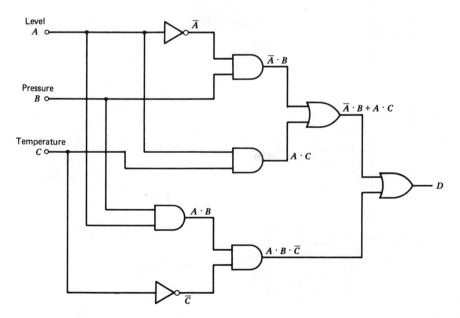

FIGURE 3.2   Solution for Example 3.5.

case, the circuit developed would look like Figure 3.2 but with an inverter after every gate. A second approach is to use the Boolean theorems to *reformulate* the equation for better implementation using NAND/NOR logic. For example, if we are to get the desired equation for D as output from a NAND gate, then the inputs must have been

$$\overline{\bar{A} \cdot B + A \cdot C}$$

and

$$\overline{A \cdot B \cdot \bar{C}}$$

because NAND between these produces

$$\overline{(\overline{\bar{A} \cdot B + A \cdot C}) \cdot (\overline{A \cdot B \cdot \bar{C}})}$$

which, by DeMorgan's theorem, becomes

$$\bar{A} \cdot B + A \cdot C + A \cdot B \cdot \bar{C}$$

that is, the desired output. Working backward from this result allows the circuit to be realized as shown in Figure 3.3. Many other correct configurations are possible.

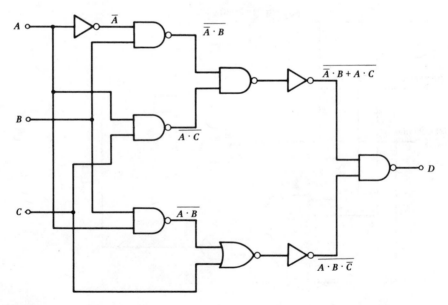

FIGURE 3.3   Solution for Example 3.6.

3.3 CONVERTERS

The most important digital tool for the process-control technologist is one that translates digital information to analog and vice versa. Most measurements of dynamic variables are performed by devices that translate information about the variable to an analog electrical signal. To interface this signal with a computer or digital logic circuit, it is necessary first to perform an analog to digital (A/D) conversion. The specifics of this conversion must be well known so that a unique, known relationship exists between the analog and digital signals. Often, the reverse situation occurs where a digital signal is required to drive an analog device. In this case, a digital to analog (D/A) converter is required.

3.3.1 Comparators

The most elementary form of communication between the analog and digital representatives is a device (usually an IC) called a *comparator*. This device, which is shown schematically in Figure 3.4, simply compares two analog voltages on its input terminals. Depending on which voltage is larger, the output will be a 1 (high) or 0 (low) digital signal. This comparator is extensively used for alarm signals to computers or digital processing systems. This element also is an integral part of the analog to digital and digital to analog converter to be discussed next.

A comparator can be constructed from an op amp provided the output is properly clamped to provide the required levels for the logic states (as +5 and 0 for TTL 1 and 0). Commercial comparators are designed to have the necessary logic levels on the output.

EXAMPLE 3.7

A process-control system specifies that temperature should never exceed 160°C if the pressure also exceeds 10 N/m² (Pa). Design an alarm system to detect this condition, using temperature and pressure transducers with transfer functions of 2.2 mV/°C and 0.2 V/N/m², respectively.

SOLUTION

The alarm conditions will be a temperature signal of $(2.2 \text{ mV/°C})(160°C) = 3.52$ V coincident with a pressure signal of $(0.2 \text{ V/Nim}^2)(10 \text{ N/m}^2) = 2$ volts. The circuit of

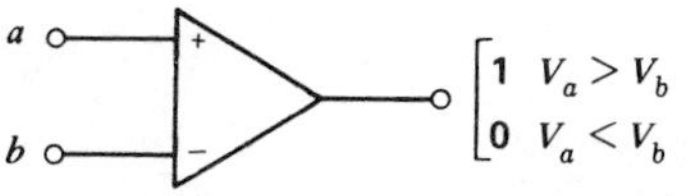

FIGURE 3.4 A comparator changes output logic state as a function of analog input voltages.

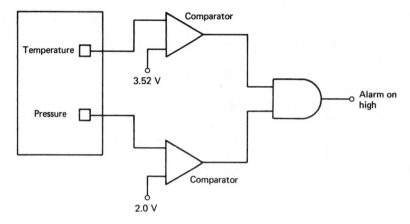

FIGURE 3.5   Diagram of circuit for Example 3.7.

Figure 3.5 shows how this alarm can be implemented with comparators and one AND gate.

3.3.2 Digital to Analog Converters (DACs)

A DAC accepts digital information and transforms it into an analog voltage. The digital information is in the form of a binary number with some fixed number of digits. Especially when used in connection with a computer, this binary number is called a binary *word* or *computer* word. The *digits* are called *bits* of the word. Thus, an 8-bit word would be a binary number having eight digits, such as 10110110_2. The D/A converter converts a digital word into an analog voltage by scaling the analog output to be zero when all bits are zero and some maximum value when all bits are one. This can be mathematically represented by treating the binary number which the word represents a *fractional* number. In this context, the output of D/A converter can be defined using Equation (3-1) as a *scaling* of some reference voltage.

$$V_x = V_R[b_1 2^{-1} + b_2 2^{-2} + \cdots + b_n 2^{-n}] \tag{3-3}$$

Where

$$V_x = \text{analog voltage output}$$
$$V_R = \text{reference voltage}$$
$$b_1 b_2 \cdots b_n = n\text{-bit binary word}$$

Note that the minimum V_x is zero, and the maximum is determined by the size of the binary word because, with all bits set to one, the decimal equivalent *approaches* V_R as the number of bits increases. Thus, a 4-bit word has a maximum of

$$V_{max} = V_R[2^{-1} + 2^{-2} + 2^{-3} + 2^{-4}] = 0.9375\ V_R$$

whereas an 8-bit word has a maximum of

$$V_{max} = V_R[2^{-1} + 2^{-2} + 2^{-3} + 2^{-4} + 2^{-5} + 2^{-6} + 2^{-7} + 2^{-8}] = 0.9961\ V_R$$

CONVERSION RESOLUTION

The conversion resolution is also a function of the *number* of bits in the word. The *more* bits the smaller change in analog output for 1-bit change in binary word and hence, the *greater* the resolution. The smallest possible change is simply given by

$$\Delta V_x = V_R 2^{-n} \qquad (3\text{-}4)$$

Where

ΔV_x = smallest output change
V_R = reference voltage
n = number of bits in the word

Thus, a 5-bit word D/A converter with a 10-volt reference will provide changes of $\Delta V_x = (10)\ (2^{-5}) = 0.3125$ volts per bit.

EXAMPLE 3.8

Determine how many bits a D/A converter must have to provide output increments of 0.04 volts or less. The reference is 10 volts.

SOLUTION

One way to find the solution would be to continually try word sizes until the resolution had fallen below 0.04 volts per bit. A more analytical procedure is to form the equation

$$\Delta V = 0.04 = (10)\ (2^{-\nu})$$

Any n larger than the integer part of the exponent of two in this equation will satisfy the requirement. Taking logarithms

$$\log (0.04) = \log ((10)\ (2^{-\nu}))$$
$$\log (0.04) = \log (10) - y \log 2$$

$$y = \frac{\log (10) - \log (0.04)}{\log 2}$$

$$y = 7.966$$

Thus, an $n = 8$ will be satisfactory. This can be proved by Equation (3-4).

$$\Delta V_x = (10)\,(2^{-8})$$
$$\Delta V_x = 0.0390625 \text{ volts}$$

DAC CHARACTERISTICS

For modern applications most DACs are integrated circuit (IC) assemblies, viewed as a black box having certain input and output characteristics. In Figure 3.6, we see the essential elements of the DAC in terms of required input and output. The associated characteristics can be summarized by reference to this figure.

1. *Digital Input.* Typically, a parallel binary word of a number of bits specified by the device specification sheet. Typically, TTL logic levels are required unless otherwise noted.
2. *Power Supply.* This is bipolar at a level of ±12 V to ±18 V as required for internal amplifiers.
3. *Reference Supply.* Required to establish the range of output voltages and resolution of the converter. This must be a stable, low-ripple source. In some units, an internal reference is provided.
4. *Output.* A voltage representing the digital input. This voltage changes in steps as the digital input changes by bits with the step determined by Equation (3-4). The actual output may be bipolar if the converter is designed to interpret negative digital inputs.
5. *Offset.* Because the DAC is usually implemented with op amps, there may be the typical

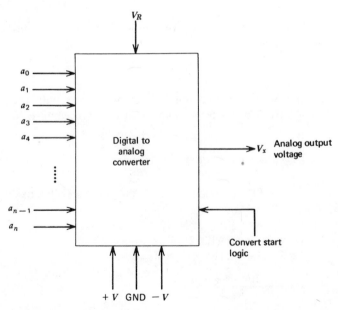

FIGURE 3.6 Diagram showing the inputs and outputs of an *n*-bit digital to analog (D/A) converter.

output offset voltage with a zero input (see Section 2.4.2). Typically, connections will be provided to facilitate a zeroing of the DAC output with a zero word input.

6. *Convert Start.* Some DAC circuits provide a logic input which prevents the conversion from occurring until a specified logic command (**1** or **0**) is received. In these cases, the input word is ignored until the proper logic input occurs.

In some cases, an input buffer is provided to hold the digital word while the conversion occurs and after, even though the word itself may appear on the input lines only briefly. These buffers are usually flip-flops (FFs) which are inserted between the input terminals of the converter and the digital lines.

DAC STRUCTURE

Generally speaking, a DAC is used as a black box and no knowledge of the internal workings is required. There is some value, however, to briefly show how such conversion can be implemented. The simplest conversion uses a series of op amps for input for which the gains have been selected to provide an output as given by

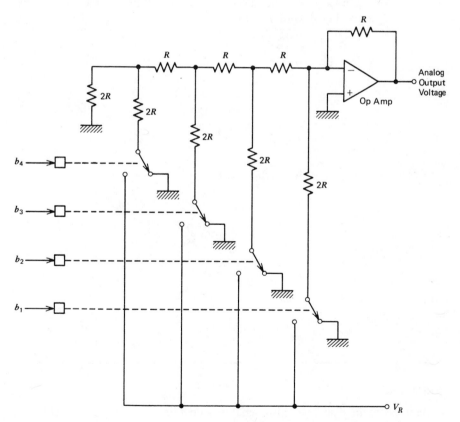

FIGURE 3.7   The ladder network is an example of a common circuit for a D/A converter.

Equation (3-3). The most common variety, however, uses a *resistive ladder network* to provide the transfer function. This is shown in Figure 3.7 for the case of a 4-bit converter. With the R-2R choice of resistors, it can be shown through network analysis that the output voltage is given by Equation (3-4). The switches are analog electronic switches.

EXAMPLE 3.9

A control valve has a linear variation of opening as the input voltage varies from 0–10 volts. A microcomputer outputs an 8-bit word to control valve opening using an 8-bit DAC to generate the valve voltage. (a) Find the reference voltage required to obtain a full open valve (10 volts); (b) Find the percentage of valve opening for a 1-bit change in the input word.

SOLUTION

(a) The full open-valve condition occurs with a 10-volt input. If a 10-volt reference is used, a full digital word **11111111** will not quite give 10 volts so we use a larger reference. Thus, we have

$$V_x = V_R(b_1 2^{-1} + b_2 2^{-2} + \cdots + b_8 2^{-8})$$

$$10 = V_R \left(\frac{1}{2} + \frac{1}{4} + \cdots + \frac{1}{256} \right) \tag{3-3}$$

$$V_R = \frac{10}{0.9961} = 10.0039 \text{ V}$$

(b) The percentage of valve change per step is found first from

$$\Delta V_x = V_R 2^{-8}$$

$$\Delta V_x = (10.0039) \frac{1}{256}$$

$$\Delta V_x = 0.0392 \text{ V}$$

Thus,

$$\text{Percent} = \frac{(0.0392)(100)}{10} = 0.392\%$$

SERIAL DACs

In some cases, the digital word is a serial type on a line instead of a parallel bit. In these cases, either a serial-to-parallel converter with a buffer output or some form of serial converter is required.

3.3.3 Analog to Digital Converters (ADCs)

Although many transducers that provide a direct digital signal output exist and are being developed, most transducers still convert the dynamic variable into an analog electrical signal. With the growing use of digital logic and computers in process control, it is necessary to employ an ADC to provide a digitally encoded output. The transfer function of the ADC can be expressed in the same manner as Equation (3-3) in that some analog voltage is provided as input, and the converter finds a binary number which, when substituted into Equation (3-3), gives the analog input. Thus,

$$V_x \simeq V_R[b_1 2^{-1} + b_2 2^{-2} + \cdots + b_n 2^{-n}] \qquad (3\text{-}5)$$

Where

$$V_x = \text{analog voltage input}$$
$$V_R = \text{reference voltage}$$
$$b_1 b_2 \cdots b_n = n\text{-bit digital outputs}$$

We use an *approximate equality* in this equation because the voltage on the right can change by a finite step size given by Equation (3-4),

$$\Delta V = V_R 2^{-n} \qquad (3\text{-}4)$$

This means that there is an inherent uncertainty of ΔV in any conversion of analog voltage to digital signal. This uncertainty must be taken into account in design applications. If the problem under consideration specifies a certain resolution in analog voltage, then the word size and reference must be selected to provide this in the converted digital number.

EXAMPLE 3.10

Temperature is to be measured by a transducer with an output of 0.02 volts/°C. Determine the required ADC reference and word size to measure 0–100°C with 0.1°C resolution.

SOLUTION

At the maximum temperature of 100°C, the voltage output is

$$(0.02 \text{ V/°C}) (100\text{°C}) = 2 \text{ V}$$

so a 2 V reference is used.
A change of 0.1°C results in a voltage change of

$$(0.1\text{°C}) (0.02 \text{ V/°C}) = 2 \text{ mV}$$

so we need a word size where

$$0.002 \text{ V} = (2)(2^{-y})$$

Choose a size n which is one integer part of y. Thus, solving with logarithms, we find

$$y = \frac{\log(2) - \log(0.002)}{\log 2}$$

$$y = 9.996 \approx 10$$

so a **10**-bit word is required for this resolution. A 10-bit word has a resolution of

$$V = (2)(2^{-10})$$
$$V = 0.00195 \text{ volts}$$

which is less than the minimum required resolution of 2 mV

EXAMPLE 3.11

Find the digital word which results from a 3.217 volt input to a 5-bit ADC with a 5 volt reference.

SOLUTION

The relationship between input and output is

$$V_x \approx V_R[a_1 2^{-1} + a_2 2^{-2} + a_3 2^{-3} + a_4 2^{-4} + a_5 2^{-5}] \qquad (3\text{-}5)$$

Thus, we are to encode a fractional number of V_x/V_R or

$$a_1 2^{-1} + a_2 2^{-2} \cdots + a_5 2^{-5} = \frac{3.127}{5} = 0.6254$$

Using the method of successive multiplication defined in Section 3.2.2, we find

$$0.6254\,(2) = 1.2508 \therefore a_1 = 1$$
$$0.2508\,(2) = 0.5016 \therefore a_4 = 0$$
$$0.5016\,(2) = 1.0032 \therefore a_3 = 1$$
$$0.0032\,(2) = 0.0064 \therefore a_4 = 0$$
$$0.0064\,(2) = 0.0128 \therefore a_5 = 0$$

So that the output is **10100$_2$**.

A/D STRUCTURE

Most ADCs are available in the form of integrated circuit (IC) assemblies which can be used as a black box in applications. To fully appreciate the characteristics of these devices, however, it is valuable to examine the standard techniques employed to perform the conversions. There are two methods in use to perform the conversion which represent very different approaches to the conversion problem.

PARALLEL-FEEDBACK ADC

The parallel-feedback A/D converter employs a feedback system to perform the conversion as shown in Figure 3.8. Essentially, a *comparator* is used to compare the input voltage V_x to a feedback voltage V_F which comes from a DAC as shown. The comparator output signal drives a logic network which steps the digital output (and hence DAC input) until the comparator indicates the two signals are the *same* within the resolution of the converter. The most popular parallel-feedback converter is the *successive approximation* device. Here, the logic circuitry is such that it

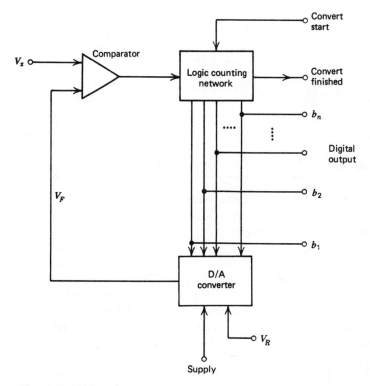

FIGURE 3.8   The successive approximation type A/D converter is very common and involves the use of a D/A converter.

successively sets and tests each bit, starting with the most significant bit of the word. We start with all bits zero. Thus, the first operation will be to set $b_1 = 1$ and test $V_F = V_R 2^{-1}$ against V_x through the comparator.

If V_x is greater, then b_1 will be 1; b_2 is set to 1 and a test is made of V_x versus $V_V = V_R(2^{-1} + 2^{-2})$, and so on.

If V_x is less than $V_R 2^{-1}$, then b_1 is reset to zero; b_2 is set to 1 and a test is made for V_x versus $V_R 2^{-2}$. This process is repeated to the least significant bit of the word. The operation can be illustrated best through an example.

EXAMPLE 3.12

Find the successive approximation ADC output for a 4-bit converter to a 3.217-volt input if the reference is 5 volts.

SOLUTION

Following the procedure outlined, we have the following operations: Let $V_x = 3.217$,

(1) Set $b_1 = 1$ $V_F = 5(2^{-1}) = 2.5$ volts
      $V_x > 2.5$ leave $b_1 = 1$
(2) Set $b_2 = 1$ $V_F = 2.5 + 5(2^{-2}) = 3.75$
      $V_x < 3.75$ reset $b_2 = 0$
(3) Set $b_3 = 1$ $V_F = 2.5 + 5(2^{-3}) = 3.125$
      $V_x > 3.125$ leave $b_3 = 1$
(4) Set $b_4 = 1$ $V_F = 3.125 + 5(2^{-4})$
      $V_x < 3.4375$ reset $b_4 = 0$

By this procedure, we find the output is a binary word of 1010_2.

In addition to the analog input, digital output, power supply, and reference inputs, most A/D converters have a *convert start logic input* and a *conversion complete logic output* as shown in Figure 3.8.

RAMP A/D

The ramp type A/D converters essentially compare the input voltage against a linearly increasing ramp voltage. A binary counter is activated which counts ramp steps until the ramp voltage equals the input. The output of the counter is then the digital word representing conversion of the analog input. The ramp itself is typically generated by an op amp integrator circuit, discussed in Section 2.5.8.

DUAL-SLOPE RAMP A/D

This ADC is the most common type of ramp converter. A simplified diagram of this device is shown in Figure 3.9. The principle of operation is based on allowing the input signal to drive the integrator for a fixed time T_1, thus generating an output of

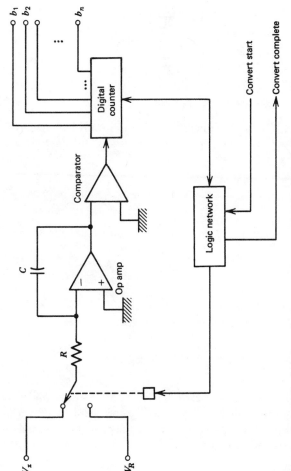

FIGURE 3.9   A dual slope A/D converter uses an op amp integrator, comparator, and associated digital circuits.

$$V_1 = \frac{1}{RC} \int V_x dt \qquad (3\text{-}6)$$

or, because V_x is constant,

$$V_1 = \frac{1}{RC} T_1 V_x \qquad (3\text{-}7)$$

After time T_1, the input of the integrator is electronically switched to the reference supply which is negative. The comparator then sees an input voltage which decreases from V_1 as

$$V_2 = V_1 - \frac{1}{RC} \int V_R dt \qquad (3\text{-}8)$$

or, since V_R is constant and V_1 is given from Equation (3-7),

$$V_2 = \frac{1}{RC} T_1 V_x - \frac{1}{RC} t V_R \qquad (3\text{-}9)$$

A counter is activated at time T_1 and counts until the comparator indicates $V_2 = 0$ at which time, t_x, Equation (3-9), indicates that V_x will be

$$V_x = \frac{t_x}{T_1} V_R \qquad (3\text{-}10)$$

Thus, the counter time t_x is linearly related to V_x and also independent of the integrator characteristics, that is, R and C. This procedure is shown in the timing diagram of Figure 3.10. Conversion *start* and *complete* digital signals are also used in these devices, and (in many cases) internal or external references may be used.

EXAMPLE 3.13

A dual slope ADC as shown in Figure 3.9 has $R = 1 \text{ k}\Omega$ and $C = 0.01 \ \mu\text{F}$. The reference is 10 volts, and the fixed integration time is 10 μs. Find the conversion time for a 6.8-volt input.

SOLUTION

We find the voltage after an integration time of 10 μs as

$$V_1 = \frac{1}{RC} T_1 V_x$$

$$V_1 = \frac{(10 \ \mu s)(6.8 \text{ V})}{(1 \text{ k}\Omega)(0.01 \ \mu F)} \qquad (3\text{-}7)$$

$$V_1 = 6.8 \text{ volts}$$

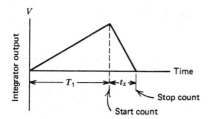

FIGURE 3.10   The dual slope A/D converter counts the time required for a zero crossing of the integrator output from a known, fixed input.

Then, we find the time required to integrate this to zero as $V_2 = 0$ in

$$V_2 = \frac{T_1 V_x}{RC} - \frac{t_x}{RC} V_R \qquad (3\text{-}9)$$

thus,

$$t_x = \frac{T_1 V_x}{V_R}$$

$$t_x = \frac{(10 \ \mu s) \ (6.8 \ V)}{10 \ V}$$

$$t_x = 6.8 \ \mu s$$

The total conversion time is then $10 \ \mu s + 6.8 \ \mu s = \textbf{16.8} \ \mu s$.

GENERAL CHARACTERISTICS

Numerous general features may be indicated for A/D converters, important in applications:

1. *Input*. Generally an analog voltage level. The most common level is 0–10 volts or −10 to +10 if bipolar conversion is possible. In some cases, the level is determined by an externally supplied reference.
2. *Output*. A parallel or serial binary word which is an encoding of the analog input.
3. *Reference*. A stable, low ripple source against which the conversion is performed.
4. *Power Supplies*. Generally, a bipolar ±12 to ±18 V supply is required for the analog amplifiers and comparators and a +5 V supply for the digital circuitry.
5. *Input Sample-and-Hold*. Errors are introduced if the input voltage changes during the conversion process. For this reason, a sample-and-hold amplifier is often employed on the input to provide a fixed voltage input for the conversion process.
6. *Digital Signals*. Most ADCs require an input logic high on a given line to initiate the conversion process. When conversion is complete, the ADC will usually provide a high level on another line as an indicator to following equipment of that status.
7. *Conversion Time*. The ADC must sequence through a set of operations before it can find the digital output as described. For this reason, an important part of the specification is the

time required for the conversion. The time is typically 10–100 μs depending upon the number of bits and the design of the converter.

EXAMPLE 3.14

A measurement of temperature using a transducer which outputs 6.5 mV/°C must measure to 100°C. A 6-bit ADC with a 10-volt reference is used. (a) Develop a circuit to interface the transducer and the ADC; (b) Find the temperature resolution.

SOLUTION

To measure to 100°C means the transducer output at 100°C will be

$$(6.5 \text{ mV/°C})(100°C) = 0.65 \text{ volts}$$

(a) The interface circuit must provide a gain so that at 100°C the ADC output is **1111111**. The input voltage which will provide this output is found from

$$V_x = V_R(a_1 2^{-1} + a_2 2^{-2} + \cdots + + a_6 2^{-6})$$

$$V_x = 10 \left(\frac{1}{2} + \frac{1}{4} + \cdots + \frac{1}{64} \right) \tag{3-5}$$

$$V_x = \textbf{9.84375 V}$$

Thus, the required gain must provide this voltage when the temperature is 100°C.

$$\text{Gain} = \frac{9.84375}{0.65}$$

$$= \textbf{15.14}$$

The op amp circuit of Figure 3.11 will provide this gain. (b) The temperature resolution can be found by working backward from the LSB voltage change of the ADC.

$$\Delta V = V_R 2^{-n}$$
$$\Delta V = (10)(2^{-6}) = 0.15625 \text{ V} \tag{3-4}$$

working back through the amplifier this corresponds to a transducer change of

$$\Delta V_T = \frac{0.15625}{15.14} = 0.01032 \text{ V}$$

or a temperature of

$$\Delta T = \frac{0.01032 \text{ V}}{0.0065 \text{ V/°C}} = \textbf{1.59°C}$$

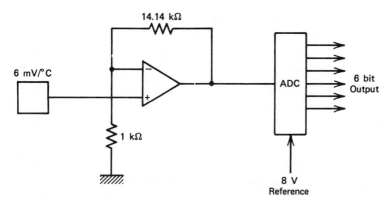

FIGURE 3.11   Figure for Example 3.14.

3.4 DATA ACQUISITION AND OUTPUT SYSTEMS

A digital computer can make a tremendous number of calculations in a second, since the typical time required to execute one instruction may be only a few microseconds. For example, a typical microprocessor can add two 8-bit binary numbers in 2 μsec. By contrast, most process-control installations involve process variable variations with a time scale on the order of minutes. For this reason and others discussed in Chapter 10, the efficient use of computers in process control means that a single computer may control many variables. To do this, the computer periodically *samples* the value of a variable, evaluates it according to programmed control operations, and outputs an appropriate controlling signal to the final control element. Under the program control, the computer then selects another of the controlled variables, samples, evaluates, and outputs, and so on for all the loops under its control. Getting a sample of a real world number into the computer is not easy. It requires a combination of hardware and software (programs) to enable the computer to read in a number which may represent some process variable, such as temperature, pressure, and so on. The overall process of doing this, and its reverse of output, is called by the general term *interface*. Now, one can take an ADC and any necessary amplifiers and write the programs required to put together an interface to some computer for a process application. If the computer is to control many loops, we would need such a system for each variable to be inputted. Instead, for input we can use a Data Acquisition System (DAS) which allows sampled variables from many sources to be inputted to the computer with appropriate programming. Similarly a Data Output Module (DOM) allows the computer to output signals to many sources under program control.

3.4.1 Data Acquisition System (DAS)

There are many different types of data acquisition systems but it is possible to generalize the essential elements as shown in Figure 3.12. The paragraphs below present general descriptions of each block of the DAS. Note that most data acquisi-

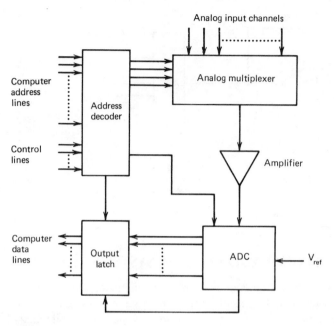

FIGURE 3.12 Data acquisition system.

tion systems are available as small modules containing the circuits shown in Figure 3.12. In general, the module accepts a number of analog inputs, called *channels*, as either differential voltage signals (two-wire) or single-ended voltage signals (referred to ground). Typically, a system may have eight differential input channels or sixteen single-ended input channels. The computer can then select any one of the channels under program control for input of data in that channel.

ADDRESS DECODER

This part of the DAS accepts an input from the computer via the address lines (16 bits for a typical 8-bit microprocessor) which serve to select a particular analog channel to be sampled. The module is often designed so that the association of a particular channel and a computer address word can be selected by the user. In some cases, this is done by making the module channel addresses appear to the computer as addresses of memory locations. Thus, if the computer executes a command to fetch the contents of some memory location, it may actually be selecting some analog input channel. In other words, the selecting of an input channel is equivalent to reading the contents of a memory location. In other systems, a binary code is sent from the computer through special input/output devices to select an analog channel and input the data on that channel. In such cases, the selection of a channel is done by what may be called a *device select* code.

ANALOG MULTIPLEXER

This element of the DAS is essentially a solid state switch which takes the decoded address signal and selects the data for the selected channel by closing a switch connected to that analog input line.

As shown in Figure 3.13 for a single-ended system, the multiplexer accepts an input from the address decoder and uses this to close the appropriate switch to allow that channel signal to be passed to the next stage of the DAS. Figure 3.13 shows that channel 2 has been selected, which would probably have been selected by a **10** on the input lines. In this sense, **00** then would select channel 0, **01** channel 1, **10** channel 2, and **11** channel 3. Thus, the address decoder must convert the computer address line to one of these four possibilities when the DAS has been addressed by the computer. The actual switch elements are usually Field Effect Transistors (FETs) which have an "on" resistance of a few hundred ohms and an "off" resistance of hundreds to thousands of megohms.

AMPLIFIER

Most data acquisition systems include a variable gain amplifier which allows the user to compensate for the input level of the signals. The integral ADC usually is designed to work from a definite unipolar or bipolar input range so that input levels must be adjusted to fall within this range. Thus, if the ADC signal input must be in the range of 0 to 5 volts, the amplifier is given a gain to assure that the inputs will be within this range. If there is a great difference between the various input signal levels, some signal conditioning may be required prior to application of the signal to the DAS.

ADC

Of course, an important part of the DAS is an analog to digital converter. The converter will accept voltages which span a specific range as provided by the preceeding signal conditioning. The converter usually can be configured to accept either unipolar or bipolar inputs. Features such as offset adjustment and full-scale adjustment are common.

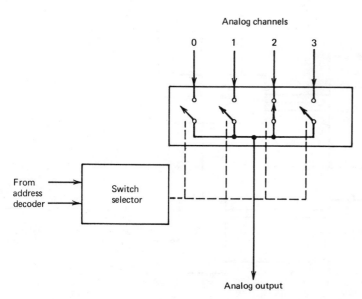

FIGURE 3.13   Four-channel analog multiplexer.

3.4.2 Data Output Module (DOM)

The preceding paragraphs described the system used to input data to the computer. Usually, this is a process-control loop controlled variable. In either supervisory control or direct digital control, it is also necessary to provide a mechanism by which the computer can output a signal to either a setpoint adjustment or to the final control element. This type of interface is provided for a many channel system by a Data Output Module (DOM). A general block diagram of this device is given in Figure 3.14. The general purpose of the address decoder is the same as the DAS, that is, to allow the computer to select a particular output channel. In this case, however, the computer is *writing* information into a memory location or output address which is converted to an analog voltage by the DAC. We use a demultiplexer which is able to switch the output of the DAC into any one of the selected channels. The latch is necessary because the computer will only allow the output data word to be present on the data lines for a few microseconds. The latch holds this data long enough for conversion and application in the process-control loop.

3.4.3 Application Notes

There are numerous factors which must be considered when a DAS or DOM is to be used. The following paragraphs discuss several of these factors.

SAMPLE AND HOLD

When using the DAS, account must be made for the fact that the signals on the input channels may be changing very rapidly. If the changes are fast enough so that the signal varies during the conversion time, a sample-and-hold must be used on

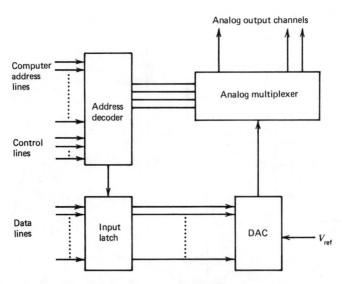

FIGURE 3.14   Data output module.

that channel to hold the value during conversion. This adds to the complexity of the software because account must be made for commands to the sample and hold module.

COMPATIBILITY TO COMPUTER

In many cases, a data module is designed to work with only one model or type of computer. This is particularly true when microprocessor-based computers where architecture varies greatly between families are used. Thus, it is necessary to select a data module (DAS or DOM) which is compatible with the input/output characteristics of the computer.

HARDWARE PROGRAMMING

Most data modules give the user many options for use in input/output operations. These options include unipolar/bipolar operation, address selection, amplifier gain, differential/single-ended operation, and others. They are typically selected by wired jumpers between pins of the module or by the attachment of resistors as specified in the specification sheet of the module.

SOFTWARE PROGRAMMING

Another important aspect of input/output interfacing is the software routines which will use the data modules. These routines must be compatible with the hardware programming and the other characteristics of the module. The programs may include delays waiting for the ADC to complete conversion, for example. This aspect is discussed further in Chapter 10.

OVERALL RESPONSE TIME

A data acquisition system does not provide the digital conversion of data on a selected channel the instant the selection occurs. Rather, there is a delay while the multiplexer acquires the system channel, while the amplifier settles to the value on the channel, and while the ADC performs the conversion operations described under the section on ADC operation. This time will be important for a determination of the maximum sampling rate from the DAS. The time may run from tens of microseconds to hundreds of microseconds, depending on the number of bits converted, gains of the amplifiers, and signal-switching speed.

SUMMARY

This chapter provides a digital electronics background to make the reader conversant with the elements of digital signal processing and able to perform simple analysis and design as associated with process control.

1. The use of digital words enables the encoding of analog information into a digital format.
2. It is possible to encode fractional decimal numbers as binary and vice versa using

$$N_{10} = b_1 2^{-1} + b_2 2^{-2} + \cdots + b_m 2^{-m} \tag{3-1}$$

3. Boolean algebra techniques can be applied to the development of process alarms and elementary control functions.

4. Digital electronic gates and comparators allow the implementation of process Boolean equations.

5. DACs are used to convert digital words into analog numbers using a fractional number representative. The resolution is

$$\Delta V = V_R 2^{-n} \tag{3-4}$$

6. An ADC of the successive approximations type determines an output digital word for an input analog voltage in as many steps as bits to the word.

7. The dual-slope ADC converts analog to digital information by a combination of integration and time counting.

8. The Data Acquisition System (DAS) is a modular device which interfaces many analog signals to a computer. Signal address decoding, multiplexing, and ADC operations are included in the device.

9. The Data Output Module (DOM) provides all the hardware necessary for a computer to output an analog signal, including addressing D/A conversion and multiplexing.

PROBLEMS

3.1 Convert the binary numbers (a) 1010_2 (b) 111011_2 (c) 010110_2 in decimal equivalents.

3.2 Convert the following binary numbers to decimal, octal, and hex equivalents.
(a) 1011010_2 (b) 0.1101_2 (c) 1011.0110_2.

3.3 Convert the decimal numbers (a) 21_{10} (b) 630_{10} (c) 427_{10} into binary, octal and hex.

3.4 Convert 27.156_{10} into a binary number with the binary part less than one having 6 bits. What actual decimal does this binary equal?

3.5 Find the 2s complement of (a) 1011_2 (b) 10101100_2.

3.6 Prove by a table of values that $\overline{A \cdot B} = \overline{A} + \overline{B}$ (De Morgan's Theorem)

3.7 Show that the Boolean equation $A \cdot B + A \cdot (\overline{A \cdot B})$ reduces to A

3.8 A process controls moving speed, load weight, and rate of loading in a conveyor system, measured and provided as high (1) or low (0) levels for digital control. An alarm should be initiated whenever any of the following occur:
(a) Speed is low, both weight and loading rate are high.
(b) Speed is high, loading rate is low.
Find a Boolean equation describing the required output to effect the alarm. Let the logic variables be S for speed, W for weight, and R for loading rate.

3.9 Implement Problem 3.8 with (a) AND/OR Logic (b) NAND/NOR Logic.

3.10 A 6-bit DAC has an input of 100101_2 and a 10-volt reference. (a) Find the output voltage produced. (b) Specify the conversion resolution.

3.11 An 8-bit ADC with a 10-volt reference has an input of 3.797 volts. Find the digital output word.

3.12 An ADC which will encode pressure data is required. The input signal is 666.6 mV/psi. (a) If a resolution of 0.5 psi is required, find the number of bits necessary in the A/D converter. The reference is 10 volts. (b) Find the maximum measurable pressure.

3.13 Prove that the ladder network of Figure 3.7 gives the output predicted by Equation (3-3) for an input of 1011.

3.14 A 4-bit DAC must have an 8-volt output when the inputs are all high. Find the required reference.

3.15   A DAS has an ADC of 8-bits with a 0 to 2.5-volt range of input for 00h to FFh output range. Inputs are temperature from 20°C to 100°C scaled at 40 mV/°C, pressure from 0 to 100 psi scaled at 100 mV/psi, and flow from 30 gal/min to 90 gal/min scaled at 150 mV/(gal/min). Develop a block diagram of the required signal conditioning so that each of these can be connected to the DAS and so that the indicated variable range corresponds to 00h to FFh output of the DAS. Specify the resolution of each in terms of the change of each variable which corresponds to a LSB change of the output.

3.16   Design op am circuits which will provide the signal conditioning blocked out in Problem 3.15.

3.17   A DAC converts 8 bits to a −5 to +5 volt range. If the conversion code is modified 2's complement, as discussed in the chapter, specify the voltage output for inputs of 00101110_2, 344_8, and B3h. What is the output increment of voltage for a LSB change?

3.18   Flow is to be measured and inputted to a computer. For maximum resolution we want the minimum flow, 30 m³/hr, to correspond to 00h and the maximum flow, 60 m³/hr, to correspond to FFh from an 8-bit ADC. Flow is available as a voltage given by the relation, $V = 0.0022Q^2$, where Q = flow in m³/hr. Develop the signal conditioning and ADC reference which will provide this specification. What is the resolution in flow when the flow is 30m³/hr and when the flow is 60 m³/hr? Explain why the resolution is different.

4

THERMAL TRANSDUCERS

INSTRUCTIONAL OBJECTIVES

The objectives of this and the following two chapters stress an understanding beyond that required for simple application of process-control transducers. After you have read it, you should be able to:

1. Define thermal energy, the relation of temperature scales to thermal energy, and temperature scale calibrations.
2. Transform a temperature reading between the Kelvin, Rankine, Celsius, and Fahrenheit temperature scales.
3. Design the application of an RTD temperature transducer to specific problems in temperature measurement.
4. Design the application of a thermistor to specific temperature measurement problems.
5. Design the application of a thermocouple to specific temperature measurement problems.
6. Explain the operation of a bimetal strip for temperature measurement.
7. Explain the operation of a gas thermometer and a vapor pressure thermometer.

4.1 INTRODUCTION

Process control is a term used to describe any condition, natural or artificial, by which a physical quantity is regulated. There is no more widespread evidence of such control than that associated with temperature and other thermal phenomena. In our natural surroundings, some of the most remarkable techniques of temperature regulation are found in the bodily functions of living creatures. On the artificial side, man has been vitally concerned with temperature control since the first fires were struck for warmth. Industrial temperature regulation has always been of paramount importance and becomes even more so with the advance of technology. In this chapter we will be concerned first with developing an understanding of the principles of thermal energy and temperature, and then developing a working knowledge of the various thermal transducers employed for temperature measurement.

4.2 DEFINITION OF TEMPERATURE

It is well known that the materials which surround us, and indeed of which we are constructed, are composed of assemblages of atoms. Each of the 92 natural elements of nature is represented by a particular type of atom. The materials that surround us typically are not pure elements, but rather combinations of atoms of several elements, forming what are referred to as molecules. Thus, helium is a natural element consisting of a particular type of atom; water, on the other hand, is composed of molecules, each molecule consisting of a combination of two hydrogen atoms and one oxygen atom. In presenting a physical picture of thermal energy we need to consider the physical relations or interaction of elements and molecules in a particular material as either solid, liquid, or gas. These statements actually refer to the manner in which the molecules of the material are interacting and to the thermal energy of the molecules.

4.2.1 Thermal Energy

SOLID

In any solid material, the individual atoms or molecules are quite strongly attracted and bonded to each other, so that no atom is able to move far from its particular location or *equilibrium position*. Each atom, however, is capable of vibration about its particular location. We can introduce the concept of thermal energy by considering the molecules' vibration.

Consider a particular solid material in which molecules are found to be exhibiting no vibration; that is, the molecules are at rest. Such a material is said to have zero thermal energy $W_{TH} = 0$. If we now add energy to this material, by placing it on a heater, for example, this energy starts the molecules vibrating about their equilibrium positions. We may say that the material now has some finite thermal energy, $W_{TH} > 0$.

LIQUID

If more and more energy is added to the above material, the vibrations become more and more violent as the thermal energy increases. Finally, a condition is reached where the bonding attractions that holds the molecules in their equilibrium positions are overcome and the molecules "break away" and move about in the material. When this occurs, we say the material has *melted* and becomes a liquid. Now, even though the molecules are still attracted to one another, the thermal energy is sufficient to cause the molecules to move about and no longer maintain the rigid structure of the solid. Instead of vibrating, one considers the molecules as randomly sliding about each other, and the average speed with which they move is a measure of the thermal energy imparted to the material.

GAS

Further increases in thermal energy of the material intensify the velocity of the molecules until finally the molecules gain sufficient energy to completely escape from the attraction of other molecules. Such a condition is manifested by *boiling* of the liquid. When the material consists of such unattached molecules moving randomly throughout a containing volume, we say the material has become a gas. The molecules still collide with each other and the walls of the container, but otherwise move freely throughout the container. Here, the average speed of the molecules is again a measure of the thermal energy imparted to the molecules of the material.

Note that all materials may not undergo the above transitions at the same thermal energy and indeed some not at all. Thus, nitrogen can be solid, liquid, and gas, but paper will experience a breakdown of its molecules before a liquid or gaseous state can occur. The whole subject of thermal transducers is associated with the measurement of the thermal energy of a material or an environment containing many different materials.

4.2.2 Temperature

If we are to measure thermal energy, we must have some sort of units by which to classify the measurement. The original units used were "hot" and "cold." These were quite satisfactory for their time but seem inadequate for modern use. The proper unit for energy measurement would be the *joules* of the sample in the metric system, but this would depend on the size of the material because it would indicate the total thermal energy contained. Thus, a measurement of the average thermal energy per molecule, expressed in joules, could be used to define thermal energy. We say "could be" because it is not traditionally used. Instead, special sets of units, whose origins are contained in the history of thermal energy measurements, are employed to define the *average energy per molecule* of a material. We will consider the four most common units. In each case, the name used to describe the thermal energy per molecule of a material is related by the statement that the material has a certain *degree of temperature*, and the different sets of units are referred to as *temperature scales*.

CALIBRATION

In order to define the temperature scales, a set of *calibration points* is used; for each, the average thermal energy per molecule is well defined through equilibrium conditions existing between solid, liquid, or gaseous states of various pure materials. Thus, for example, a state of equilibrium exists between the solid and liquid phase of a pure substance when the *rate of phase change* is the same in *either* direction, liquid to solid, and solid to liquid. Some of the standard calibration points are:

1. Oxygen:liquid/gas equilibrium.
2. Water:solid/liquid equilibrium.
3. Water:liquid/gas equilibrium.
4. Gold:solid/liquid equilibrium.

The various temperature scales are defined by the assignment of numerical values of temperatures to the above, and additional calibration points. Essentially, the scales differ in two respects: (1) the location of the 0 of temperature, and (2) the size of one unit of measure, that is, the average thermal energy per molecules represented by one unit of the scale.

ABSOLUTE TEMPERATURE SCALES

An absolute temperature scale is one that assigns a 0-unit temperature to a material which has no thermal energy, that is, no molecular vibration. There are two such scales in common use, the Kelvin scale in Kelvin (K) and the Rankine scale in degrees Rankine (°R). These temperature scales differ only by the quantity of energy represented by one unit of measure; hence, a simple proportionality relates the temperature in °R to the temperature in K. Table 4.1 shows the values of temperature in Kelvin and degrees Rankine at the calibration points introduced earlier. From this table we can determine the transformation of temperature in K to °R and vice versa. To do this, we note that the difference in temperature between the water liquid/solid point and water liquid/gas point is 100 K and 180°R, respectively. Because these two numbers represent the same difference of thermal energy, it is clear that 1 K must be larger than 1 degree R by the ratio of the two numbers.

$$(1 \text{ K}) = \tfrac{180}{100} (1°\text{R}) = \tfrac{9}{5} (1°\text{R})$$

Thus, the transformation between scales is given by,

$$T(\text{K}) = \tfrac{5}{9} T(°\text{R}) \qquad (4\text{-}1)$$

Where

$T(\text{K})$ = temperature in K
$T(°\text{R})$ = temperature in °R

TABLE 4.1   TEMPERATURE SCALE CALIBRATION POINTS

Calibration Point	Temperature			
	K	°R	°F	°C
Zero thermal energy	0	0	−459.6	−273.15
Oxygen: liquid/gas	90.18	162.3	−297.3	−182.97
Water: solid/liquid	273.15	491.6	32	0
Water: liquid/gas	373.15	671.6	212	100
Gold: solid/liquid	1336.15	2405	1945.5	1063

EXAMPLE 4.1

A material has a temperature of 335 K. Find the temperature in °R.

SOLUTION

$$T(°R) = \tfrac{9}{5} T(K)$$
$$T(°R) = \tfrac{9}{5} (335 \text{ K}) = 603°R \tag{4.1}$$

RELATIVE TEMPERATURE SCALES

The *relative* temperature scales differ from the *absolute* scales only in a shift of the zero axis. Thus, when these scales indicate a zero of temperature, the thermal energy of the sample is not zero. These two scales are the Celsius (related to the Kelvin) and the Fahrenheit (related to the Rankine) with temperature indicated by °C and °F, respectively. Table 4.1 shows various calibration points of these scales. Note that the quantity of energy represented by 1°C is the same as indicated by 1 K, but the zero has been shifted in the Celsius scale, so that

$$T(°C) = T(K) - 273.15 \tag{4-2}$$

Similarly, the size of 1°F is the same as the size of 1°R but with a scale shift so that:

$$T(°F) = T(°R) - 459.6 \tag{4-3}$$

To transform from Celsius to Fahrenheit we simply note that the two scales differ by the size of the degree just as in the K and °R, and a scale shift of 32 separates the two, thus,

$$T(°F) = \tfrac{9}{5} T(°C) + 32 \tag{4-4}$$

RELATION TO THERMAL ENERGY

It is possible to relate temperature to actual thermal energy in joules by using a constant called *Boltzmann's constant*. Although not true in all cases, it is a good approximation to state that the thermal energy W_{TH} of a molecule can be found from the absolute temperature in K from

$$W_{TH} = \tfrac{3}{2} kT \tag{4-5}$$

where $k = 1.38 \times 10^{-23}$ J/K is Boltzmann's constant. Thus, it is possible to determine the average thermal speed, v_{TH}, of a gas molecule by equating the kinetic energy of the molecule to its thermal energy,

$$\tfrac{1}{2} m v_{TH}^2 = W_{TH} = \tfrac{3}{2} kT$$

and

$$v_{TH} = \sqrt{\frac{3kT}{m}}$$ (4-6)

where m is the molecule mass in kilograms.

EXAMPLE 4.2

Given temperature of 144.5°C, express this temperature in (a) K and (b) °F.

SOLUTION

a. $T(K) = T(°C) = 273.15$
    $T(K) = 144.5 + 273.15$ (4-2)
    $T(K) = \textbf{417.65 K}$

b. $T(°F) = \frac{9}{5} T(°C) + 32$
    $T(°F) = \frac{9}{5} (144.5°C) + 32$ (4-4)
    $T(°F) = \textbf{292.1°F}$

EXAMPLE 4.3

A sample of oxygen gas has a temperature of 90°F. If its molecular mass is 5.3×10^{-26} kg, find the average thermal speed of a molecule.

SOLUTION

We first convert the 90°F into K, and then use Equation (4-6) to find the speed.

$T(°R) = T(°F) + 459.6$
$T(°R) = (90°F + 459.6)$ (4-3)
$T(°R) = 549°R$

$T(K) = \frac{5}{9} T(°R) = \frac{5}{9} (549.6 K)$ (4-1)
$T(K) = 305.33 K$

Then the velocity is

$$v_{TH} = \sqrt{\frac{3kT}{m}}$$

$$v_{TH} = \left[\frac{(3)(1.38 \times 10^{-23} \text{ J/K})(305.33°K)}{5.3 \times 10^{-26} \text{ kg}} \left(1 \frac{kg - m^2}{s^2 J} \right) \right]^{1/2}$$ (4-6)

$$v_{TH} = \textbf{488.37 m/s}$$

4.3 METAL RESISTANCE VERSUS TEMPERATURE DEVICES

One of the primary methods for electrical measurement of temperature involves changes in the electrical resistance of certain materials. In this, as well as other cases, the principal measurement technique is to place the temperature-sensing device in contact with the environment whose temperature is to be measured. The sensing device then takes on the temperature of the environment. Thus, a measure of its resistance indicates the temperature of the device and also the environment. Time response becomes very important in these cases because the measurement must wait until the device comes into thermal equilibrium with the environment. The two basic devices used are the *resistance-temperature detector* (RTD), based on the variation of metal resistance with temperature and the *thermistor*, based on the variation of semiconductor resistance with temperature.

4.3.1 Metal Resistance Versus Temperature

A metal is an assemblage of atoms in the solid state in which the individual atoms are in an equilibrium position with superimposed vibration induced by the thermal energy. The chief characteristic of a metal is the fact that each atom gives up one electron, called its valence electron, which can move freely throughout the material; that is, it becomes a conduction electron. We say then, for the whole material, that the valence band of electrons and conduction band of electrons in the material overlap in energy as shown in Figure 4.1a. Contrast this with a semiconductor where a small gap exists between the top electron energy of the valence band and the bottom electron energy of the conduction band, as shown in Figure 4.1b. In this same scheme, Figure 4.1c shows that an insulator has a very large gap between valence and conduction electrons. Now, when a current is passed through a material, the conduction band electrons are current carriers (or better yet) the current.

When we stated earlier that the electrons of a metal "move freely throughout the material," there should have been a condition imposed—that is, that this can be true at absolute zero temperature. In fact, when a certain thermal energy is present in the material and the atoms vibrate, the conduction electrons tend to collide with the vibrating atoms. This impedes the free movement of electrons and absorbs some of their energy; that is, the material exhibits a *resistance* to electrical current flow. Thus, metallic resistance is a function of the vibration of the atoms and thus of the temperature. As the temperature is raised, the atoms vibrate with greater amplitude and frequency which causes even more collision with electrons, further impeding their flow and absorbing more energy. From this argument, we can see that metallic resistance should increase with temperature, and it does.

The graph in Figure 4.2 shows the effect of increasing resistance with temperature for several metals. In order to compare the different materials, the graph shows the relative resistance versus temperature. For a specific metal of high purity, the curve of relative resistance versus temperature is highly repeatable, and thus, either tables or graphs can be used to determine the temperature from a resistance measurement using that material. It is possible to express the resistance of a particular metal sample at a constant temperature (T) analytically using the equation

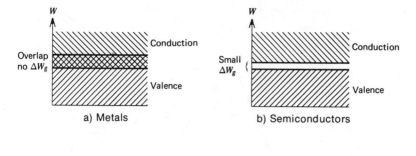

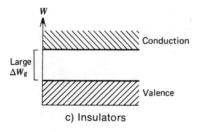

FIGURE 4.1   Energy bands for solids. Only conduction band electrons are free to carry current.

$$R = \rho \frac{l}{A} \quad (T = \text{constant}) \tag{4-7}$$

Where

R = sample resistance (Ω)
l = length (m)
A = cross-sectional area (m²)
ρ = resistivity ($\Omega - $m)

In Equation (4-7) the principal increase in resistance with temperature is due to changes in the resistivity (ρ) of the metal with temperature. If the resistivity of some metal is known as a function of temperature, then Equation (4-7) can be used to determine the resistance of any particular sample of that material at the same temperature. In fact, curves such as those in Figure 4.2 are curves of resistivity versus temperature because, for example,

$$\frac{R(T)}{R(75°)} = \frac{\rho(T)\, l/A}{\rho(75°)\, l/A} = \frac{\rho(T)}{\rho(75°)} \tag{4-8}$$

The use of either Equation (4-7), resistance versus temperature graphs, or resistance versus temperature tables is only practical when high accuracy is desired. For many applications we can use an analytical approximation of the curves for which we simply insert the temperature and quickly calculate the resistance, described in Section 4.3.2.

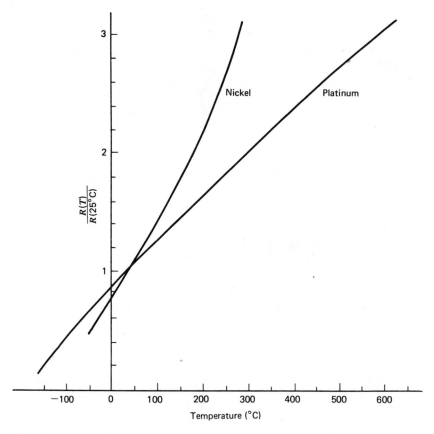

FIGURE 4.2   Metal resistance increases almost linearly with temperature.

4.3.2 Resistance Versus Temperature Approximations

An examination of the resistance versus temperature curves of Figure 4.2 shows that the curves are very nearly linear, that is, a straight line. In fact, when only short temperature spans are considered, the linearity is even more evident. This fact is employed to develop approximate analytical equations for resistance versus temperature of a particular metal.

LINEAR APPROXIMATION

A linear approximation means that we may develop an equation for a stright line which approximates the resistance versus temperature $(R - T)$ curve over some specified span. In Figure 4.3, we see a typical $R - T$ curve of some material. Here a straight line has been drawn between the points of the curve which represent temperatures T_1 and T_2 as shown, and T_0 represents the midpoint temperature. The equation of this straight line is the linear approximation to the curve over the span T_1 to T_2. The equation for this line is typically written as

$$R(T) = R(T_0) [1 + \alpha_0 \, \Delta T] \qquad T_1 < T < T_2 \tag{4-9}$$

Where

$R(T)$ = approximation of resistance at temperature T
$R(T_0)$ = resistance at temperature T_0
$\Delta T = T - T_0$
α_0 = fractional change in resistance per degree of temperature at T_0

The reason for using α_0 as the fractional slope of the $R - T$ curve is that this same constant can be used for cases of other physical dimensions (length and cross-sectional area) of the same kind of wire. Note that α_0 depends on the midpoint temperature T_0, which simply says that a straight-line approximation over some other span of the curve would have a different slope.

The value of α_0 can be found from values of resistance and temperature taken either from a graph, as given in Figure 4.2, or a table of resistance versus temperature as given in Problem 4.8. In general, then,

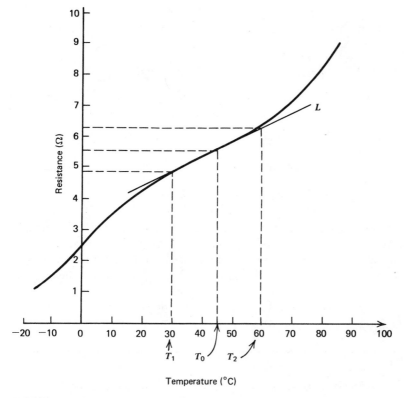

FIGURE 4.3   Line L represents a linear approximation of resistance versus temperature between T_1 and T_2.

$$\alpha_0 = \frac{1}{R(T_0)} \cdot (\text{slope at } T_0) \qquad (4\text{-}10)$$

or, for example, from Figure 4.3

$$\alpha_0 = \frac{1}{R(T_0)} \cdot \left[\frac{R_2 - R_1}{T_2 - T_1} \right] \qquad (4\text{-}11)$$

Where

R_2 = resistance at T_2
R_1 = resistance at T_1

Note that α has units of *inverse* temperature degrees and therefore depends on the temperature scale being used. Thus, the units of α are typically $1/°C$ or $1/°F$.

QUADRATIC APPROXIMATION

A quadratic approximation to the $R - T$ curve is a more accurate representation of the $R - T$ curve over some span of temperatures. It includes both a linear term, as before, and a term that varies as the square of the temperature. Such an analytical approximation is usually written as

$$R(T) = R(T_0) \left[1 + \alpha_1 \Delta T + \alpha_2 (\Delta T)^2 \right] \qquad (4\text{-}12)$$

Where

$R(T)$ = quadratic approximation of the resistance at T
$R(T_0)$ = resistance at T_0
α_1 = linear fractional change in resistance with temperature
$\Delta T = T - T_0$
α_2 = quadratic fractional change in resistance with temperature

Values of α_1 and α_2 are found from tables or graphs as indicated in the example below, using values of resistance and temperature at three points. As before, both α_1 and α_2 depend on the temperature scale being used and have units of $1/°C$ and $(1/°C)^2$ if Celsius temperature is used, and $1/°F$ and $(1/°F)^2$ if the Fahrenheit scale is used.

The following examples show how these linear approximations are formed.

EXAMPLE 4.4

Find a linear approximation for resistance between 30°C and 60°C using the $R - T$ curve of Figure 4.3.

SOLUTION

Using Equations (4-9) and (4-11), we first identify: $T_1 = 30°C$, $T_0 = 45°C$ and $T_2 = 60°C$. Then, from Figure 4.3, we find $R(30°C) = 4.8 \ \Omega$, $R(45°C) = 5.5 \ \Omega$ and $R(60°C) = 6.2 \ \Omega$.

$$\alpha_0 = \frac{1}{5.5 \ \Omega} \frac{6.2 \ \Omega - 4.8 \ \Omega}{60°C - 30°C} \tag{4-11}$$

$$\alpha_0 = 0.0085/°C$$

Thus,

$$R(T) = 5.5[1 + .0085(T - 45)]\Omega$$

with T in °C

The following typical values of resistance versus temperature are to be used for Examples (4.5) and (4.6).

T(°F)	R(Ω)
60	106.06
65	107.14
70	108.22
75	109.30
80	110.38
85	111.46
90	112.53

EXAMPLE 4.5

The following equation is a linear approximation for the resistance between 65°F and 75°F from the above data.

$$R(T) = 108.22[1 + 0.002(T - 70°F)]\Omega$$

Find the resistance at 20°C.

SOLUTION

We must first convert the 20°C to its equivalent °F to keep the units consistent.

$$T(°F) = \tfrac{9}{5}(20°C) + 32$$
$$T(°F) = 68°F \tag{4-4}$$

Then we use the stated linear equation to find

$$R(68°F) = 108.22[1 + 0.002(68 - 70)]$$
$$R(68°F) = 108.65 \ \Omega$$

EXAMPLE 4.6

Find a quadratic approximation for resistance versus temperature about 75°F between 60°F and 90°F, using the table of values given above.

SOLUTION

We can find the quadratic terms in Equation (4-12) by forming two equations using two points about 75°F. Thus, at 75°F the resistance is 109.3 Ω; therefore,

$$R(T_0) = 109.3 \ \Omega, \ T_0 = 75°F$$

Now, using 60°F and 90°F as the two points, we get two equations, using

$$R(T) = R(T_0)[1 + \alpha_1 \Delta T + \alpha_2(\Delta T)^2]$$
$$106.06 = 109.3[1 + \alpha_1(60 - 75) + \alpha_2(60 - 75)^2] \qquad (4\text{-}12)$$
$$112.53 = 109.3[1 + \alpha_1(90 - 75) + \alpha_2(90 - 75)^2]$$

Solving these equations for α_1 and α_2, we find:

$$\alpha_1 = 1.973 \times 10^{-3}/°F$$
$$\alpha_2 = -2.033 \times 10^{-7}/(°F)^2$$

4.3.3 Resistance-Temperature Detectors

A *resistance-temperature detector* (RTD) is a temperature transducer that is based on the principles discussed in the preceding sections, that is, metal resistance increasing with temperature. Metals used in these devices vary from *platinum*, which is very repeatable, quite sensitive, and very expensive, to *nickel*, which is not quite as repeatable, more sensitive, and less expensive.

SENSITIVITY

An estimation of RTD *sensitivity* can be noted from typical values of the linear fractional change in resistance with temperature. For platinum, this number is typically on the order of 0.004/°C whereas for nickel a typical value is 0.005/°C. Thus, with platinum, for example, a change of only 0.4 Ω would be expected for a 100 Ω RTD if the temperature is changed by 1°C. Usually, a specification will provide calibration information either as a graph of resistance versus temperature or as a table of values from which the sensitivity can be determined. For the same materials, however, this number is relatively constant because it is a function of resistivity.

RESPONSE TIME

In general, RTD has a response time of 0.5 to 5 seconds or more. The slowness of response is due principally to the slowness of thermal conductivity in bringing the device into thermal equilibrium with its environment. Generally, time constants are specified either for a "free air" condition (or its equivalent) or an "oil bath" condition (or its equivalent). In the former case, there is poor thermal contact and hence, slow response, and in the latter, good thermal contact and fast response. These numbers yield a range of response times to be expected depending on the application.

CONSTRUCTION

An RTD, of course, is simply a length of wire whose resistance is to be monitored as a function of temperature. The construction is typically such that the wire is wound on a form (in a coil) to achieve small size and improve thermal conductivity to decrease response time. In many cases, the coil is protected from the environment by a *sheath* or protecting tube which inevitably increases response time but may be necessary in hostile environments. A loosely applied standard sets the resistance at multiples of 100 Ω for a temperature of 0°C.

SIGNAL CONDITIONING

In view of the very small fractional changes of resistance with temperature (0.4%), the RTD is generally used in a *bridge* circuit in which a null condition is accurately detected. For process-control applications, the bridge then requires a *self-nulling* property. The output of the nulling circuit then provides the controller output of 4 to 20 mA or 10 to 50 mA. Figure 4.4 illustrates the essential features of such a system. The *compensation line* in the R_3 leg of the bridge is required when the lead lengths are so long that thermal gradients along the RTD leg may cause changes in line resistance. These changes would show up as false information, suggesting changes in RTD resistance. By using the *compensation* line, the same resistance changes also appear on the R_3 side of the bridge and cause no net shift in the bridge null.

The feedback of the controller may occur in many forms, ranging from an electromechanical change of the R_2 setting to a current source providing a nulling current as in the current balance bridge (Section 2.3) or even to a person watching a galvanometer and turning the R_2 knob. Note that because the RTD is a resistance, there is an I^2R power dissipated by the device itself which causes a slight heating effect, a *self-heating*. This may also cause an erroneous reading or even upset the environment in delicate measurement conditions. Thus, the current through the RTD must be kept sufficiently low and constant to avoid self-heating. Typically, a *dissipation constant* is provided in RTD specifications. This number relates the power required to raise the RTD temperature by 1 degree of temperature, usually in

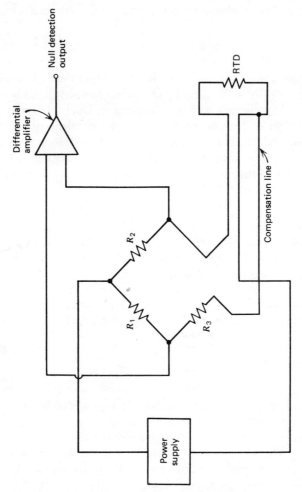

FIGURE 4.4 Note the compensation line in this typical RTD signal conditioning circuit. Any of the resistances in the bridge could be used to renull the bridge.

still air. Thus, a 25 mW/°C dissipation constant shows that if I^2R power losses in the RTD equal 25 mW, then the RTD will be heated by 1°C.

The dissipation constant is usually specified under two conditions, free air and a well-stirred oil bath. This is because of the difference in capacity of the medium to carry heat away from the device. The self-heating temperature rise can be found from the power dissipated by the RTD and the dissipation constant from

$$\Delta T = \frac{P}{P_D} \qquad (4\text{-}13)$$

Where

ΔT = temperature rise due to self-heating in °C
P = power dissipated in the RTD from the circuit in W
P_D = dissipation constant of the RTD in W/°C

EXAMPLE 4.7

An RTD has $\alpha_0 = 0.005/°C$, $R = 500\ \Omega$, and a dissipation constant of $P_D = 30$ mW/°C at 20°C. The RTD is used in a bridge circuit so that as in Figure 4.4 with $R_1 = R_2 = 500\ \Omega$ and R_3 a variable resistor used to null the bridge. If the supply is 10 volts and the RTD is placed in a bath at 0°C, find the value of R_3 to null the bridge.

SOLUTION

First we find the value of the RTD resistance at 0°C without including the effects of dissipation. From Equation (4-9) we get

$$R = 500[1 + 0.005(0 - 20)]\ \Omega$$
$$R = 450\ \Omega$$

Now, except for the effects of self heating, we would expect the bridge to null with R_3 equal to 450 Ω also. Let's see what self heating does to this problem. First, we find the power dissipated in the RTD from the circuit assuming the resistance is still 450 Ω. The power is

$$P = I^2R$$

and the current, I, is found from

$$I = \frac{10}{500 + 450} = 0.011 \text{ amps}$$

so that the power is

$$P = (0.011)^2 (450) = 0.054 \text{ W}$$

Now we get the temperature rise from Equation (4-13)

$$\Delta T = \frac{0.054}{0.030} = 1.8°C$$

Thus, the RTD is not actually at the bath temperature of 0°C but at a temperature of 1.8°C. This means we must find the RTD resistance from Equation (4-9) as

$$R = 500[1 + 0.005(1.8 - 20)] \ \Omega$$
$$R = 454.5 \ \Omega$$

Thus, the bridge will null with $R_3 = 454.5 \ \Omega$.

RANGE

The effective range of RTDs depends principally on the type of wire used as the active element. Thus, a typical platinum RTD will have a range of −100°C to 650°C, whereas an RTD constructed from nickel might typically have a specified range of −180°C to 300°C.

4.4 THERMISTORS

The thermistor represents another class of temperature transducer which measures temperature through changes of material resistance. The characteristics of these devices are very different from RTDs discussed in the previous section and depend on the peculiar behavior of semiconductor resistance versus temperature.

4.4.1 Semiconductor Resistance Versus Temperature

In contrast to metals, electrons in semiconductor materials are bound to each molecule with sufficient strength so that no conduction electrons are contributed from the valence band to the conduction band. We say that a gap of energy ΔW_g exists between valence and conduction electrons as shown in Figure 4.1b. Such a material behaves as an insulator because there are no conduction electrons to carry current through the material. This is true only when there is no thermal energy present in the sample, that is, at a temperature of 0°K. When the temperature of the material is increased, the molecules begin to vibrate. In the case of a semiconductor, such vibration provides additional energy to the valence electrons. When such energy equals or exceeds the gap energy ΔW_g, these electrons become free of the molecules. Thus, the electron is now in the conduction band and is free to carry current through the bulk of the material. As the temperature is increased still further, more and more electrons gain sufficient energy to enter the conduction band. It is then clear that the semiconductor becomes a *better* conductor of current as its temperature is *increased*, that is, as its resistance decreases, From this discussion we form a picture of the resistance of a semiconductor material decreasing from very large values at low temperature to smaller resistance at high temperature. This is just the *opposite* of a metal. An important distinction, however, is that the change in semiconductor resistance is highly nonlinear as shown in Figure 4.5. The reason semiconductors, but not insulators and other materials, behave this way is that the energy gap between conduction and valence bands is small enough to allow thermal excitation of electrons across the gap.

It is important to note that the effect just described requires that the thermal energy provide sufficient energy to overcome the band gap energy ΔW_g. In general, a material is classified as a semiconductor when the gap energy is typically 0.01–4 eV (1 eV = 1.6×10^{-19} J). That this is true is exemplified by a consideration of silicon, a semiconductor that has a band gap of $\Delta W_g = 1.107$ eV. When heated,

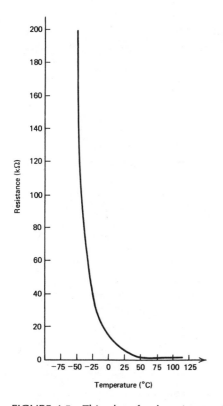

FIGURE 4.5   This plot of a thermistor resistance versus temperature shows the highly nonlinear behavior characteristic of semiconductor materials.

this material passes from insulator to conductor. The corresponding thermal energies that bring this about can be found using Equation (4-5) and the joules to eV conversion, thus:

$$\text{for } T = 0°\text{K} \qquad W_{TH} = 0.0 \text{ eV}$$
$$\text{for } T = 100°\text{K} \qquad W_{TH} = 0.013 \text{ eV}$$
$$\text{for } T = 300°\text{K} \qquad W_{TH} = 0.039 \text{ eV}$$

With average thermal energies as high as 0.039 eV, sufficient numbers of electrons are raised to the conduction level for the material to become a conductor. In *true* insulators, the gap energy is so large that temperatures less than destructive to the material cannot provide sufficient energy to overcome the gap energy.

4.4.2 Thermistors

A thermistor is a temperature transducer which has been developed from the principles just discussed regarding semiconductor resistance change with temperature. The particular semiconductor material used varies widely to accommodate temperature ranges, sensitivity, resistance ranges, and other factors. The devices are usually mass produced for a particular configuration, and tables or graphs of resistance versus temperature are provided for calibration purposes. Variation of individual units from these nominal values is indicated as a net percentage deviation or a percentage deviation as a function of temperature.

SENSITIVITY

The sensitivity of the thermistors is a significant factor in their application. Changes in resistance of 10% per °C are not uncommon. Thus, a thermistor with a nominal resistance of 10 kΩ at some temperature may change by 1 kΩ for a 1°C change in temperature. When used in null detecting bridge circuits, sensitivity this large can provide for control, in principle, to less than 1°C in temperature.

CONSTRUCTION

Because the thermistor is a bulk semiconductor, it can be fabricated in many forms. Thus, common forms include discs, beads, and rods varying in size from a bead 1 mm in diameter to a disc several centimeters in diameter and several centimeters thick. By variation of doping and use of different semiconducting materials, a manufacturer can provide a wide range of resistance values at any particular temperature.

RESPONSE TIME

The response time of a thermistor depends principally on the quantity of material present and the environment. Thus, for the smallest bead thermistors in an oil bath (good thermal contact), a response of 1/2 second is typical. The same thermistor in still air will respond with a typical response time of 10 seconds. When encapsulated, as in Teflon or other materials for protection against a hostile environment, the time response is increased due to the poor thermal contact with the environment. Large disc or rod thermistors may have response times of 10 seconds or more even with good thermal contact.

SIGNAL CONDITIONING

Because a thermistor exhibits such a large change in resistance with temperature, there are many possible circuit applications. In many cases, however, a bridge circuit with null detection (to maintain a particular temperature) is used because the nonlinear features of the thermistor make its use difficult as an actual measurement device. Because these devices are resistances, care must be taken to ensure that power dissipation in the thermistor does not exceed limits specified or possibly interfere with the environment for which the temperature is being measured. *Dissipation constants* are quoted for thermistors as the power in milliwatts required to raise a thermistor's temperature 1°C above its environment. Typical values vary from 1 mW/°C in free air to 10 mW/°C or more in an oil bath.

4.5 THERMOCOUPLES

In previous sections, we have considered the change in material resistance as a function of temperature. Such a resistance change is considered a variable parameter property in the sense that the measurement of resistance, and thereby temperature, requires external power sources. There exists another dependence of electrical behavior of materials on temperature which forms the basis of a large percentage of all temperature measurement. This effect is characterized by a voltage-generating transducer in which an emf is produced that is proportional to temperature. Such an emf is found to be almost linear with temperature and very repeatable for constant materials. Devices that measure temperature on the basis of this thermoelectric principle are called thermocouples.

4.5.1 Thermoelectric Effects

The basic theory of the thermocouple effect is found from a consideration of the electrical and thermal transport properties of different metals. In particular, when a temperature differential is maintained across a given metal, the vibration of atoms and motion of electrons is affected in such a manner that a difference in potential exists across the material. This potential difference is related to the fact that electrons in the hotter end of the material have more thermal energy than those in the cooler end and thus tend to drift toward the cooler end. This drift varies for different metals at the same temperature due to differences in their thermal conductivities. If a circuit is closed by connecting the ends through another conductor, a current is found to flow in the closed loop.

The proper description of such an effect is to say that an emf has been established in the circuit which is causing the current to flow. In Figure 4.6a, we see a pictorial representation of this effect where two different metals A and B are used to close the loop with the connecting junctions at temperature T_1 and T_2. We could not close the loop with the same metal because the potential differences across each leg would be the same, and thus no net emf would be present. Note that the emf produced is proportional to the difference in temperature between the two junctions. Theoretical treatments of this problem are found to involve the thermal activities of the two metals.

SEEBECK EFFECT

Using solid state theory the situation above may be analyzed to show that this emf can be given by an integral over temperature

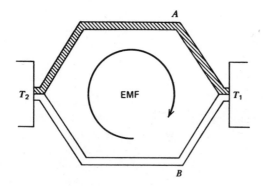

a) Seebeck effect

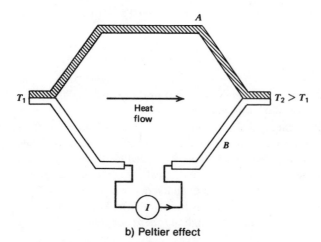

b) Peltier effect

FIGURE 4.6   The Seebeck and Peltier effects refer to the relation between emfs and temperature differences in a two-wire system.

$$\epsilon = \int_{T_1}^{T_2} (Q_A - Q_B)dT$$

Where

ϵ = emf produced in volts
T_1, T_2 = junction temperatures in °K
Q_A, Q_B = thermal transport constants of the two metals

This equation, which describes the *Seebeck effect*, shows that the emf produced is proportional to the *difference* in temperature and, further, to the difference in the metallic thermal transport constants. Thus, if the metals are the same, the emf is zero, and if the temperatures are the same, the emf is also zero.

In practice, it is found that the two constants Q_A and Q_B, are *nearly* independent of temperature and that an approximate linear relationship exists as

$$\epsilon = \alpha(T_2 - T_1)$$

Where

α = constant in volts/°K
T_1, T_2 = junction temperatures in °K

However, the small but finite temperature dependence of Q_A and Q_B is necessary for accurate considerations.

EXAMPLE 4.8

Find the Seebeck emf for a material with $\alpha = 50 \ \mu V/°C$ if the junction temperatures are 20°C and 100°C.

SOLUTION

The emf can be found from

$$\epsilon = \alpha(T_2 - T_1)$$
$$\epsilon = (50 \ \mu V/°C) (100°C - 20°C)$$
$$\epsilon = 4 \ mV$$

PELTIER EFFECT

An interesting and sometimes useful extension of the same thermoelectric properties discussed above occurs when the reverse of the Seebeck effect is considered. In this case, we construct a closed loop of two different metals, A and B as before. Now, however, an external voltage is applied to the system to cause a current to flow in the circuit as shown in Figure 4.6b. Because of the different electrothermal transport properties of the metals, it is found that one of the junctions will be *heated* and the other *cooled*, that is, the device is a refrigerator! This process is referred to as the Peltier effect. Some practical applications of such a device, such as cooling small electronic parts, for example, have been employed.

4.5.2 Thermocouples

In order to use the Seebeck effect as the basis of a temperature transducer, we need to establish a definite relationship between the measured emf of the thermocouple and the unknown temperature. We see first that one temperature must already be known because the Seebeck voltage is proportional to the *difference* between junction temperatures. Furthermore, every connection of different metals made in the thermocouple loop for measuring devices, extension leads, and so on will contribute an emf depending on the difference in metals and various junction temperatures. In order to provide an output that is definite with respect to the temperature to be measured, an arrangement such as that shown in Figure 4.7a is used. This shows that the measurement junction T_M is exposed to the environment whose temperature is to be measured. This junction is formed of metals A and B as shown. Two other junctions are then formed to a common metal C, which then connects to the measurement apparatus. The "reference" junctions are held at a common, known temperature T_R, the reference junction temperature. When an emf is measured, such problems as voltage drops across resistive elements in the loop must be considered. In this arrangement, an open-circuit voltage is measured (at high impedance) which is then a function of only the temperature difference $(T_M - T_R)$ and the type of metals A and B. Note that the voltage produced has a magnitude dependent on the absolute magnitude of the temperature *difference* and a *polarity* dependent on *which* temperature is larger, reference or measurement junction. Thus, it is *not* necessary that the measurement junction have a higher temperature than the reference junctions, but both magnitude and sign of the measured voltage must be noted.

THERMOCOUPLE TYPES

Certain standard configurations of thermocouples using specific metals (or alloys of metals) have been adopted and given letter designations; examples are shown in Table 4.2. Each type has its particular features, such as range, linearity, inertness to hostile environments, sensitivity, and so on, and is chosen for specific applications accordingly. In each type, various sizes of conductors may be employed for specific cases, such as oven measurements, highly localized measurements, and so on. The curves of voltage versus temperature in Figure 4.8 are shown for a reference temperature of 25°C and for several types of thermocouples. We wish to note several important features from these curves.

First, we see that the type J and K thermocouples are noted for their rather large slope,—that is, high sensitivity—making measurements easier for a given change in temperature. We note that the type R and S thermocouples have much less slope

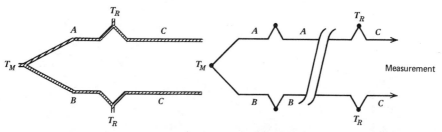

a) Three-wire thermocouple system

b) Use of extension wires in a thermocouple system

FIGURE 4.7   Practical measurements with a thermocouple system employ a third
wire to carry the emf to the measurement device.

and are appropriately less sensitive. They have the significant advantages of a much
larger possible *range* of measurement, including very high temperatures, and they
are very inert materials. Another important feature is the observation that these
curves are not exactly linear. To take advantage of the inherent accuracy possible
with these devices, comprehensive tables of voltage versus temperature have been
determined for many types of thermocouples. Such tables are found in Appendix A.3.

THERMOCOUPLE TABLES

The thermocouple tables simply give the voltage that results for a particular type of
thermocouple when the reference junctions are at a particular reference temperature, and the measurement junction is at a temperature of interest. Referring to the
tables, for example, we see that for a type J thermocouple at 210°C with a 0°C
reference the voltage is

$$V(210°C) = 11.34 \text{ mV} \qquad (\text{type J, 0°C ref.})$$

TABLE 4.2   STANDARD THERMOCOUPLES

Type	Materials[a]	Normal Range
J	Iron-constantan[b]	−190°C to 760°C
T	Copper-constantan	−200°C to 371°C
K	Chromel-alumel	−190°C to 1260°C
E	Chromel-constantan	−100°C to 1260°C
S	90% Platinum + 10% rhodium − platinum	0°C to 1482°C
R	87% Platinum + 13% rhodium − platinum	0°C to 1482°C

[a] First material is more positive when the measurement temperature is more than the reference
temperature.
[b] Constantan, chromel, and alumel are registered trade names of alloys.

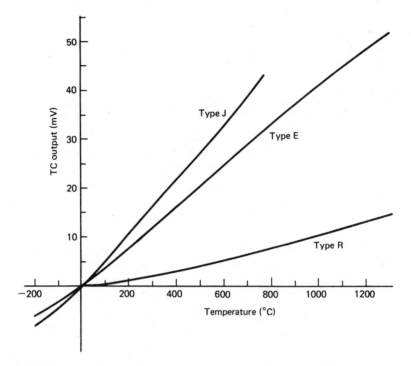

FIGURE 4.8   These curves of TC voltage versus temperature for a 0°C reference show the different sensitivities and nonlinearities of types of thermocouples.

Conversely, if we measured a voltage of 4.768 mV with a type S and a 0°C reference, we find from the table

$$T(4.768 \text{ mV}) = 560°C \qquad \text{(type S, 0°C ref.)}$$

In most cases, the measured voltage does not fall exactly on a table value as in the above case. When this happens, it is necessary to *interpolate* between table values which bracket the desired value. In general, the value of temperature can be found using the following interpolation equation

$$T_M = T_L + \left[\frac{T_H - T_L}{V_H - V_L}\right](V_M - V_L) \qquad (4\text{-}14)$$

Here, the measured voltage V_M lies between a higher voltage V_H and a lower voltage V_L, which are in the tables. The temperatures corresponding to these voltages are T_H and T_L, respectively, as shown in Example 4.9.

EXAMPLE 4.9

A voltage of 23.72 mV is measured with a type K TC at a 0°C reference. Find the temperature of the measurement junction.

SOLUTION

From the table we find $V_M = 23.72$ lies between $V_L = 23.63$ mV and $V_H = 23.84$ mV with corresponding temperatures of $T_L = 570°C$ and $T_H = 575°C$, respectively. The junction temperature is found from Equation (4-14).

$$T_M = 570°C + \frac{(575°C - 570°C)}{(23.84 - 23.63 \text{ mV})} (23.72 \text{ mV} - 23.63 \text{ mV})$$

$$T_M = 570°C + \frac{5°C}{0.21} (0.09 \text{ mV}) \qquad (4\text{-}14)$$

$$T_M = 572.1°C$$

The reverse situation, although not so common in practice, may occur when the voltage for a particular temperature T_M, which is not in the table, is desired. Again, an interpolation equation can be used, such as

$$V_M = V_L + \left[\frac{V_H - V_L}{T_H - T_L}\right] (T_M - T_L) \qquad (4\text{-}15)$$

where all terms are as defined for Equation (4-14).

EXAMPLE 4.10

Find the voltage of a type-J TC with a 0°C reference if the junction temperature is −172°C.

SOLUTION

We do not let the signs bother us but merely apply the interpolation relation directly. From the tables, we see that the junction temperature lies between a high (algebraically) $T_H = -170°C$ and a low of $T_L = -175°C$. The corresponding voltages are $V_H = -7.12$ mV $V_L = -7.27$. The TC voltage will be

$$V_M + -7.27 \text{ mV} + \frac{-7.12 + 7.27}{-170 + 180} (-172°C + 175°C)$$

$$V_M = -7.27 \text{ mV} + \frac{0.15 \text{ mV}}{5°C} (3°C) \qquad (4\text{-}15)$$

$$V_M = -7.18 \text{ mV}$$

CHANGE OF TABLE REFERENCE

It has already been pointed out that thermocouple tables are prepared for a particular junction temperature. It is possible to use these tables with a thermocouple (TC) which has a different reference temperature by an appropriate shift in the table scale. The key point to remember is that the voltage is proportional to the difference between the reference and measurement junction temperature. Thus, if a new reference is greater than the table reference, all voltages of the table will be less for this thermocouple (TC). The amount less will be just the voltage of the new reference as found on the table. Perhaps a few examples are in order here. Suppose we have a type-J TC with a 30°C reference. On the 0°C reference table, a type-J at 30°C will produce 1.54 mV. This means that any temperature with this TC will generate a voltage 1.54 mV less than those in the table. Thus, referring to the table,

$$400°C \text{ results in } V = 21.85 - 1.54 = 20.13 \text{ mV (type-J, 30°C)}$$
$$150°C \text{ results in } V = 8.00 - 1.54 = 6.28 \text{ mV (type-J, 30°C)}$$
$$-90°C \text{ results in } V = -4.21 - 1.54 = -5.75 \text{ mV (type-J, 30°C)}$$

In a similar fashion, if the new reference is lower than the reference, all of the table voltages will be larger. For example, consider a type-K thermocouple with a reference at $-26°C$. First, by interpolation, we find the voltage that this corresponds to on the 0°C reference tables

$$V(-26°C) = -1.14 + \frac{-0.95 + 1.14}{-25 + 30}(-26 + 30)$$

$$V(-26°C) = -0.98 \text{ mV (type-K, 0°C ref)}$$

Thus, every voltage on the table must be increased by 0.98 mV, so

$$400°C \text{ results in } V = 16.40 + 0.98 = 17.38 \text{ mV}$$
$$150°C \text{ results in } V = 6.13 + 0.98 = 7.11 \text{ mV}$$
$$-90°C \text{ results in } V = -3.19 + 0.98 = -2.21 \text{ mV}$$

In effect, we are sliding the curves of TC voltage versus temperature along the temperature axis to give a zero voltage at the reference being used. This is shown in Figure 4.9. Note that the shifts made, as in the examples above, are not exact because of the dependence on temperature of the metallic thermoelectric constants. If a very large difference in temperature exists between the table reference temperature and the reference being used, inaccuracies exist.

4.5.3 Thermocouple Transducers

The use of a thermocouple for a temperature transducer has evolved from an elementary process with crudely prepared thermocouple constituents into a very precise and exacting technique.

SENSITIVITY

A review of the tables shows that the range of thermocouple voltages is typically less than 100 mV. The actual sensitivity strongly depends on the type of signal conditioning employed and on the TC itself. We see from Figure 4.8 the following worst and best case of sensitivity:

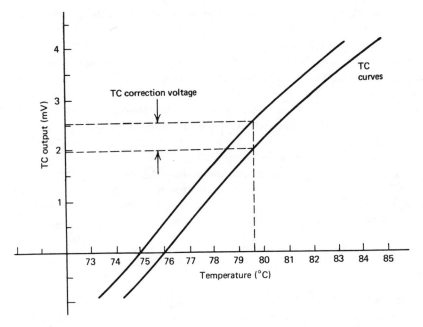

FIGURE 4.9   When the reference temperature is changed, a correction voltage must be applied. Thus, if the tables are for 75°C and the actual reference is 76°C, a correction of 0.5 mV is needed.

- ▫ Type J: 0.05 mV/°C (typical)
- ▫ Type R: 0.006 mV/°C (typical)

CONSTRUCTION

A thermocouple (TC) by itself is, of course, simply a welded or even twisted junction between two metals and, in many cases, that is the construction. There are cases, however, where the TC is sheathed in a protective covering or even sealed in glass to protect the unit from a hostile environment. The size of the TC wire is determined by the application and can range from #10 wire in rugged environments to fine #30 AWG wires or 0.02-mm microwire in refined biological measurements of temperature.

SIGNAL CONDITIONING

In general, the most critical element in TC signal conditioning is the need to make the measurements at high impedance. Even though the internal d-c resistance of a TC is very small (mΩ), the voltage generated is also very small. Thus, if high current is drawn from the TC, an error of several percent can occur in a reading. Historically, TC voltages are measured using the potentiometer circuit, described in Chapter 2, which effectively makes the measurement at an infinite impedance. Modern techniques have evolved which allow electronic means of measurement, such as using an electrometer containing a field-effect transistor with its inherently high input impedance or an appropriate operational amplifier configuration with high input impedance. Another factor of importance in the use of TC is the requirement of known reference temperature of the reference junctions. In many applications, particularly field usage, a thermometer is used to determine local ambient temperature, which is also the *reference* temperature. A correcting factor, such as was discussed in a previous section, is then used to provide a corrected, measured TC voltage from which to determine the temperature. In some cases, it is desirable to place the reference junctions at a point far removed from the measurement junction as, for example, when the temperature in the vicinity of the measurement junction varies over wide limits. In these cases, *extension wires* are used which are constructed of the *same* material as the TC themselves. The actual reference junctions are not formed until a junction is made to the third wire as shown in Figure 4.7b. The extension wires must be of the same material as the TC, but they can be stranded making them more flexible than the solid thermocouple wires.

NOISE

Perhaps the biggest obstacle to the use of thermocouples for temperature measurement in industry is their susceptibility to electrical noise. First, the voltages generated generally are less than 50 mV and often are only 2 or 3 millivolts, and in the industrial environment it is common to have hundreds of millivolts of electrical noise generated by large electrical machines in any electrical system. Second, a thermocouple constitutes an excellent antenna for pickup of noise from electromagnetic radiation in the radio, TV, and microwave bands. In short then, a bare thermocouple may have many times more noise than temperature signal at a given time.

In order to use thermocouples effectively in industry, a number of noise reduction techniques are employed. The following three are the most popular:

1. The extension or lead wires from the thermocouple to the reference junction or measurement system are twisted and then wrapped with a grounded foil sheath.
2. The measurement junction itself is grounded at the point of measurement. The grounding is typically to the inside of the stainless steel sheath which covers the actual thermocouple. This means that that voltage generated must be measured by a differential system, which cancels noise which appears in equal magnitude on each wire of the thermocouple.
3. An instrumentation amplifier which has excellent common mode rejection is employed for measurement. This goes hand-in-hand with the use of a grounded junction.

4.6 OTHER THERMAL TRANSDUCERS

The transducers discussed in the previous sections cover a large fraction of the temperature-measurement techniques used in process control. There remain, however, numerous other devices or methods of temperature measurement which may be encountered. Pyrometric methods involve measurement of temperature by the electromagnetic radiation which is emitted in proportion to temperature. This technique is discussed in detail in Chapter 6. Several other techniques are discussed in this section.

4.6.1 Bimetal Strips

This type of temperature transducer has the characteristics of being relatively inaccurate, having hysteresis, having relatively slow time response, and being low cost. Such devices are used in numerous applications, particularly where an on/off cycle rather than smooth or continuous control is desired.

THERMAL EXPANSION

We have seen that greater thermal energy causes the molecules of a solid to execute greater amplitude and higher frequency vibrations about their average positions. It is natural to expect that an expansion of the volume of a solid would accompany this effect, since the molecules tend to occupy more volume on the average with their vibrations. This effect varies in degree from material to material due to many factors, including molecule size and weight, lattice structure, and others. Although one can speak of a volume expansion, as described, it is more common to consider a length expansion when dealing with solids, particularly in the configuration of a rod or beam. Thus, if we have a rod of length l_0 at temperature T_0 as shown in Figure 4.10, and the temperature is raised to a new value T, then the rod will be found to have a new length l given by

$$l = l_0[1 + \gamma \, \Delta T] \tag{4-16}$$

where $\Delta T = T - T_0$ and γ is the linear thermal expansion coefficient appropriate to the material of which the rod is produced. Several different expansion coefficients are given in Table 4.3.

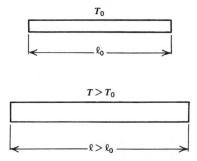

FIGURE 4.10   A solid object experiences a physical expansion in proportion to temperature. The effect is highly exaggerated in this representation.

TABLE 4.3 THERMAL EXPANSION
COEFFICIENTS

Material	Expansion Coefficient
Aluminum	$25 \times 10^{-6}/°C$
Copper	$16.6 \times 10^{-6}/°C$
Steel	$6.7 \times 10^{-6}/°C$
Beryllium/copper	$9.3 \times 10^{-6}/°C$

BIMETALLIC TRANSDUCER

The thermal transducer exploiting the effect discussed above occurs when two materials with grossly different thermal expansion coefficients are bonded together. Thus, when heated, the different expansion rates cause the assembly curve shown in Figure 4.11. This effect can be used to close switch contacts, or to actuate an on/off mechanism when the temperature increases to some appropriate setpoint. The effect also is used for temperature indicators, by means of assemblages, to convert the curvature into dial rotation.

EXAMPLE 4.11

How much will an aluminum rod of 10 m length at 20°C expand when the temperature is changed from 0°C to 100°C?

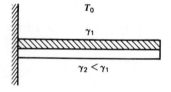

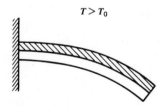

FIGURE 4.11 A bimetal strip will curve when exposed to a temperature change because of the different thermal expansion coefficients of the metals. The thickness of the metal strips has been exaggerated.

SOLUTION

First find the length of 0°C and at 100°C, then find the difference. Using Equation (4-16) at 0°C, we get

$$l_1 = (10 \text{ m}) [1 + (2.5 \times 10^{-5}/°C) (0°C - 20°C)] \qquad (4-16)$$
$$l_1 = 9.995 \text{ m}$$

and at 100°C

$$l_2 = 10 \text{ m } 1 + (2.5 \times 10^{-5}/°C) (100°C - 20°C)$$
$$l_2 = 10.02 \text{ m}$$

Thus, the expansion is

$$l_2 - l_1 = 0.025 \text{ m} = \textbf{25mm}$$

4.6.2 Gas Thermometers

The operational principle of the gas thermometer is based on a basic law of gases. In particular, if a gas is kept in a container at constant volume and the pressure and temperature vary, then the ratio of gas pressure and temperature is a constant,

$$\frac{p_1}{T_1} = \frac{p_2}{T_2} \qquad (4-17)$$

Where

p_1, T_1 = absolute pressure and temperature (in K) in state 1
p_2, T_2 = absolute pressure and temperature in state 2

EXAMPLE 4.12

A gas in a closed volume has a pressure of 120 psi at a temperature of 20°C. What will the pressure be at 100°C?

SOLUTION

First, we convert the temperature to the absolute scale of Kelvin using Equation (4-2). We get 293.15 K and 373.15 K. Then, we use Equation (4-17) to find the pressure in state 2,

$$p_2 = \frac{T_1}{T_2} p_1$$

$$p_2 = \frac{373.15}{293.15}(120 \text{ psi})$$

$$p_2 = 153 \text{ psi}$$

Since the gas thermometer converts temperature information directly into pressure signals, it is particularly useful in pneumatic systems. Such transducers are also advantageous because there are no moving parts, and no electrical stimulation is necessary. For electronic analog or digital process control applications, however, it is necessary to devise systems for converting the pressure to electrical signals. This type of transducer is often used with bourdon tubes (see Chapter 5) to produce direct indicating temperature meters and recorders. The gas most commonly employed is nitrogen. Time response is quite slow in relation to electrical devices because of the greater mass which must be heated.

4.6.3 Vapor Pressure Thermometers

A vapor pressure thermometer converts temperature information into pressure as does the gas thermometer, but it operates by a different process. If a closed vessel is partially filled with liquid, then the space above the liquid will consist of evaporated vapor of the liquid at a pressure which depends on the temperature. If the temperature is raised, more liquid will vaporize and the pressure will increase. A decrease in temperature also will result in condensation of some of the vapor, and the pressure will decrease. Thus, vapor pressure depends on temperature. Different materials will have different curves of pressure versus temperature, and there is no simple equation like that for a gas thermometer. Figure 4.12 shows a curve of vapor pressure versus temperature for methyl chloride, which is often employed in these transducers. Note that the pressure available is substantial as the temperature rises. As in the case of gas thermometers, the range is not great, and response time is slow (20 seconds and more) because the liquid and the vessel must be heated.

EXAMPLE 4.13

Two methyl chloride vapor pressure temperature transducers will be used to measure the temperature difference between two reaction vessels. The nominal temperature is 85°C. Find the pressure difference per degree Celcius at 85°C from the graph in Figure 4.12.

SOLUTION

We are just estimating the slope of the graph in the vicinity of 85°C. To do this let us find the slope between 80°C and 90°C,

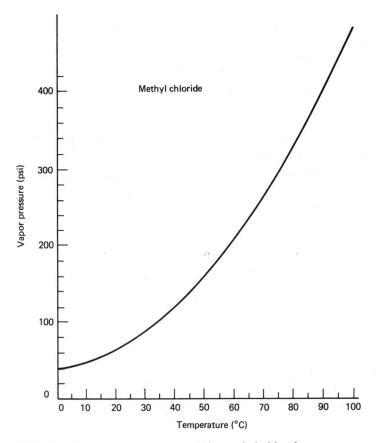

FIGURE 4.12   Vapor pressure curve for methyl chloride.

$$\frac{\Delta p}{\Delta T} = \frac{400 - 320 \text{ psi}}{90 - 80°C}$$

$$\frac{\Delta p}{\Delta T} = 8 \text{ psi/°C}$$

4.6.4 Liquid-Expansion Thermometers

Just as a solid experiences an expansion in dimension with temperature so does a liquid show an expansion in volume with temperature. This effect forms the basis for the traditional liquid-in-glass thermometers which are so common in temperature measurement. The relationship that governs the operation of this device is,

$$V(T) = V(T_0) [1 + \beta \, \Delta T] \tag{4-18}$$

Where

$V(T)$ = volume at temperature T
$V(T_0)$ = volume at temperature T_0
$\Delta T = T - T_0$
β = volume thermal expansion coefficient

In actual practice the expansion effects of the glass container must be accounted for high accuracy in temperature indications. This type of temperature transducer is not commonly used in process-control work because further transduction is necessary to convert the indicated temperature into an electrical signal.

4.7 DESIGN CONSIDERATIONS

In the design of overall process-control systems, specific requirements are set up for each element of the system. The design of the elements themselves, which constitutes subsystems, involves careful matching of the elemental characteristics to the overall system design requirements. Even in the design of monitoring systems, where no integration of subsystems is required, it is necessary to match the transducer to the measurement environment and to the required output signal. In keeping with these considerations, we can treat temperature transducer design procedures by the following steps.

1. *Identify the Nature of the Measurement.* This includes the nominal value and range of the temperature measurement, the physical conditions of the environment where the measurement is to be made, required speed of measurement, and any other features that must be considered.

2. *Identify the Required Output Signal.* In most applications the output will be either a standard 4–20 mA current or a voltage which is scaled to represent the range of temperature in the measurement. There may be further requirements related to isolation, output impedance, or other factors. In some cases, a specific digital encoding of the output may be specified.

3. *Select an Appropriate Temperature Transducer.* Based primarily on the results of the first step, a transducer which matches the specifications of range, environment, and so forth is selected. To some extent, factors such as cost and availability will be important in the selection of a transducer. The requirements of output signals also enter into this selection but with lower significance because signal conditioning generally provides the required signal transformations.

4. *Design the Required Signal Conditioning.* Using the signal conditioning techniques treated in Chapters 2 and 3, the direct transduction of temperature by the transducer is converted into the required output signal. The specific type of signal conditioning depends, of course, on the type of transducer employed as well as the nature of the specified output signal characteristics.

In one form or another these steps are required of any temperature transducer application, although they are not necessarily performed in the sequence indicated.

The primary concern of this text is to enable the reader to accomplish transducer selection and signal conditioning associated with a particular requirement. Note that in a particular situation where a thermal transducer is required there is no unique design to fit the application. There are, in fact, so many different designs possible that one must adjust his or her thinking from searching for *the* solution to searching for a solution to *a* design problem. A solution is *any* arrangement that satisfies *all* of the specified requirements of the problem. The following examples illustrate several problems in thermal design and typical solutions.

EXAMPLE 4.14

Design a system which will turn on an alarm light when an oven temperature exceeds 140°C. The light is actuated by a relay with a 5.5 mA pull-in and a 4.1 mA release having a 1000 Ω resistance. Specify the light turn-off temperature.

SOLUTION

This is a simple example that could be solved using many approaches. Let us first identify the various parts of the design criteria.

1. *Identify Measurement.* All that is specified here is a 140°C oven.
2. *Output Signal.* This can be found from Ohm's law applied to the relay. Thus, an output is required at 140°C of 5.5 mA into 1000 Ω or

$$V = IR = (5.5 \text{ mA}) (1000 \text{ }\Omega)$$
$$V = 5.5 \text{ volts}$$

3. *Select Transducer.* Here many transducers would work quite well. Let us use a thermistor specified such that the resistance at 140°C is 701.2 Ω.
4. *Signal Conditioning.* Again, many configurations would work but for the sake of illustration, let us use a simple op amp d-c amplifier with the thermistor in the circuit as shown in Figure 4.13. Thus, we have,

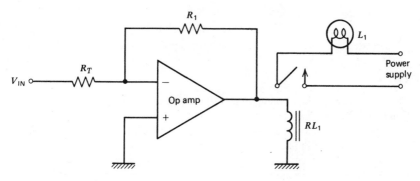

FIGURE 4.13   This circuit represents one possible solution to Example 4.14.

$$V_{OUT} = -\frac{R_1}{R_T} V_{IN}$$

We can select V_{IN} to be any convenient value, let us say $V_{IN} = 1$ volt. Thus, pull-in requires

$$V_{OUT} = -5.5 \text{ volts} = -\frac{R_1}{701.2 \ \Omega} 1 \text{ volt}$$

or

$$R_1 = 3856.6\Omega$$

This completes the design except for specifying the temperature at which the relay will open. This occurs when the voltage drops to $V_0 = 4.1$ volts. Then, we simply find the thermistor resistance at which this will occur and determine the temperature corresponding to this resistance. Thus,

$$V_{OUT} = 4.1 \text{ volts} = \frac{3856.6 \ \Omega}{R_T} 1 \text{ volt}$$

$$R_T = 940.66 \ \Omega$$

From the specs we find that $R = 951.1 \ \Omega$ at 128°C and 926.7 Ω at 129°C. Interpolating between these we find the light-off temperature to be

$$T(OFF) = 128°C + \frac{129°C - 128°C}{951.1 - 926.7}(951.1 - 940.6) \tag{4-14}$$

$$T(OFF) = 128.43°C$$

EXAMPLE 4.15

Design a temperature alarm that will use an RTD with $\alpha(30°C) = 0.002/°C$ and $R(30°C) = 140 \ \Omega$ and $\tau = 5$ s. If the temperature changes suddenly from 30°C to 35°C, the system should alarm after 1.5 seconds by triggering a +1 volt trigger.

SOLUTION

1. *Environment.* The specified condition is a 30°C nominal temperature which may suddenly change to 35°C. An alarm must be set within 1.5 seconds.
2. *Output.* Here we need a voltage at the input of a trigger which reaches +1 volts after 1.5 s from the 30°C to 35°C jump.
3. *Transducer.* The transducer has been specified in the problem statement.
4. *Signal Conditioning.* One solution to this problem is shown in Figure 4.14. Here a bridge is used to detect the RTD resistance change and a differential amplifier provides the

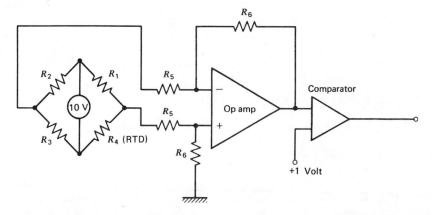

FIGURE 4.14   This circuit represents one possible solution to Example 4.15.

gain to assure the change is detected by providing a +1 volt signal to the trigger at 1.5 s. The components necessary to accomplish this are found from the following analysis.

The RTD resistance at 35°C is found from Equation (4-9) to be

$$R(35°C) = 140 [1 + (.002/°C) (5°C)]$$
$$R(35°C) = 141.4 \ \Omega$$

(4-9)

The time variation is found from the response time equation.

$$R(t) = 140 \ \Omega + 1.4(1 - e^{-t/5}) \ \Omega$$

Now, from Section 2.3.1, we know the bridge offset voltage will be

$$\Delta V = \left[\frac{R_4}{R_1 + R_4} - \frac{R_3}{R_2 + R_3} \right] V_0$$

(2-7)

We choose R_4 to be the RTD and R_1, R_2, R_3 so that the bridge is balanced at 30°C. This can be accomplished by setting $R_3 = 140 \ \Omega$, $R_1 = R_2 = 100 \ \Omega$. Now we can write the offset voltage as a function of time by using $R(t)$ and noting that $R_1 + R_4 \cong 240 \ \Omega$, that is, ignoring the 1.4 Ω change.

$$\Delta V(t) = \frac{(1.4 \ \Omega) (10 \ V)}{240 \ \Omega} [1 - e^{-t/5}]$$

$$\Delta V(t) = 0.06[1 - e^{-t/5}] \text{volts}$$

At a time of 1.5 seconds the voltage will be

$$\Delta V(1.5 \ s) = 0.06[1 - e^{-1.5/5}]$$
$$\Delta V(1.5 \ s) = 0.016 \text{ volts}$$

Thus, to get the required 1-volt output we need a gain of

$$G = \frac{1}{0.015} = 67$$

This can be provided for by making $R_5 = 1 \text{ k}\Omega$ and $R_6 = 67 \text{ k}\Omega$, which completes the design.

SUMMARY

1.   The measurement and control of temperature plays a very important role in the process-control industry. The class of transducers that performs this measurement consists primarily of three types: (1) the resistance temperature detector (RTD), (2) the thermistor, and (3) the thermocouple. In this chapter the basic operating principles and application information have been provided for these transducers. Several other transducers have been briefly described.

2.   The concept of temperature is contained in the representation of a body's thermal energy as the average thermal energy per molecule expressed in units of degrees of temperature. Four units of temperature are in common use. Two of these scales are called absolute because an indication of zero units corresponds to zero thermal energy. These scales, which are designated by Kelvin (K) and degrees Rankine (°R), differ only in that the size of each degree is different; that is, the amount of energy represented by 1 K is different. The amount of energy represented by 1 K would correspond to 9/5°R. Thus, we can transform temperatures using

$$T(\text{K}) = \tfrac{5}{9} T(\text{°R}) \tag{4-1}$$

3.   The other two scales are called relative because their zero does not occur at a zero of thermal energy. The Celsius scale (°C) corresponds in degree size to the Kelvin but has a shift of the zero so that

$$T(\text{°C}) = T(\text{K}) - 273.15 \tag{4-2}$$

4.   In a similar fashion the Fahrenheit (°F) and Rankine scales are related by

$$T(\text{°F}) = T(\text{°R}) - 459.6 \tag{4-3}$$

5.   The RTD is a transducer which depends on the increase in metallic resistance with temperature. This increase is very nearly linear, and analytical approximations are used to express the resistance versus temperature as either a linear equation

$$R(T) = R(T_0)[1 + \alpha_0 \Delta T] \tag{4-9}$$

or quadratic relationship

$$R(T) = R(T_0)[1 + \alpha_1 \Delta T = \alpha_2 (\Delta T)^2] \tag{4-12}$$

6.   When greater accuracy is desired, tables or graphs of resistance versus temperature are

used. Because of the small fractional change in resistance with temperature, the RTD is usually used in a bridge circuit with a high-gain null detector.

7.   The thermistor is based on the decrease of semiconductor resistance with temperature. This device has a highly nonlinear resistance versus temperature curve and is not typically used with any analytical approximations. Such transducers can exhibit a very large change in resistance with temperature and hence make very sensitive temperature change detectors. Many circuit configurations are used, including bridges and operational amplifiers.

8.   A thermocouple is a junction of two dissimilar metal wires usually joined to a third metal wire through two reference junctions. A voltage is developed across the common metal wires which is proportional, almost linearly, to the difference in temperature between the measurement and reference junctions. Extensive tables of temperature versus voltage for numerous types of TCs using standard metals and alloys allow an accurate determination of temperature at the reference junctions. The voltage must be measured at high impedance to avoid loading effects on the measured voltage. Thus, potentiometric, operational amplifiers, or other high-impedance techniques are employed in signal conditioning.

9.   A bimetal strip converts temperature into a physical motion of metal elements. This flexing can be used to close switches or cause dial indications.

10.   Gas and vapor pressure temperature transducers convert temperature into gas pressure which then is converted to an electrical signal or is used directly in pneumatic systems.

PROBLEMS

4.1   Convert 453.1°R into K, °F, and °C.

4.2   Convert −222°F into °C, °R, and K.

4.3   A control system is used to change a process temperature by 33.4°F. Calculate the change in °C. (*Hint:* Note that a change in temperature does not involve a scale shift.)

4.4   A sample of hydrogen gas has a temperature of 500°C. Calculate the average molecular speed in m/s. Express this also in ft/s. (mass of H_2 molecule is 3.3×10^{-27} kg).

4.5   An RTD has an $\alpha(20°C) = 0.004/°C$, if $R = 106 \; \Omega$ at 20°C, find the resistance at 25°C.

4.6   The RTD of Problem 4.5 is used in the bridge circuit of Figure 4.4. If $R_1 = R_2 = R_3 = 100 \; \Omega$ and the supply voltage is 10 volts, calculate the voltage the detector must resolve to define a 1°C change in temperature.

4.7   In Problem 4.6 the RTD is replaced by a thermistor with $R(20°C) = 100 \; \Omega$ and a change rate of $-10 \; \Omega/°C$ about 20°C. Calculate the voltage resolution of the detector to resolve a 1°C change in temperature.

4.8   Use the values of resistance versus temperature shown below to find the linear and the quadratic approximations of resistance between 100°C and 130°C about 115°C.

T(°C)	R(Ω)
90	562.66
95	568.03
100	573.40
105	578.77
110	584.13
115	589.48
120	594.84
125	600.18
130	605.52

4.9   A type-J thermocouple measures 22.5 mV with a 0°C reference. Calculate the junction temperature.

4.10   A type-S TC with a 21°C reference measures 12.120 mV. Calculate the junction temperature.

4.11   If a type-J TC is to measure 500°C with a −10°C reference, what voltage is produced?

4.12   A type-K TC with 0°C reference will monitor an oven temperature at 300°C. Extension wires of 1000-ft length and 0.01 Ω/ft will be used. Calculate the input impedance which the voltage measurement system must have to make measurements within 0.2%.

4.13   Using op amps, comparator, and a type-J TC, design a system which will signal an alarm when a process temperature exceeds 700°C.

4.14   Using an RTD with $\alpha(20°C) = 0.0034/°C$, $R(20°C) = 100\ \Omega$ design a bridge and op-amp arrangement to provide a 0- to 10-volt signal for a temperature span of 20°C to 100°C.

4.15   The RTD of Problems 4.5 and 4.6 has a dissipation constant of 25 mW/°C. Used as in Problem 4.6, the RTD is placed in a bath at 20°C. Find the bridge offset voltage.

4.16   A copper rod 10 cm long at 20°C is to touch a contact when the temperature reaches 150°C. How far from the end of the rod should the contact be placed?

4.17   A gas thermometer has a pressure of 125 kPa at 0°C and 215 kPa at some unknown temperature. Find this temperature in °C.

4.18   A welding gas bottle has a pressure of 500 psi at 0°F. What will the pressure be at 110°F?

4.19   A methyl chloride vapor-pressure thermometer will be used between 70°F and 200°F. What pressure range does this correspond to?

4.20   Find a linear approximation of methyl chloride pressure versus temperature from 70°C to 90°C. Find the maximum difference between the approximation and the actual value, expressed as a percentage error.

4.21   A type-K TC with a 0°C reference will be used to measure temperature between 200°C and 350°C. Devise a system which will convert this temperature range into an 8-bit digital word with 00h meaning 200°C and FFh meaning 350°C.

5

MECHANICAL
TRANSDUCERS

INSTRUCTIONAL OBJECTIVES

The objectives of this chapter are confined to the subject of mechanical transducers. After you have read it, you should be able to:

1. Define the relationship between acceleration, velocity, and position.
2. Define the characteristics of vibration and shock.
3. Draw and label a typical stress/strain curve.
4. Design the application of an LVDT to a displacement measurement problem.
5. Describe the types of accelerometers and the characteristics of each.
6. Design a system of strain measurement using metal foil strain gages.
7. Define two types of pressure measurement with electrical signal output.
8. Diagram a system of flow measurement using differential pressure measurement.

5.1 INTRODUCTION

The class of transducers used for the measurement of *mechanical* phenomena is of special significance because of the extensive use of these devices throughout the process-control industry. In many instances, an interrelation exists by which a transducer designed to measure some mechanical variable is used to measure another variable. To learn to use mechanical transducers, it is important to understand the mechanical phenomenon itself, the operating principles of the transducer, and the application details of the transducer. All of these will be discussed for the measurement devices presented in this chapter.

The objectives of this chapter are confined to the subject of mechanical transducers. Each of the mechanical dynamic variables presented is itself a complex field of study. Our purpose here is to give an overview of the essential features associated with each to make the reader conversant with the principal transducers used to measure mechanical variables, the characteristics of each, and appropriate application notes. As in previous chapters, an expert understanding of the phenomenon is not required in order to effectively employ transducers for its measurement.

5.2 DISPLACEMENT, LOCATION, OR POSITION TRANSDUCERS

The measurement of displacement, position, or location is an important topic in the process industries. Examples of industrial requirements to measure these variables are many and varied, and the required transducers are also of greatly varied designs. To give a few examples: (1) location and position of objects on a conveyor system, (2) orientation of steel plates in a rolling mill, (3) liquid/solid level measurements, (4) location and position of work piece in automatic milling operations, and (5) conversion of pressure to a physical displacement which is measured to indicate pressure. In the sections to follow, the basic principles of several common types of displacement, position, and location transducers are given.

5.2.1 Potentiometric

The simplest type of displacement transducer involves the action of displacement in moving the wiper of a potentiometer. This device then converts linear or angular motion into a changing resistance which may be converted directly to voltage and/or current signals. Such potentiometric devices often suffer from the obvious problems of mechanical wear, friction in the wiper action, limited resolution in wirewound units, and high electronic noise. (See Figure 5.1.)

5.2.2 Capacitive and Inductive

A second class of transducers for displacement measurement involves changes in capacity or inductance.

CAPACITIVE

The basic idea of a capacitive displacement transducer is that the capacity of a two-plate assembly is proportional to the distance between the plates and the common area of the plates ($C \propto A/d$). Thus, by moving the plates closer, the capacity is increased, and by reducing the exposed area, the capacity is reduced. In either case, the linear or angular displacement can be converted to a changing capacity. An a-c

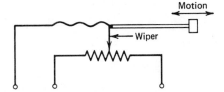

FIGURE 5.1   Potentiometric displacement transducer.

bridge circuit or other active electronic circuit is employed to convert the capacity change to a current or voltage signal. Figure 5.2 shows two common types of capacitive transducers.

INDUCTIVE

If a permeable core is inserted into an inductor as shown in Figure 5.3, the net inductance is increased. Every new position of the core produces a different inductance. In this fashion, the inductor and movable core assembly may be used as a displacement transducer. An a-c bridge or other active electronic circuit sensitive to inductance may then be employed for signal conditioning.

5.2.3 Variable Reluctance

The class of variable reluctance displacement transducers differs from the inductive in that a moving core is used to vary the magnetic flux coupling between two or more coils, rather than changing an individual inductance. Such devices find application in many circumstances for the measure of both translational and angular displacements. Many configurations of this device exist but the most common and extensively used is called a *linear variable differential transformer* (LVDT).

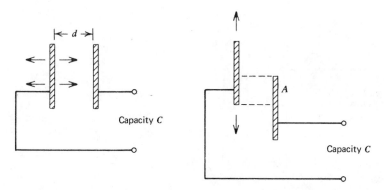

FIGURE 5.2   The capacity varies with the distance between the plates and the exposed area. Both effects are employed in transducers.

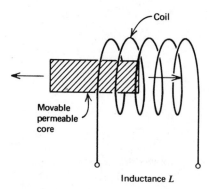

FIGURE 5.3   Different core positions in the coil produce changes in inductance which can be employed for displacement measurement.

LVDT

The basic structure of the LVDT is a movable core of permeable material and three coils, shown in Figure 5.4. The inner core is the primary which provides magnetic flux through its excitation by some a-c source. The two secondary coils have voltages induced due to the flux linkages with the primary. When the core is centrally located, the voltage induced in each secondary is the same. But when the core is displaced, the change in flux linkage causes one secondary voltage to increase and the other to decrease. The secondary windings are generally connected in series opposition so that the voltages induced in each are out of phase with the other. In this case, as shown in Figure 5.5, the output voltage amplitude is zero when the core is centrally located and increases as the core is moved either in or out. It happens that the voltage amplitude is *linear* with core displacement over

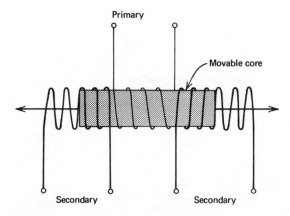

FIGURE 5.4   A linear variable differential transformer (LVDT) has a primary excitation coil and two series-connected secondary coils with a movable core.

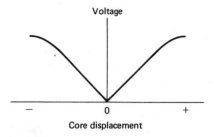

FIGURE 5.5   Over some range of core motion the LVDT voltage amplitude versus core displacement is very linear. This is for opposition connected secondaries.

some range of core travel. Furthermore, there is a phase shift as the core moves both to and from the central location, so that phase measurement relates the direction of core motion.

A signal conditioning circuit providing a d-c level proportional to amplitude and a polarity proportional to direction is easily designed. The LVDT is a device that outputs a *bipolar* voltage *linearly proportional* to displacement. A simple circuit is shown in Figure 5.6.

Commercial LVDTs which can resolve displacements as small as 0.002 mm or 2 μm are available. Some have builtin oscillator excitation, phase sensitive detector, and demodulation in an IC so that only a power supply is required. The d-c voltage output is linear with displacement. Many variations of this device have been designed for special applications, including angular displacement measurements.

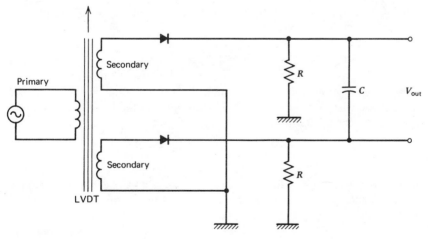

FIGURE 5.6   This LVDT and circuit provide an output voltage relating the direction of core motion by the polarity and the extent of motion by the amplitude of the output dc voltage.

5.3 STRAIN TRANSDUCERS

Although not obvious at first, the measurement of strain in solid objects is very common in process control. The reason it is not obvious is that the strain measurement is used as a secondary step in the measurement of many other process variables, including flow, pressure, weight, and acceleration. Strain measurements have been used to measure pressures from over a million pounds per square inch to those within living biological systems. We will first review the concept of strain and how it is related to the forces which produce it, and then discuss the transducers used to measure strain.

5.3.1 Strain and Stress

Strain is the result of the applications of forces to solid objects. The forces are defined in a special way described by the general term, *stress*. For those readers needing a review of force principles, Appendix A-4 discusses elementary mechanical principles, including force. In this section we will define stress and the resulting strain.

DEFINITION

A special case exists for the relation between force applied to a solid object and the resulting deformation of that object. Solids are assemblages of atoms in which the atomic spacing has been adjusted to render the solid in equilibrium with all external forces acting on the object. This spacing determines the physical dimensions of the solid. If the applied forces are changed, the object atoms rearrange themselves again to come into equilibrium with the *new* set of forces. This rearrangement results in a change in physical dimensions which is referred to as a *deformation* of the solid.

The study of this phenomena has evolved into a very exact technology. The effect of applied force is referred to as a *stress* and the resulting deformation as a *strain*. In order to facilitate a proper analytical treatment of the subject, stress and strain are carefully defined to emphasize the physical properties of the material being stressed and the specific type of stress applied. We delineate here the three most common types of *stress-strain* relationships.

TENSILE STRESS-STRAIN

In Figure 5.7a, the nature of a tensile force is shown as a force applied to a sample of material in such a way as to elongate or pull apart the sample. In this case, the stress is defined as

$$\text{Tensile Stress} = \frac{F}{A} \tag{5-1}$$

Where

    F = applied force in N
    A = cross-sectional area of the sample in m²

We see that the units of stress are N/m² in the SI units (or lb/in² in the English units) and are thus like a pressure.

The *strain* in this case is defined as the *fractional change in length* of the sample

$$\text{Tensile Strain} = \frac{\Delta\ell}{\ell} \tag{5-2}$$

Where

    $\Delta\ell$ = change in length in m (in.)
    ℓ = original length in m (in.)

Strain is thus a unitless quantity.

COMPRESSIONAL STRESS-STRAIN

The only differences between *compressional* and *tensile* stress are the direction of the applied force and the polarity of the change in length. Thus, in a compressional stress, the force is such as to *compress* the sample, shown in Figure 5.7b. The compressional stress is defined as in Equation (5-1).

$$\text{Compressional Stress} = \frac{F}{A} \tag{5-3}$$

The resulting strain is also defined as the fractional change in length but where the sample will now decrease in length.

$$\text{Compressional Strain} = \frac{\Delta\ell}{\ell} \tag{5-2}$$

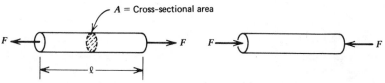

a) Tensile stress applied to a rod   b) Compressional stress applied to a rod

FIGURE 5.7   Tensile and compressional stress can be defined in terms of forces applied to a rod.

SHEAR STRESS-STRAIN

Figure 5.8*a* shows the nature of the shear stress. In this case, the force is applied as a *couple* (that is, *not* along the same line), tending to shear off the solid object which separates the force arms. In this case, the stress is again

$$\text{Shear Stress} = \frac{F}{A} \tag{5-4}$$

Where

F = force in N
A = cross-sectional area of sheared member in m²

Now the strain in this case is defined as the fractional change in dimension of the sheared member. This is shown in the cross-sectional view of Figure 5.8*b*.

$$\text{Shear Strain} = \frac{\Delta x}{\ell} \tag{5-5}$$

Where

Δx = deformation in m (as shown in Figure 5.8*b*)
ℓ = width of a sample in m

STRESS-STRAIN CURVE

If a specific sample is exposed to a range of applied stress, and the resulting strain is measured, a graph similar to Figure 5.9 results. This shows that the relationship between stress and strain is *linear* over some range of stress. If the stress is kept *within* the *linear* region, the material is essentially *elastic* in that if the stress is removed, the deformation is also gone. But if the elastic limit is exceeded, *permanent* deformation results. The material may begin to "neck" at some location and finally break. Within the *linear* region, it is found that a specific type of material will always follow the same curves despite different physical dimensions. Thus, we can say that the linearity and slope are a constant of the *type of material* only. In

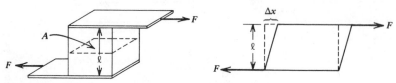

a) **Shear stress results from a force couple**   b) **Shear stress tends to deform an object as shown**

FIGURE 5.8   Shear stress is defined through the elements of this figure.

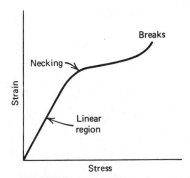

FIGURE 5.9   A typical stress-strain curve showing the linear region, necking, and eventual breaking.

tensile and compressional stress this constant is called the Modulus of Elasticity or Young's Modulus as given by

$$E = \frac{\text{stress}}{\text{strain}} = \frac{F/A}{\Delta \ell / \ell} \qquad (5\text{-}6)$$

Where

stress = F/A in N/m²
strain = $\Delta \ell / \ell$ unitless
    E = Modulus of Elasticity in N/m²

Note that the modulus of elasticity has units of stress, that is, N/m². Table 5.1 gives the modulus of elasticity for several materials. In an exactly similar fashion the shear modulus is defined for shear stress-strain as

$$M = \frac{\text{stress}}{\text{strain}} = \frac{F/A}{\Delta x / \ell} \qquad (5\text{-}7)$$

Where all units have been defined in Equation (5-6).

TABLE 5.1   MODULUS OF ELASTICITY

Material	Modulus (N/m²)
Aluminum	6.89×10^{10}
Copper	11.73×10^{10}
Steel	20.70×10^{10}
Polyethylene (plastic)	3.45×10^{8}

EXAMPLE 5.1

Find the strain that results from a tensile force of 1000 N applied to a 10 m aluminum beam having a 4×10^{-4} m^2 cross-sectional area.

SOLUTION

The modulus of elasticity of aluminum is found from Table 5.1 to be $E = 6.89 \times 10^{10}$ N/m^2. Now we have

$$E = \frac{F/A}{\Delta \ell / \ell} \tag{5-6}$$

so that

$$\text{Strain} = \frac{F}{EA}$$

$$= \frac{10^3 \text{ N}}{(4 \times 10^{-4} \text{ m}^2)\,(6.89 \times 10^{10} \text{ N/m}^2)}$$

$$= 3.62 \times 10^{-5} \text{ or } 36.3 \ \mu\text{m/m (see below)}$$

STRAIN UNITS

Although strain is a unitless quantity, it is common practice to express the strain as the ratio of two length units, as m/m or in./in.; also, since the strain is usually a very small number a micro-(μ) prefix is often included. In this sense, a strain of 0.001 would be expressed as 1000 μin./in. or 1000 μm/m. In the example above the solution is stated as 36.3 μm/m. In general, the smallest value of strain encountered in most applications is 1 μm/m. Note that strain is a unitless quantity so that it is not necessary to do unit conversions. A strain of 153 μm/m could also be written in the form of 153 μin./in. or even 153 μfurlongs/furlong.

5.3.2 Strain Gage Principles

In Section 4.2.1, we saw that the resistance of a metal sample is given by

$$R_0 = \rho \frac{\ell_0}{A_0} \tag{4-7}$$

Where

R_0 = sample resistance Ω
ρ = sample resistivity Ω-m
ℓ_0 = length in m
A_0 = cross-sectional area in m^2

Suppose this sample is now stressed by the application of a force F as shown in Figure 5.7a. Then we know that the material elongates by some amount $\Delta\ell$ so that the new length is $\ell = \ell + \Delta\ell$. It is also true that in such a stress-strain condition, although the sample lengthens, its volume will remain nearly constant. Because the volume unstressed is $V = \ell_0 A_0$ it follows that if the volume remains constant and the length increases, then the area must *decrease* by some amount ΔA

$$V = \ell_0 A_0 = (\ell_0 + \Delta\ell)(A_0 - \Delta A) \qquad (5\text{-}8)$$

Now, because *both* length and area have changed, we find that the resistance of the sample will have also changed

$$R = \rho\,\frac{\ell_0 + \Delta\ell}{A_0 - \Delta A} \qquad (5\text{-}9)$$

Using Equations (5-8) and (5-9), the reader can verify the new resistance is approximately given by

$$R \simeq \rho\,\frac{\ell_0}{A_0}\left(1 + 2\frac{\Delta\ell}{\ell_0}\right) \qquad (5\text{-}10)$$

from which we conclude that the change in resistance is

$$\Delta R \simeq 2R_0\,\frac{\Delta\ell}{\ell_0} \qquad (5\text{-}11)$$

Equation (5-11) is the basic equation that underlies the use of metal strain gages because it shows that the strain $\Delta\ell/\ell$ converts directly into a *resistance change*.

EXAMPLE 5.2

Find the change in a nominal wire resistance of 120 Ω which results from a strain of 1000 μm/m.

SOLUTION

We can find the change in gage resistance from

$$\Delta R \simeq 2R_0\,\frac{\Delta\ell}{\ell}$$

$$\Delta R \simeq (2)(120)(10^{-3}) \qquad (5\text{-}11)$$
$$\Delta R = 0.24\ \Omega$$

Example 5.2 shows a very significant factor regarding strain gages. The change in resistance is very small for typical strain values. For this reason, resistance change measurement methods used with SGs must be very sophisticated.

MEASUREMENT PRINCIPLES

The basic technique of strain gage (SG) measurement involves attaching (gluing) a metal wire or foil to the element whose strain is to be measured. As stress is applied and the element deforms, the SG material experiences the *same* deformation if it is securely attached. Because strain is a *fractional* change in length, the change in SG resistance reflects the strain of both the gage and the element to which it is secured.

TEMPERATURE EFFECTS

If not for temperature compensation effects, the method of SG measurement outlined above would be useless. To see this we need only note that the metals used in SG construction have linear temperature coefficients of $\alpha \cong 0.004/°C$, typical for most metals. Temperature changes of 1°C are not uncommon in measurement conditions in the industrial environment. If the temperature change in Example 5.2 had been 1°C, substantial change in resistance would have resulted. Thus, from Chapter 4,

$$R(T) = R(T_0)[1 + \alpha_0 \, \Delta T] \qquad (4\text{-}9)$$

or

$$\Delta R_T = R_0 \alpha \, \Delta T$$

Where

ΔR_T = resistance change due to temperature change
$\alpha_0 \simeq 0.004/°C$ in this case
$\Delta T \simeq 1°C$ in this case
$R(T_0) = 120 \, \Omega$ nominal resistance

Then, we find $\Delta R_T = 0.48 \, \Omega$ which is *twice the change* due to strain! Obviously temperature effects can mask the strain effects we are trying to measure! Fortunately, we are able to compensate for temperature and other effects as shown in the signal conditioning methods in the section below.

5.3.3 Metal Strain Gages (SGs)

Metal SGs are devices that operate on the principles discussed in the previous section. The following items are important to understanding SG applications.

GAGE FACTOR

The relation between strain and resistance change of Equation (5-11) is only approximately true. Impurities in the metal, the type of metal, and other factors lead to slight corrections. A SG specification always indicates the correct relation through statement of a *gage factor* (GF) which is defined as

$$GF = \frac{\Delta R/R}{strain} \tag{5-12}$$

Where

$\Delta R/R$ = fractional change in gage resistance due to strain
strain = $\Delta \ell/\ell$ = fractional change in length

For metal gages this number is always close to 2. For some special alloys and carbon gages, the GF may be as large as 10. Note that a high gage factor is desirable because it indicates a large change in resistance for a given strain and is easier to measure.

CONSTRUCTION

Strain gages (SGs) are used in two forms, wire and foil. The basic characteristics of each are the same in terms of resistance change for a given strain. The design of the SG itself is such as to make it very long to give a large enough nominal resistance (to be practical) and to make the gage of sufficiently fine wire or foil so as to not resist strain effects. Finally, it is necessary to make the gage sensitivity *unidirectional* so that it responds to strain *only in one direction*. In Figure 5.10, we see the common *pattern* of SGs that provides these characteristics. Note that by folding the material back and forth as shown, we achieve a long length to provide high resistance. Further, if a strain is applied transversely to the SG length, the pattern will tend to *unfold* rather than stretch with no change in resistance. These gages are usually

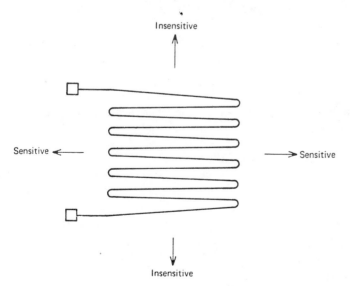

FIGURE 5.10 The most common foil and wire strain gage pattern. The device is sensitive to strain in one direction only.

mounted on a paper backing which is bonded (using epoxy) to the element whose strain is to be measured. The nominal SG resistances available are typically 60, 120, 240, 350, 500, and 1000 ohms.

SIGNAL CONDITIONING

Two effects are critical in the signal conditioning techniques used for SGs. The first is the small, fractional changes in resistance that require carefully designed resistance *measurement* circuits. A good SG system might require a resolution of 2 μm/m strain. From Equation (5-11), this would result in a ΔR of only 4.8 $\times$ 10^{-4} Ω for a nominal gage resistance of 120 Ω.

The second effect is the need to provide some compensation for temperature effects, to eliminate masking changes in strain.

The bridge circuit provides the answer to both areas. The sensitivity of the bridge circuit for detecting small changes in resistance is well known. Furthermore, by using a dummy gage as shown in Figure 5.11a, we can provide the required temperature compensation. In particular, the dummy is mounted in an insensitive orientation (Figure 5.11b), but in the same proximity as the active SG. Then, both gages change in resistance from temperature effects, but the detector does not respond to a change in both. Only the active SG responds to strain effects. Other configurations are used where more than one gage is active, but the basic idea of temperature compensation is retained.

EXAMPLE 5.3

A strain gage with a GF = 2.03 and R = 350 Ω is used in the bridge of Figure 5.11a. The bridge resistors are $R_1 = R_2 = 350$ Ω and the dummy gage has R = 350 Ω. If a strain of 1450 μm/m is applied, find the bridge offset voltage if V_S = 10.0 volts.

SOLUTION

Note that with no strain, the bridge is balanced. Now when the strain is applied, the gage resistance will change by a value given from

$$GF = \frac{\Delta R/R}{\text{strain}} \qquad (5\text{-}12)$$

thus,

$$\Delta R = (GF)\,(\text{strain})\,(R)$$
$$\Delta R = (2.03)(1.45 \times 10^{-3})(350\ \Omega)$$
$$\Delta R = 1.03\ \Omega$$

Thus, the new resistance R = 351 Ω. The bridge offset voltage is

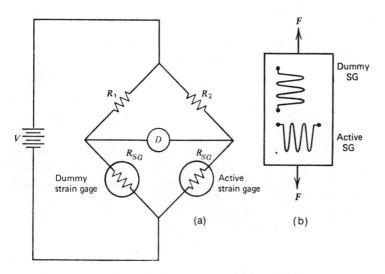

FIGURE 5.11 It is necessary to employ temperature compensation in strain gage bridges. This is usually accomplished by use of a dummy gage as shown.

$$\Delta V = \frac{RV}{R_1 + R} - \frac{R_{SG}V}{R_{SG} + R_2}$$

thus,

$$\Delta V = 5 - \frac{(351)(10)}{701}$$
$$\Delta V = -0.007 \text{ V}$$

so that a 7 mV offset results.

5.3.4 Semiconductor Strain Gages (SGs)

The use of semiconductor material, notably silicon, for SG application has increased over the past few years. There are presently several disadvantages to these devices compared to the metal variety, but also numerous advantages for their use.

PRINCIPLES

As in the case of the metal SGs, the basic effect is a change of resistance with strain. In the case of a semiconductor, the resistivity also changes with strain along with the physical dimensions. This is due to changes in electron and hole mobility with changes in crystal structure as strain is applied. The net result is a much *larger* gage factor than is possible with metal gages.

GAGE FACTOR

The semiconductor device gage factor (GF) is still given by

$$GF = \frac{\Delta R/R}{\text{strain}} \tag{5-12}$$

The value of the semiconductor gage factor varies between -50 and -200. Thus, resistance changes will be factors of from 25 to 100 times those available with metal SGs. It must also be noted, however, that these devices are highly *nonlinear* in resistance versus strain. In other words, the gage factor is not a constant as the strain takes place. Thus, the gage factor may be -150 with no strain but drop (nonlinearly) to -50 at 5000 μm/m. This means the resistance change will be nonlinear with respect to strain. In order to use the semiconductor strain gage to measure strain, we must have a curve or table of values of gage factor versus resistance.

CONSTRUCTION

The semiconductor strain gage physically appears as a band or strip of material with electrical connection as shown in Figure 5.12. The gage is either bonded directly onto the test element or, if encapsulated, is attached by the encapsulation material. These SGs also appear as IC assemblies in configurations used for other measurements.

SIGNAL CONDITIONING

The signal conditioning is still typically a bridge circuit with temperature compensation. An added problem is the need for linearization of the output because the basic resistance versus strain characteristic is nonlinear. This is usually provided by *active* linearization circuits.

EXAMPLE 5.4

(a) Contrast the resistance change produced by a 150 μm/m strain in a metal gage with GF = 2.13 with,

(b) A semiconductor SG with GF = -151. Nominal resistances are *both* 120 Ω.

SOLUTION

From the basic equation

$$GF = \frac{\Delta R/R}{\text{strain}}$$

(a) We find for the *metal* gage SG

$$\Delta R = (120\ \Omega)\,(2.13)\,(0.15 \times 10^{-3})$$
$$\Delta R = 0.038\ \Omega$$

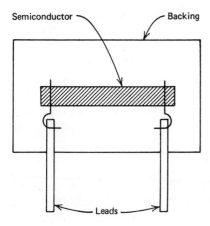

FIGURE 5.12   Typical semiconductor strain gage configuration.

(b) Whereas for the semiconductor gage, the change is

$$\Delta R = (120\ \Omega)\,(-151)\,(0.15 \times 10^{-3})$$
$$\Delta R = -2.72\ \Omega$$

5.3.5 Load Cells

One important direct application of SGs is for the measurement of force or weight. These transducer devices, called *load cells*, measure deformations produced by the force or weight. In general, a beam or yoke assembly is used which has several strain gages mounted so that the application of a force causes a strain in the assembly which is measured by the gages. A common application uses one of these devices in support of a hopper or feed of dry or liquid materials. A measure of the weight through a load cell yields a measure of the quantity of material in the hopper. Generally, these devices are calibrated so that the force (weight) is directly related to the resistance change. Forces as high as 5 MN (approximately 10^6 lb) can be measured with an appropriate load cell.

5.4 MOTION TRANSDUCERS

A special class of transducers is used to measure the *velocity* and *acceleration* of objects in industrial processes and testing. Often, these variables are not under specific control but are used to evaluate the performance, durability, and failure modes of manufactured products and the processes that produce them. The basic physical concepts of motion are discussed in Appendix A.4. In the following sections several special types of motion are discussed, and then the transducers used for measurement of these variables are studied.

5.4.1 Types of Motion

The design of a transducer to measure motion is often tailored to the type of motion which is to be measured. It will help you understand these transducers if you have a clear understanding of the types of motion considered.

RECTILINEAR

This type of motion is characterized by velocity and acceleration which is composed of straight-line segments. Thus, objects may accelerate forward to a certain velocity, deaccelerate to a stop, reverse, and so on. There are many types of transducers deisgned to handle this type of motion. Typically, maximum accelerations are less than 10 gs, and no angular motion (in a curved line) is allowed. It is perhaps too strong to say that no angular motion is allowed. Rather, if there is angular motion, then several rectilinear motion transducers must be used, each sensitive to only one line of motion. Thus, if vehicle motion is to be measured, two transducers may be used, one to measure motion in the forward direction of vehicle motion and the other perpendicular to the forward axis of the vehicle.

ANGULAR

Some transducers are designed to measure *only* rotations about some axis, such as the angular motion of the shaft of a motor. Such devices cannot be used to measure the physical displacement of the whole shaft, but only its rotation.

VIBRATION

In the normal experiences of daily living, a person rarely experiences accelerations which vary from 1 **g** by more than a few percent. Even the severe environments of rocket launchings involve accelerations of only 1 **g** to 10 **g**. If an object is placed in periodic motion about some equilibrium position as in Figure 5.13, it is found that very large *peak* accelerations may result, reaching to 100 **g** or more. This motion is called *vibration*. Clearly, the measurement of acceleration of this magnitude is very important to industrial environments, where vibrations are often encountered from machinery operations. In general, vibrations are somewhat random in both the frequency of periodic motion and the magnitude of displacements from equilib-

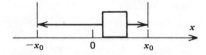

FIGURE 5.13   An object in periodic motion about an equilibrium at $x = 0$. The peak motion is x_0.

rium. For analytical treatments, vibration is defined in terms of a regular periodic motion where the position of an object in time is given by

$$x(t) = x_0 \sin \omega t \qquad (5\text{-}13)$$

Where

$x(t)$ = object position in m
x_0 = peak displacement from equilibrium in m
ω = angular frequency in rad/s

Note that the definition of ω as angular frequency is not inconsistent with the reference to ω as angular speed because they are the same. If an object rotates, we define the time to complete one rotation as a period T which corresponds to a frequency $f = 1/T$. The frequency represents the number of revolutions per second and is measured in hertz (Hz) where 1 Hz = 1 revolution per second. An angular rate of one revolution per second corresponds to an angular velocity of 2π rad/s because one revolution sweeps out 2π radians. From this argument, we see that f and ω are related by

$$\omega = 2\pi f \qquad (5\text{-}14)$$

Because f and ω are related by a constant, we refer to ω as both angular frequency as well as angular velocity.

Now we can find the vibration velocity as a derivative of Equation (5-13)

$$v(t) = -\omega x_0 \cos \omega t \qquad (5\text{-}15)$$

and we can get the vibration acceleration from a derivative of (5-15)

$$a(t) = -\omega^2 x_0 \sin \omega t \qquad (5\text{-}16)$$

Note that vibration position, velocity, and acceleration are all periodic functions having the *same* frequency and period. Of particular interest is the *peak* acceleration:

$$a_{\text{peak}} = \omega^2 x_0 \qquad (5\text{-}17)$$

We see that the peak acceleration is dependent on ω^2, the angular frequency squared. This may result in very large acceleration values, even with *modest* peak displacements as Example 5.5 shows.

EXAMPLE 5.5

A water pipe vibrates at a frequency of 10 Hz with a displacement of 0.5 cm. Find (a) the peak acceleration in m/s², and (b) **g** acceleration.

SOLUTION

The peak acceleration will be given by

(a)
$$a_{peak} = \omega^2 x_0 \qquad\qquad (5\text{-}17)$$

Where

$\omega = 2\pi f = 20\pi$ rad/s and $x_0 = 0.5$ cm $= 0.005$ m
$a_{peak} = (20\pi)^2 (0.005)$
$a_{peak} = 19.7$ m/s²

(b) Noting that 1 g $= 9.8$ m/s² we get

$$a_{peak} = (19.7 \text{ m/s}^2)\left(\frac{1 \text{ g}}{9.8 \text{ m/s}^2}\right)$$

$$a_{peak} = 2.0 \text{ g}$$

A 2 **g** vibrating excitation of any mechanical element can be very destructive, yet this is generated under the modest conditions of Example 5.5. A special class of transducers has been developed for measuring vibration acceleration.

SHOCK

A very special type of acceleration occurs when an object which may be in uniform motion or modestly accelerating, is suddenly brought to rest, such as from a collision. Such phenomena are the result of very large accelerations or actually *deaccelerations* such as an object being dropped from some height onto a hard surface. The name *shock* is given to such deaccelerations which are characterized by very short times, typically in the order of milliseconds with peak accelerations over 500 g. In Figure 5.14 we have a typical acceleration graph as a function of time for a shock experiment. This graph is characterized by a maximum or peak deacceleration a_{peak}, a shock duration T_d, and bouncing as indicated. We can find an average shock by knowing the velocity with which the object is moving and the shock duration as considered in Example 5.6.

EXAMPLE 5.6

A TV set is dropped from a 2 m height. If the shock duration is 5 ms, find the average shock in **g**.

SOLUTION

The TV accelerates at 9.8 m/s² for 2 m. We find the velocity as

$$v^2 = 2gx$$
$$v^2 = (2)(9.8 \text{ m/s}^2)(2 \text{ m})$$
$$v = 6.2 \text{ m/s}$$

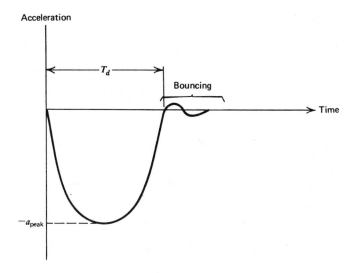

FIGURE 5.14   A typical shock graph showing acceleration versus time. Shock duration is T_d.

Then, if the duration is 5 ms, we have

$$\bar{a} = \frac{6.2 \text{ m/s}}{5 \times 10^{-3} \text{ s}}$$

$$\bar{a} = 1200 \text{ m/s}^2$$

or 122 **g**! No wonder that the TV breaks apart when it hits ground.

5.4.2 Accelerometer Principles

Many accelerometers operate according to the one basic principle. The variations in design involve only the manner by which this principle is implemented. The basic principle involves Newton's law ($F = ma$).

SPRING-MASS SYSTEM

The principle of acceleration measurement is based upon a marriage of Newton's law (relating force and acceleration) and Hooke's law (relating force and spring action). In Figure 5.15a we see the combination of a mass free to move and a spring attached to some base. If the entire assembly is accelerated to the right, Newton's law states that the mass must be under the influence of a force, $F = ma$. This force is provided by the spring that extends until the force provided by Hooke's law matches that required by the acceleration. As long as the system accelerates in this manner, the spring-mass system remains in this state, as shown in Figure 5.15b. We can then form an equality

$$k \, \Delta x = ma \tag{5-18}$$

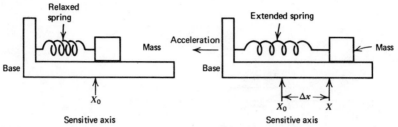

a) Spring-mass system with no acceleration   b) Spring-mass system with acceleration

FIGURE 5.15   The basic accelerometer is a spring-mass system.

Where

k = spring constant in N/m
Δx = spring extension in m
m = mass in kg
a = acceleration in m/s²

Now Equation (5-18) allows the measurement of acceleration to be reduced to a measurement of spring extension (linear displacement) because

$$a = \frac{k}{m}\Delta x \qquad (5\text{-}19)$$

Note that if the acceleration is reversed, the same physical argument would apply except that the spring is compressed instead of extended. Equation (5-19) describes the relationship between spring displacement and acceleration.

The *spring-mass principle* stated above applies to all common accelerometer designs. The mass that converts the acceleration to spring displacement is referred to as the *test mass* or *seismic mass*. We see then that acceleration measurement reduces to linear displacement measurement; most designs differ in how this displacement measurement is made.

NATURAL FREQUENCY AND DAMPING

On closer examination of the simple principle described above, we find another characteristic of spring-mass systems which complicates the analysis. In particular, a system consisting of a spring and attached mass always exhibits oscillations at some characteristic *natural frequency*. Experience tells us that if we pull a mass back and then release it (in the absence of acceleration), it will be pulled back by the spring, overshoot the equilibrium, and oscillate back and forth. Only friction associated with the mass and base eventually brings the mass to rest. Any displacement measuring system must respond to this oscillation as if an actual acceleration occurred; in fact, none has been applied. This natural frequency is given by

$$f_N = \frac{1}{2\pi}\sqrt{\frac{k}{m}} \qquad (5\text{-}20)$$

Where

f_N = natural frequency in Hz
k = spring constant in N/m
m = seismic mass in kg

The friction that eventually brings the mass to rest is called a *damping coefficient* α, which has the units of s^{-1}. In general, the effect of oscillation is called *transient response*, described by a periodic damped signal, as shown in Figure 5.16, whose equation is

$$X_T(t) = X_0\, e^{-\alpha t}\sin(2\pi f_N t) \qquad (5\text{-}21)$$

Where

$X_T(t)$ = transient mass position
X_0 = peak position, initially
α = damping coefficient
f_N = natural frequency

The parameters, natural frequency, and damping coefficient in Equation (5-21), have a profound effect on the application of accelerometers.

VIBRATION EFFECTS

The effect of natural frequency and damping on the behavior of spring-mass accelerometers is best described in terms of an applied vibration. If the spring mass system is exposed to a vibration, then the resultant acceleration is given by Equation (5-16)

$$a(t) = -\omega^2 x_0 \sin \omega t \qquad (5\text{-}16)$$

If this is used in Equation (5-18), we can show that the mass motion is given by

$$\Delta x = -\frac{m x_0}{k}\, \omega^2 \sin \omega t \qquad (5\text{-}22)$$

where all terms were previously defined and $\omega = 2\pi f$, with f the applied frequency.

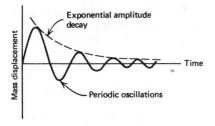

FIGURE 5.16   A spring-mass system exhibits a natural oscillation with damping as transient response to an impulse.

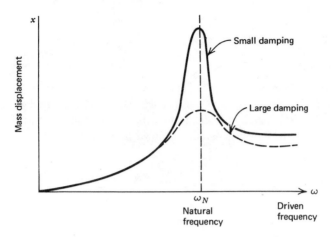

FIGURE 5.17 A spring-mass accelerometer driven by a periodic acceleration of varying frequency shows a peak mass displacement or resonance at the natural frequency.

The derivation of Equation (5-22), which has ignored the natural frequency, indicates that the mass displacement amplitude varies as the *applied* frequency is squared. To see the effect of the natural frequency and damping on this result we plot the actual displacement amplitude versus applied frequency in Figure 5.17. This figure shows that the behavior predicted by Equation (5-22) is correct for frequencies lower than the natural frequency but deviates as the natural frequency is reached. A peak, or resonance, occurs at the natural frequency. After further increases in applied frequency, we find the amplitude is *independent* of frequency.

The effect of damping is seen as a reduction of the resonance peak and constant displacement at higher frequency. Thus, we can make the following observations concerning the effects of natural frequency and an applied vibration frequency.

1. $f < f_N$—For an applied frequency less than the natural frequency, the natural frequency has little effect on the basic spring-mass response given by Equations (5-18) and (5-22). A rule of thumb states that a safe maximum applied frequency is $f < 1/2.5f_N$.
2. $f > f_N$—For an applied frequency much larger than the natural frequency, the accelerometer output is independent of the applied frequency. Although not shown in Figure 5.17, the accelerometer becomes a measure of *vibration displacement* X_0 of Equation (5-13) under these circumstances. It is interesting to note that the seismic mass is stationary in space in this case and the housing, which is driven by the vibration, moves about the mass. A general rule sets $f > 2.5 f_N$ for this case.

Generally, accelerometers are *not* used near their natural frequency because of high nonlinearities in output.

EXAMPLE 5.7

An accelerometer has a seismic mass of 0.05 kg and a spring constant of 3.0 × 10^3 N/m. Maximum mass displacement is ±0.02 m (before the mass hits the stops).

Calculate (a) the maximum measurable acceleration in **g**, and (b) the natural frequency.

SOLUTION

We find the maximum acceleration when the maximum displacement occurs

(a)
$$a = \frac{k}{m} \Delta x$$

$$a = \left(\frac{3.0 \times 10^3 \text{ N/m}}{0.05 \text{ kg}} \right) (0.02 \text{ m}) \qquad (5\text{-}19)$$

$$a = 1200 \text{ m/s}^2$$

or, since

$$1 \text{ g} = 9.8 \text{ m/s}^2$$

$$a = (1200 \text{ m/s}^2) \left(\frac{1 \text{ g}}{9.8 \text{ m/s}^2} \right)$$

$$a = 120 \text{ g}$$

(b) The natural frequency is

$$f_N = \frac{1}{2\pi} \sqrt{\frac{k}{m}}$$

$$f_N = \frac{1}{2\pi} \sqrt{\frac{3.0 \times 10^3 \text{ N/m}}{0.05 \text{ kg}}}$$

$$f_N = 39 \text{ Hz}$$

5.4.3 Types of Accelerometers

The variety of accelerometers used results from different applications with requirements of range, natural frequency, and damping. In this section, various accelerometers with their special characteristics are reviewed. The basic difference is in the method of mass displacement *measurement*. In general, the specification sheets for an accelerometer will give the natural frequency, damping coefficient, and a scale factor which relates the output to acceleration input. The values of test mass and spring constant are seldom known or required.

POTENTIOMETRIC

This simplest accelerometer type measures mass motion by attaching the mass to the wiper arm of a potentiometer. In this manner, the mass position is conveyed as a changing resistance. The natural frequency of these devices is generally less than 30 Hz, limiting its application to *steady-state* acceleration or *low-frequency* vibration measurement. Numerous signal conditioning schemes are employed to convert the resistance variation into a voltage or current signal.

LVDT

A second type of accelerometer takes advantage of the natural linear displacement measurement of the LVDT (Section 5.2.3) to measure mass displacement. In these instruments, the LVDT core itself is the seismic mass. Displacements of the core are converted directly into a linearly proportional ac voltage. These accelerometers generally have a natural frequency less than 80 Hz and are also commonly used for steady-state and low-frequency vibration.

VARIABLE RELUCTANCE

This accelerometer type falls in the same general category as the LVDT in that an inductive principle is employed. Here, the test mass is usually a permanent magnet. The measurement is made from the voltage induced in a surrounding coil as the magnetic mass moves under the influence of an acceleration. This accelerometer is used in vibration studies only, because it has an output *only* when the mass is in motion. Its natural frequency is typically less than 100 Hz. This type of accelerometer often is used in oil exploration to pick up vibrations reflected from underground rock strata. In this form, it is commonly referred to as a *geophone*.

PIEZOELECTRIC

The piezoelectric accelerometer is based on a property exhibited by certain crystals where a voltage is generated across the crystal when stressed. This property is also the basis for such familiar transducers as crystal phonograph cartridges and crystal microphones. In accelerometers, the principle is shown in Figure 5.18. Here, a

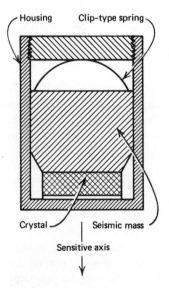

FIGURE 5.18 A piezoelectric accelerometer.

piezoelectric crystal is spring-loaded with a test mass in contact with the crystal. When exposed to an acceleration, the test mass stresses the crystal by a force ($F = ma$), resulting in a voltage generated across the crystal. A measure of this voltage is then a measure of the acceleration. The crystal per se is a very high-impedance source and thus requires a high-input impedance low-noise detector. Output levels are typically in the millivolt range. The natural frequency of these devices may exceed 5 kHz, so that they can be used for vibration and shock measurements.

5.4.4 Applications

A few notes about the application of accelerometers are of aid in understanding how the selection of a transducer is made in a particular case.

STEADY-STATE ACCELERATION

In steady-state accelerations we are interested in a measure of acceleration which may vary in time but which is *nonperiodic*. Thus, the stop-go motion of an automobile is an example of a steady-state acceleration. For these steady-state accelerations, we select a transducer having (1) adequate range to cover expected acceleration *magnitudes* and (2) a natural frequency sufficiently high that its period is shorter than the characteristic time span over which the measured acceleration changes. Note that by using electronic integrators, the basic accelerometer can provide both velocity (first integration) and position (second integration) information.

EXAMPLE 5.8

An accelerometer outputs 14 mV per **g**. Design a signal conditioning system that provides (a) a velocity signal scaled at 0.25 volt for every m/s, and (b) determine the gain of the system and the feedback resistance ratio.

SOLUTION

First we note that 14 mV/**g** becomes

$$\left(14 \ \frac{mV}{g}\right)\left(\frac{1 \ g}{9.8 \ m/s^2}\right) = 1.43 \ \frac{mV}{m/s^2}$$

Now we need an integrator to get the velocity and amplifier to provide the proper scale. Such a circuit is shown in Figure 5.19. We chose $T = RC = 1$ so that the integrator output is scaled at

$$\left(1.43 \ \frac{mV}{m/s^2}\right)\left(\frac{1}{1 \ s}\right) = 1.43 \ \frac{mV}{m/s}$$

in units of output voltage/speed.

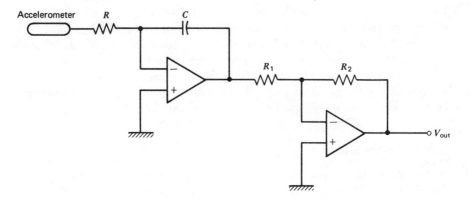

FIGURE 5.19 An integrator and inverter can be used to obtain velocity information from an accelerometer.

Then, the required gain is found from

$$G = \frac{0.25 \text{ V/m/s}}{1.43 \text{ mV/m/s}} = 175$$

Thus, we can make $R_2 = 174 \text{ k}\Omega$ and $R_1 = 1 \text{ k}\Omega$ so that

$$G = 1 + \frac{R_2}{R_1} = 175$$

VIBRATION

The application of accelerometers for vibration first requires that the applied frequency is less than the natural frequency of the accelerometer. Second, one must be sure the stated range of acceleration measured will never exceed that of the specification for the device. This assurance must come from a consideration of Equation (5-20) under circumstances of maximum frequency and vibration displacement.

SHOCK

The primary elements of importance in shock measurements are that the device have a natural frequency which is greater than 1 kHz and a range typically greater than 500 **g**. The only accelerometer that can usually satisfy these requirements is the piezoelectric type (Section 5.4.3).

5.5 PRESSURE TRANSDUCERS

The measurement and control of fluid (liquid and gas) pressure has to be one of the most common of all the process industries. Because of the great variety of conditions, ranges, and materials for which pressure must be measured, there are many different types of pressure transducer designs. In the paragraphs below, the basic concepts of pressure are presented, and a brief description is given of the most common types of pressure transducers. You will see that pressure measurement is often accomplished by conversion of the pressure information to some intermediate form, such as displacement, which is then measured to determine the pressure. One of the reasons that this section is near the end of the chapter is that the intermediate measurement is often displacement or strain.

5.5.1 Pressure Principles

Pressure is simply the force per unit area which a fluid exerts on its surroundings. If it is a gas, then the pressure of the gas is the force per unit area which the gas exerts on the walls of the container which holds it. If the fluid is a liquid, then the pressure is the force per unit area which the liquid exerts on the container in which it is contained. Obviously, the pressure of a gas will be uniform on all the walls which must enclose the gas completely. In a liquid, the pressure will vary, being greatest on the bottom of the vessel and zero on the top surface, which need not then be enclosed.

STATIC PRESSURE

The statements made in the previous paragraph are explicitly true for a fluid which is not moving in space, that is, which is not being pumped through pipes or flowing through a channel. The pressure in cases where no motion is occurring is referred to as *static* pressure.

DYNAMIC PRESSURE

If a fluid is in motion, the pressure which it exerts on its surroundings *depends* on the motion. Thus, if we measure the pressure of water in a hose with the nozzle closed, we may find a pressure of, say, 40 pounds per square inch (note: force per unit area). If the nozzle is opened, the pressure in the hose will drop to a different value, say, 30 pounds per square inch. For this reason, a thorough description of pressure must note the circumstances under which it is measured. Pressure can depend on flow, compressibility of the fluid, external forces, and numerous other factors.

UNITS

Since pressure is force per unit area, we describe it in the SI system of units by newtons per square meter. This unit has been named the *pascal* (Pa), so that $1 \text{ Pa} = 1 \text{ N/m}^2$. As will be seen later, this is not a very convenient unit, and it is often used in conjunction with the SI standard prefixes, as kPa or MPa. You will see the combination N/cm^2 used, but use of this combination should be avoided in favor of Pa with the appropriate prefix. In the English system of units, the most common designation is the pound per square inch, $lb/in.^2$. This is usually just written *psi*.

The conversion is that one psi is approximately 6.895 kPa. For very low pressures, such as may be found in vacuum systems, the unit *Torr* is often used. One Torr is approximately 133.3 Pa. Again, use of the pascal with appropriate prefix is preferred. Other units which you may encounter in the pressure description are the *atmosphere* (at) which is 101.325 kPa or $\simeq$ 14.7 psi, and the *bar* which is 100 kPa. The use of inches or feet of water and mm of mercury will be discussed later.

GAUGE PRESSURE

In many cases the absolute pressure is not the quantity of major interest in describing the pressure. The atmosphere of gas which surrounds the earth exerts a pressure, due to its weight, at the surface of the earth of approximately 14.7 psi, as noted in the atmosphere unit above. If a closed vessel at the earth's surface contained a gas at an absolute pressure of 14.7 psi, then it would exert *no effective pressure* on the walls of the container because the atmospheric gas exerts the same pressure from the outside. In cases like this it is more appropriate to describe pressure in a relative sense, that is, compared to atmospheric pressure. This is called *gauge pressure* and is given by

$$p_g = p_{abs} - p_{at} \qquad (5\text{-}23)$$

Where

p_g = gauge pressure
p_{abs} = absolute pressure
p_{at} = atmospheric pressure

In the English system of units the abbreviation *psig* is used to represent the gauge pressure.

HEAD PRESSURE

For liquids, the expression *head pressure* or *pressure head* is often used to describe the pressure of the liquid in a tank or pipe. This refers to the *static* pressure produced by the weight of the liquid above the point at which the pressure is being described. This pressure depends *only* on the height of the liquid above that point and the liquid density (mass per unit volume). In terms of an equation, if a liquid is contained in a tank, then the pressure at the bottom of the tank is given by

$$p = \rho gh \qquad (5\text{-}24)$$

Where

p = pressure in Pa
ρ = density in kg/m^3
g = acceleration due to gravity (9.8 m/s^2)
h = depth of liquid in m

This same equation could be used to find the pressure in the English system, but it is common practice to express the density in this system as the weight density, ρ_w, in lb/ft³, which includes the gravity term of Equation (5-24). In this case, the relation between pressure and depth becomes

$$p = \rho_w h \qquad (5\text{-}25)$$

Where

p = pressure in lb/ft²
ρ_w = weight density in lb/ft³
h = depth in ft

If the pressure was desired in psi, then the ft² would be expressed as 144 in.² Because of the common occurrence of liquid tanks and the necessity to express the pressure of such systems, it has become common practice to describe the pressure directly in terms of the *equivalent* depth of a particular liquid. Thus, the term "mm of mercury" means that the pressure is equivalent to that produced by so many mm of mercury depth, which could be calculated from Equation (5-24) using the density of mercury. In the same sense, the expression "inches of water" or "feet of water" mean the pressure which is equivalent to some particular depth of water, using its weight density.

EXAMPLE 5.9

A tank holds water with a depth of 7.0 feet. What is the pressure at the tank bottom in psi and Pa? (density = 10^3 kg/m³)

SOLUTION

We can find the pressure in Pa directly by converting the 7.0 ft into meters, thus, (7.0 ft)(0.3048 m/ft) = 2.1 m. Then, from Equation (5-24),

$$p = (10^3 \text{ kg/m}^3)(9.8 \text{ m/s}^2)(2.1 \text{ m})$$
$$p = 21 \text{ kPa} \qquad \text{(note significant figures)}$$

To find the pressure in psi we can convert the pressure in Pa to psi or use Equation (5-25). Let's use the latter. The weight density is found from

$$\rho_w = (10^3 \text{ kg/m}^3)(9.8 \text{ m/s}^2) = 9.8 \times 10^3 \text{ N/m}^3$$

or

$$= (9.8 \times 10^3 \text{ N/m}^3)(0.3048 \text{ m/ft})^3(0.2248 \text{ lb/N})$$
$$= 62.4 \text{ lb/ft}^3$$

Now, the pressure is

$$p = (62.4 \text{ lb/ft}^3)(7.0 \text{ ft}) = 440 \text{ lb/ft}^2$$
$$p = 3 \text{ psi}$$

5.5.2 Pressure Transducers (p > one atmosphere)

In general, the design of pressure transducers employed for measurement of pressure higher than one atmosphere differs from those employed for pressure less than one atmosphere. In this section the basic operating principles of many types of pressure transducers used for the higher pressures are considered. You should be aware that this is not a rigid separation, since you will find many of these same principles employed in the lower (vacuum) pressure measurements.

Most pressure transducers used in process control require the transduction of pressure information into a physical displacement. Measurement of pressure requires techniques for producing the displacement and means for converting such displacement into a proportional electrical signal. This is *not* true, however, in the *very low* pressure region ($p < 10^{-3}$ atm) where many purely electronic means of pressure measurement may be used.

DIAPHRAGM

One common element used to convert pressure information into a physical displacement is the diaphragm shown in Figure 5.20. Here we note that if a pressure p_1 exists on one side of the diaphragm and p_2 on the other, then a net force is exerted given by

$$F = (p_2 - p_1)A \tag{5-26}$$

Where

$$A = \text{diaphragm area in m}^2$$
$$p_1, p_2 = \text{pressure in N/m}^2$$

A diaphragm is like a spring and therefore extends or contracts until a Hooke's law force is developed which balances the pressure difference force. This is shown in Figure 5.20 for p_1 greater than p_2. A *bellows* shown in Figure 5.21 is another device much like the diaphragm which converts a pressure differential into a physical displacement, except that here the displacement is much more a straight-line expansion.

BOURDON TUBE

A very special and common pressure-to-displacement conversion is accomplished by a specially constructed tube, shown in Figure 5.22. If a section of tubing is

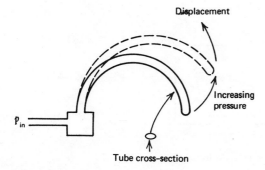

FIGURE 5.20   The Bourdon tube is a very common device for converting a pressure to a displacement.

partially flattened and coiled as shown, then the application of pressure inside the tube causes the tube to uncoil. This then provides a displacement which is proportional to pressure.

ELECTRONIC CONVERSIONS

Many techniques are used to convert the displacements generated in the previous examples into electronic signals. The simplest is to use a mechanical linkage connected to a potentiometer. In this fashion, pressure is related to a resistance change. Other methods of conversion employ strain gages directly on a diaphragm. LVDTs and other inductive devices are used to convert bellows or Bourdon tube motions into proportional electrical signals.

Often pressure measurement is accomplished using a diaphragm in a special feedback configuration, shown in Figure 5.23. The feedback system keeps the diaphragm from moving, using an induction motor. The error signal in the feedback system provides an electrical measurement of the pressure.

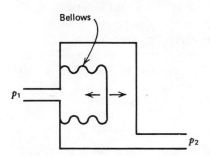

FIGURE 5.21   The bellows transforms a pressure difference into a proportional and unidirectional displacement.

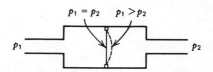

FIGURE 5.22   A diaphragm is used in many pressure measurement systems. The displacement of the diaphragm is proportional to the pressure difference.

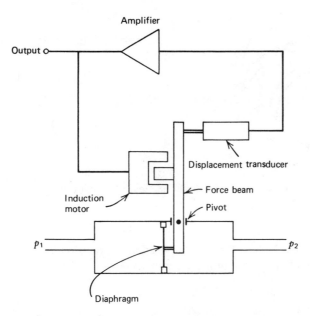

FIGURE 5.23   A differential pressure (D/P) cell measures a pressure difference with a diaphragm, but the diaphragm is effectively kept from moving by an electronic feedback system.

5.5.3 Pressure Transducers (p < one atmosphere)

Measurements of pressure less than 1 atm are most conveniently made using purely electronic methods. There are three common methods of electronic pressure measurements.

The first two devices (given below) are useful for pressure less than 1 atm down to about 10^{-3} atm. They are both based on the rate at which heat is conducted and radiated away from a heated filament placed in the low pressure environment. The heat loss is proportional to the number of gas molecules per unit volume, and thus, under constant filament current, the filament temperature is proportional to gas pressure. We have thus transduced a pressure measurement to a temperature measurement.

PIRANI GAUGE

This gauge determines the filament temperature through a measure of filament *resistance* in accordance with the principles established in Section 4.3. Filament excitation and resistance measurement are both performed with a bridge circuit. The response of resistance versus pressure is highly nonlinear.

THERMOCOUPLE

A second pressure transducer or gauge measures filament temperature using a thermocouple directly attached to the heated filament. In this case, ambient room temperature serves as a reference for the thermocouple, and the voltage output, which is proportional to pressure, is highly nonlinear. Calibration of both Pirani and Thermocouple gauges depends on the type of gas for which the pressure is being measured.

IONIZATION GAUGE

This device is useful for the measurement of very low pressures from about 10^{-3} atm to 10^{-13} atm. This gauge employs electrons, usually from a heated filament, to ionize the gas whose pressure is to be measured, and then measures the current flowing between two electrodes in the ionized environment, as shown in Figure 5.24. The number of ions per unit volume depends on the gas pressure, and hence the current also depends on gas pressure. This current is then monitored as an approximately *linear* indication of pressure.

5.6 FLOW TRANSDUCERS

The measurement and control of flow can be said to be the very heart of the process industries. Continuously operating manufacturing processes involve the movement of raw materials, products, and waste throughout the process. All such functions can be considered flow, whether automobiles through an assembly line or methyl chloride through a pipe. The methods of measurement of flow are at least as varied as the industry. It would be unreasonable to try to present every type of flow transducer, and in this section we will consider flow on three broad fronts—solid,

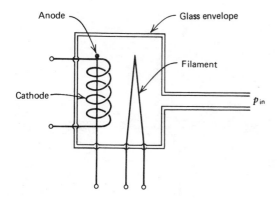

FIGURE 5.24   The ionization gauge is employed to measure very low pressures (down to 10^{-13} atm).

liquid and gas. Like pressure, we will find that flow information is often translated into an intermediate form which is then measured using techniques developed for that form.

5.6.1 Solid Flow Measurement

The most common solid flow measurement occurs when material in the form of small particles, such as crushed material or powder, is carried by a conveyor belt system or by some other host material. For example, if solid material is suspended in a liquid host, the combination is called a *slurry*, which is pumped through pipes like a liquid. We will consider the conveyor system and leave the slurry to be treated as liquid flow.

CONVEYOR FLOW CONCEPTS

For solid objects, the flow usually is described by a specification of the mass or weight per unit time which is being transported by the conveyor system. The units will be in many forms, for example, kg/min or lb/min. To make a measurement of flow it is only necessary to weight the quantity of material on some fixed length of the conveyor system. Knowing the speed at which the conveyor is moving allows calculation of the material flow rate.

In Figure 5.25, a typical conveyor system is shown where material is drawn from a hopper and transported by the conveyor system. Assuming that the material can flow freely from the hopper, the faster the conveyor is moved, the faster material will flow from the hopper, and the greater the material flow rate on the conveyor. In this case, flow rate can be calculated from

$$Q = \frac{W\,R}{L} \tag{5-27}$$

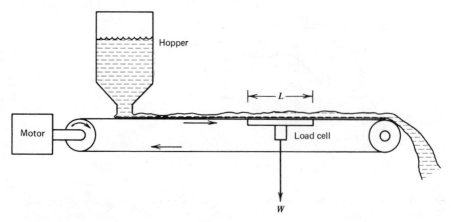

FIGURE 5.25   Conveyor system for illustrating solid flow measurement.

Where

Q = flow in kg/min
W = weight of material on section of length L
R = conveyor speed in m/min
L = length of weighing platform in m

FLOW TRANSDUCER

In the example with which we are working in Figure 5.25, it is evident that the flow transducer is actually the assembly of conveyor, hopper opening, and weighing platform. It is the actual weighing platform that performs the measurement from which flow rate is determined, however. We see that flow measurement becomes weight measurement. In this case, we have suggested that this weight is measured by means of a load cell, which is then a strain gage measurement. Another popular device for weight measurement of moving systems like this is an LVDT which measures the droop of the conveyor at the point of measurement due to the material which it carries.

EXAMPLE 5.10

A coal conveyor system moves at 100 ft/min. A weighing platform is 5.0 ft in length, and a particular weighing shows that 75 lb of coal are on the platform. Find the coal delivery in lb/hr.

SOLUTION

We can use Equation (5-27) directly to find the flow

$$Q = \frac{(75 \text{ lb})(100 \text{ ft/min})}{5 \text{ ft}}$$

$$Q = 1500 \text{ lb/min}$$

Then, converting to lb/hr by multiplying by 60 min/hr

$$Q = 90000 \text{ lb/hr}$$

5.6.2 Liquid Flow

The measurement of liquid flow is involved in nearly every facet of the process industry. The conditions under which the flow occurs and the vastly different types of material which flow result in a great many types of flow measurement methods. Indeed, entire books are written devoted to the problems of how to measure liquid flow and how to interpret the results of flow measurements. It would be impractical and not within the scope of this book to present a comprehensive study of liquid flow, and only the basic ideas of liquid flow measurement will be presented.

FLOW UNITS

The units used to describe the flow measured can be of several types depending on how the specific process needs the information. The most common descriptions are:

1. *Volume Flow Rate.* Expressed as a volume delivered per unit time. Typical units are: gallons/min, m^3/hr, ft^3/hr.
2. *Flow Velocity.* Expressed as the distance the liquid travels in the carrier per unit time. Typical units are m/min, ft/min. This is related to the volume flow rate by

$$V = \frac{Q}{A}$$ (5-28)

Where

V = flow velocity
Q = volume flow rate
A = cross-sectional area of flow carrier (pipe, and so on)

3. *Mass* or *Weight Flow Rate.* Expressed as mass or weight flowing per unit time. Typical units are kg/hr, lb/hr. This is related to the volume flow rate by

$$F = \rho Q$$ (5-29)

Where

F = mass or weight flow rate
ρ = mass density or weight density
Q = volume flow rate

EXAMPLE 5.11

Water is pumped through a 1.5-inch diameter pipe with a flow velocity of 2.5 ft per second. Find the volume flow rate and weight flow rate. The weight density is 62.4 lb/ft^3.

SOLUTION

The flow velocity is given as 2.5 ft/s so the volume flow rate can be found from Equation (5-28), $Q = VA$. The area is given by

$$A = \pi d^2/4$$

Where

the diameter $d = (1.5\ \text{in})(1/12\ \text{ft/in}) = 0.125\ \text{ft}$
so that $A = (3.14)(0.125)^2/4 = 0.0122\ \text{ft}^2$. Then, the volume flow rate is

$$Q = (2.5\ \text{ft/s})(0.0122\ \text{ft}^2)(60\ \text{s/min})$$
$$Q = 1.8\ \text{ft}^3/\text{min}$$

The weight flow rate is found from Equation (5-29)

$$F = (62.4\ \text{lb/ft}^3)(1.8\ \text{ft}^3/\text{min})$$
$$\mathbf{F = 112\ lb/min}$$

PIPE FLOW PRINCIPLES

The flow rate of liquids in pipes is determined primarily by the pressure which is forcing the liquid through the pipe. The concept of pressure head or simply *head* introduced in the previous sections often is used to describe this pressure because it is easy to relate the forcing pressure to that produced by a depth of liquid in a tank from which the pipe exits. In Figure 5.26, flow through pipe, P, is driven by the pressure in the pipe, but this pressure is caused by the weight of liquid in the tank of height, h (head). The pressure is found from Equation (5-24) or (5-25). Many other factors affect the actual flow rate produced by this pressure, including liquid viscosity, pipe size, pipe roughness (friction), turbulence of flowing liquid, and others. It is beyond the scope of this book to detail exactly how these factors determine the flow. Instead, it is our objective to discuss how such flow is measured, regardless of those features which may determine exactly what the flow is relative to the conditions mentioned.

RESTRICTION FLOW TRANSDUCERS

One of the most common methods of measuring the flow of liquids in pipes is made by introducing a restriction in the pipe and measuring the pressure drop which

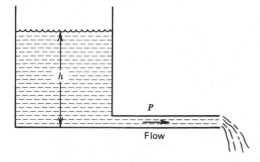

FIGURE 5.26   Flow through pipe, P, is determined in part of the head pressure, h.

results across the restriction. When such a restriction is placed in the pipe, the velocity of the fluid through the restriction *increases,* and the pressure in the restriction *decreases.* We find that there is a relationship between the pressure drop and the rate of flow such that as the flow increases, the pressure drops. In particular, one can find an equation of the form

$$Q = K \sqrt{\Delta p} \qquad (5\text{-}30)$$

Where

    Q = volume flow rate
    K = a constant for the pipe and liquid type
    Δp = drop in pressure across the restriction

The constant, K, depends on many factors, including the type of liquid, size of pipe, velocity of flow, temperature, and others. The type of restriction employed also will change the value of the constant used in this equation. Note that the flow rate is not linearly dependent on the pressure drop but rather on the square root. Thus, if one found that the pressure drop in a pipe increased by a factor of two when the flow rate was increased, the flow rate will have only increased by a favor of 1.4 (the square root of two). Certain standard types of restrictions are employed in exploiting the pressure-drop method of measuring flow.

    Figure 5.27 shows the three most common methods. It is interesting to note that having converted flow information to pressure, we now employ one of the methods of measuring pressure, which may translate to displacement, which is measured by a displacement transducer before finally getting a signal which will be used in the process-control loop. The most common method of measuring the pressure drop is to use a differential pressure transducer similiar to that shown in Figure 5.23. These are often described by the name *DP cell.*

OBSTRUCTION FLOW TRANSDUCER

Another type of flow transducer operates by the effect of flow on an obstruction placed in the flow stream. In a *rotameter,* the obstruction is a float which rises in a vertical tapered column. The lifting force, and thus distance to which the float rises in the column, is proportional to the flow rate. The lifting force is produced by the differential pressure which exists across the float, since it is a restriction in the flow. This type of flow transducer is used for both liquids and gases. A *moving vane* flow meter has a vane target immersed in the flow region, which will be rotated out of the flow as the flow velocity increases. The angle of the vane is a measure of the flow rate. If the rotating vane shaft is attached to an angle-measuring transducer, the flow rate can be measured for use in a process-control application. A *turbine* type of flow meter is composed of a freely spinning turbine blade assembly in the flow path. The rate of rotation of the turbine is proportional to the flow rate. If the turbine is

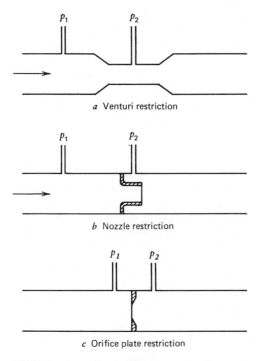

a Venturi restriction

b Nozzle restriction

c Orifice plate restriction

FIGURE 5.27   Three different types of restrictions commonly used to convert flow rate to a pressure difference, $p_1 - p_2$.

attached to a tachometer, a convient electrical signal can be produced. In all of the above methods of flow measurement, it is necessary to present a substantial obstruction into the flow path in order to measure the flow. For this reason, these devices are used only in special circumstances where an obstruction would not cause any unwanted reaction on the flow system. These devices are illustrated in Figure 5.28.

MAGNETIC FLOW METER

It can be shown that if charged particles move across a magnetic field, a potential is established across the flow, perpendicular to the magnetic field. Thus, if the flowing liquid is also a conductor, though not necessarily a good conductor, of electricity, the flow can be measured by allowing the liquid to flow through a magnetic field and measuring the transverse potential produced. The pipe section in which this measurement is made must be insulated and a nonconductor itself, or the potential produced will be cancelled by currents in the pipe. A diagram of this type of flow meter is presented in Figure 5.29. This type of transducer produces an electrical signal directly and is convenient for process-control applications involving conducting fluid flow.

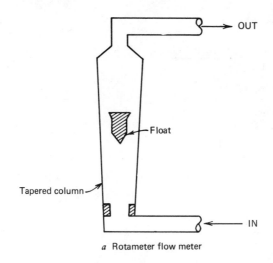

a Rotameter flow meter

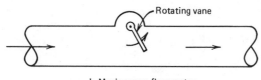

b Moving vane flow meter

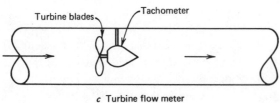

c Turbine flow meter

FIGURE 5.28   Three different types of obstruction flow meters.

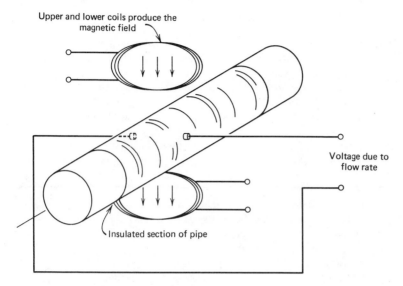

FIGURE 5.29   Magnetic flow meter.

SUMMARY

In this chapter, an assortment of measurement systems, which fall under the general description of mechanical transducers, has been studied. The objective was to gain familiarity with the essential features of the variables themselves and the typical measurement methods.

Topics covered were:

1.   Position, location, and displacement transducers, including the potentiometric, capacitive, and LVDT. The LVDT converts displacement linearly into a voltage.

2.   The strain gage measures deformation of solid objects resulting from applied forces called stress. The strain gage converts strain into a change of resistance.

3.   Accelerometers are used to measure the acceleration of objects due to rectilinear motion, vibration, and shock. Most of them operate by the spring-mass principle, which converts acceleration information into a displacement.

4.   Pressure is the force per unit area which a fluid exerts on the walls of a container. Pressure transducers often convert pressure information into a displacement. Examples include diaphragms, bellows, and the Bourdon tube. Electronic measures are often used for low pressures.

5.   For gas pressures less than 1 atmosphere purely electrical techniques are used. In some cases, the temperature of a heated wire is used to indicate pressure.

6.   Flow transducers are very important in the manufacturing world. Typically, solid flow is mass or weight per unit time.

7.   Fluid flow through pipes or channels typically is measured by converting the flow information into pressure by means of a restriction in the flow system.

PROBLEMS

5.1   A capacitive displacement transducer is used to measure rotating shaft wobble as shown in Figure 5.30. The capacity is 880 pF with no wobble. Find the change in capacity for a $+0.02$ to -0.02 mm shaft wobble.

5.2   An aluminum beam supports a 550-kg mass. If the beam diameter is 6.2 cm, calculate (a) the stress, (b) the strain of the beam.

5.3   A strain gage has GF = 2.14 and a nominal resistance of 120 Ω. Calculate the resistance change resulting from a strain of 144 μin./in.

5.4   A strain gage with GF = 2.03 and 120 Ω nominal resistance is used to measure strain with a resolution of 5μm/m. Design a bridge and detector that provides this, together with a smooth range of five switched 1000 μm/m spans: 0 to 1000 μm/m, 1000 to 2000 μm/m, and so on.

5.5   We will weigh objects by a strain gage of R = 120 Ω, GF = 2.02 mounted on a copper column of 6-inch diameter. Find the change in resistance per pound placed on the column. Is this change a decrease or increase? Draw a diagram of the system showing how the active and dummy gage should be mounted.

5.6   Calculate the rotation rate of a 10,000 rpm motor in rad/second.

5.7   An object falls from rest near the earth's surface, accelerating downward at 1g. After 5 s, find (a) speed and (b) distance moved.

5.8   An automobile fender vibrates at 16 Hz with a peak-to-peak amplitude of 5 mm. Calculate the peak acceleration in **g**.

5.9   A force of 2.7 lb is applied to a 5.5-kg mass. Find the resulting acceleration in m/s².

5.10   Calculate the average shock in **g**s experienced by a transistor which falls 1.5 m from a table top if it takes 2.7 ms to deaccelerate to zero when striking the floor.

5.11   A spring-mass system has a mass of 0.02 kg and a spring constant of 140 N/m. Calculate the natural frequency.

5.12   An LVDT is used in an accelerometer to measure seismic mass displacements. The LVDT and signal conditioning outputs 0.31 mV/mm with a $\pm$2 cm maximum core displacement. The spring constant is 240 N/m and the core mass is 0.05 kg. Find (a) relation between acceleration in m/s² and output voltage, (b) maximum acceleration measurable, (c) natural frequency.

5.13   For the accelerometer of Problem 5.12, design a signal conditioning system which provides velocity at 2mV/(m/s) and position information at 0.5 V/m.

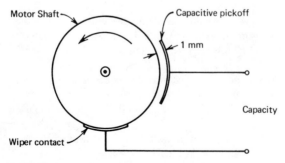

FIGURE 5.30   Figure for Problem 5.1.

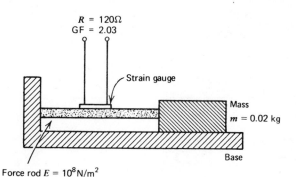

$R = 120\Omega$
GF = 2.03

Strain gauge

Mass
$m = 0.02$ kg

Base

Force rod $E = 10^8$ N/m^2

FIGURE 5.31   Figure for Problem 5.15.

5.14   A piezoelectric accelerometer has a transfer function of 61 mV/g and a natural frequency of 4500 Hz. In a vibration test at 110 Hz, a reading of 3.6 volts peak results. Find the vibration peak displacement.

5.15   An accelerometer for shock is designed as shown in Figure 5.31. Find a relation between strain gage resistance change and shock in **g**, that is, change in resistance per **g**.

5.16   Design a signal conditioning scheme for the accelerometer of Problem 5.15 using a bridge circuit. Plot the bridge offset voltage versus shock in **g** from 0 to 5000 **g**.

5.17   Calculate (a) the pressure in atmospheres which a water column 3.3 m high exerts on its base, (b) the pressure if the liquid is mercury, (c) conversions to pascals for (a) and (b).

5.18   A welding tank holds oxygen at 1500 psi. What is the tank pressure expressed in Pa? What is the pressure in atmospheres?

5.19   A diaphragm has an effective area of 25 cm^2. If the pressure difference is 5 psi, what force is exerted on the diaphragm?

5.20   A grain conveyor system finds the weight on a 1.0-meter platform to be 258 N. What conveyor speed is needed to get a flow of 5200 kg/hr?

5.21   Convert a water flow of 52.2 gal/hr into kg/hr and velocity in m/s through a 2-inch diameter pipe.

5.22   For an orifice plate in a system pumping alcohol, we find $K = 40 \ (\text{m}^3/\text{min})(\text{kPa})^{\frac{1}{2}}$. Plot the pressure versus flow rate from 0 to 100 m^3/min.

6

OPTICAL
TRANSDUCERS

INSTRUCTIONAL OBJECTIVES

After a study of this chapter the reader should be able to:

1. Describe EM radiation in terms of frequency, wavelength, speed of propagation, and spectrum.
2. Define the energy of EM radiation in terms of power, intensity, and the effects of divergence.
3. Describe the luminous energy descriptors of EM radiation.
4. Compare photoconductive, photovoltaic, and photoemissive type photodetectors.
5. Describe the principles and structure of both total radiation and optical pyrometers.
6. Distinguish incandescent, atomic, and laser light sources by the characteristics of the light they produce.
7. Design the application of optical techniques to process-control measurement applications.

6.1 INTRODUCTION

A previously emphasized characteristic of all transducers is that a *linear* relation is desired between transducer output and variable input. Such a linear characteristic simplifies both analysis and signal conditioning and thereby provides more confidence that the output signal represents the dynamic variable.

A second desirable characteristic of a transducer is negligible effect on the measured environment, that is, the process. Thus, if a Resistance Temperature Detector (RTD) *heats up* its own temperature environment, there is less confidence that the RTD resistance truly represents the environment temperature. Much effort is made in transducer design to reduce such backlash from the measuring instrument on its environment.

When electromagnetic (EM) radiation is used to perform dynamic variable measurements, transducers that virtually do not affect the system measured emerge. Such systems of measurement are called *nonlocal* or *noncontact* because no physical contact is made with the environment of the variable. Noncontact characteristic measurements often can be made from a distance.

In process control, EM radiation in either the visible or infrared light band is used in measurement applications. The techniques of such application are called *optical* because such radiation is close to visible light.

A common example of optical transduction is measurement of an object's temperature by the EM radiation it emits. Another example involves radiation reflected off the surface to yield a measurement of level or displacement.

In general, there are several levels of approach to a study of transducer technology. The first and most elementary level is the simple substitution of a black-box instrument for the measurement of some variable. A second level is the knowledge and training necessary to repair such an instrument, using manuals. This requires rudimentary knowledge of the variable being measured and the manner by which such measurement is made. A third level might be the knowledge and understanding necessary to select an appropriate transducer for the specific measurement requirement. This involves even more comprehension of the variable, the transduction methods, and measurement limitations. The fourth and highest requires all the knowledge to *design* the actual transducer, given the variation of one physical variable with another. In this text, sufficient background is provided to reach the third level.

Optical technology is a vast subject covering a span from geometrical optics, including lenses, prisms, gratings, and the like, to physical optics with lasers, parametric frequency conversion, and nonlinear phenomena. These subjects are all very interesting, but all that is required for our purposes is a familiarity with optical principles and a knowledge of specific transduction and measurement methods.

6.2 FUNDAMENTALS OF EM RADIATION

We are all familiar with EM radiation in the form of *visible light*. It appears all around us and underlies an important part of our lives. Some are familiar with it in other forms, such as radio or TV signals and ultraviolet and infrared light. Most of us falter, however, if asked to give a general technical description of such radiation including criteria for measurement and units.

This section covers a general method of characterizing EM radiation. Although much of what follows is valid for the complete range of radiation, particular attention is given to the infrared, visible, and ultraviolet, because most transducer applications are concerned with these ranges.

6.2.1 Nature of EM Radiation

We begin by noting that EM radiation is a form of *energy* which is always in motion, that is, propagation through space. An object which releases or *emits* such radiation *loses* energy. One that *absorbs* radiation *gains* energy. Thus, we must describe how this energy appears in the form of EM radiation.

FREQUENCY AND WAVELENGTH

Because we use the term *electromagnetic* radiation to name this form of energy, it is no surprise that it is intimately tied to electricity and magnetism. Indeed, careful study has shown that as a result of electrical and magnetic phenomena, EM radiation is produced. It is found that the radiation propagates through space in a manner similar to waves in water propagating from some disturbance. As such, we can define both a *frequency* and *wavelength* of the radiation. The *frequency* represents the oscillation per second as the radiation passes some fixed point in space. The *wavelength* is the spatial distance between two successive maxima or minima of the wave in the direction of propagation.

SPEED OF PROPAGATION

It has been found that EM radiation propagates through a vacuum at a speed *independent* of both the wavelength and frequency. In this case, the velocity is

$$c = \lambda f \tag{6-1}$$

Where

$c = 2.998 \times 10^8$ m/s $\approx 3 \times 10^8$ m/s $=$ speed of EM radiation in a vacuum
$\lambda =$ wavelength in meters
$f =$ frequency in hertz (Hz) or cycles per second (s^{-1})

EXAMPLE 6.1

Given an EM radiation frequency of 10^6 Hz, find the wavelength.

SOLUTION

We have

$$c = \lambda f \tag{6-1}$$

and

$$\lambda = \frac{c}{f} = \frac{3 \times 10^8 \text{ m/s}}{10^6 \text{ s}^{-1}}$$

$$\lambda = 300 \text{ m}$$

Note: This EM radiation is used to carry AM radio signals.

When such radiation moves through a nonvacuum environment, the propagation velocity is *reduced* to a value *less* than c. In general, the new velocity is indicated by the *index of refraction* of the medium. The index of refraction is a ratio defined by

$$n = \frac{c}{v} \qquad (6\text{-}2)$$

Where

n = index of refraction
v = velocity of EM radiation in the material (m/s)

The index of refraction often varies with the radiation wavelength for some sample of material.

EXAMPLE 6.2

Find the velocity of EM radiation in glass for which the index of refraction is n = 1.57.

SOLUTION

We know that

$$n = \frac{c}{v} \qquad (6\text{-}2)$$

so that

$$v = \frac{c}{n} = \frac{3 \times 10^8 \text{ m/s}}{1.57}$$

$$v = 1.91 \times 10^8 \text{ m/s}$$

WAVELENGTH UNITS

The most consistent description of EM radiation is via the *frequency* or *wavelength*. For many applications, this specification is made through the frequency of the radiation, as in a 1 MHz radio signal or a 1 GHz microwave signal. By convention, however, it has become more common to describe EM radiation by the wavelength, as given by Equation (6-1). This is particularly true near the visible light band. The proper unit of measurement is, of course, the length in meters with associated prefixes. Thus, a 10 GHz signal is described by a 30-mm wavelength. Red light is emitted at about 0.7 μm wavelength.

Another unit often employed is the *Angstrom* (Å), defined as 10^{-10} m, or 10^{-10} m/Å. Thus, the red light previously described has a wavelength of 7000Å. The conversion is left as an exercise for the reader. (See also Problem 6.3.)

EM RADIATION SPECTRUM

We have seen that EM radiation is a type of energy that propagates through space at a constant speed or velocity if we specify the direction. The oscillating nature of this radiation gives rise to a different interpretation of this radiation in relation to our enviroment, however. In categorizing radiation by wavelength or frequency, we are describing its relation in the *spectrum* of radiation. Figure 6.1 shows the range of EM radiation from very low frequency to very high frequency, together with the

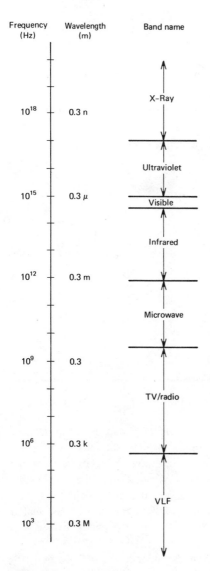

FIGURE 6.1 The electromagnetic radiation spectrum covers everything from very low frequency (VLF) to X-rays and beyond.

associated wavelength in meters, from Equation (6-1), and how the bands of frequency relate to our world.

Note that this one type of energy ranges from radio signals and visible light to x-rays and penetrating cosmic rays and all through the smooth variation of frequency. In process-control instrumentation we are particularly interested in two of the bands, infrared and visible light.

VISIBLE LIGHT

The small band of radiation between approximately 400 nm and 760 nm represents *visible light* (Figure 6.1). This radiation band covers those wavelengths to which our eyes (or radiation detectors in our heads) are sensitive.

INFRARED LIGHT

The longer wave radiation band that extends from the limit of eye sensitivity at 0.76 μm to approximately 100 μm is called *infrared* (IR) radiation. In some cases, the band is further subdivided so that radiation of wavelength 3 μm to 100 μm is called *far-infrared*.

Note that the limits of the above-defined bands are not distinct but serve only to separate roughly the described radiation into broad categories. Our treatment for the rest of this chapter refers simply to *light*, meaning either IR or visible, because our concern is with these bands, exclusively.

EXAMPLE 6.3

Describe (a) the wavelength (in m and Å units), and (b) nature of EM radiation of 5.4 × 10¹³ Hz frequency.

SOLUTION

The wavelength is given by

a. $$\lambda = \frac{c}{f} \tag{6-1}$$

$\lambda = (3 \times 10^8 \text{ m/s})/(5.4 \times 10^{+13} \text{ Hz})$
$\lambda = 5.56 \ \mu\text{m}$

$$\lambda = 5.56 \times 10^{-6} \text{ m} \times \frac{1 \text{ Å}}{10^{-10} \text{ m}} = 55,600 \text{ Å}$$

b. From Figure 6.1 we see that this radiation lies in the *infrared* band, generally, and is designated as far-infrared, specifically.

6.2.2 Characteristics of Light

Since light has been described as a source of energy, it is natural to inquire about the energy content and its relation to the spectrum.

PHOTON

No description of EM radiation is complete without a discussion of this elementary energy unit. It has been found that EM radiation at a particular frequency can propagate only in *discrete* quantities of energy. Thus, if some source is emitting radiation of one frequency, then in fact it is emitting this energy as a large number of discrete units or *quanta*. These quanta are called *photons*. The actual energy of one photon is related to the frequency by

$$W_p = hf = \frac{hc}{\lambda} \tag{6-3}$$

Where

W_p = photon energy (J)
h = 6.63×10^{-34} J $-$ s (known as Planck's constant)
f = frequency (s^{-1})
λ = wavelength (m)

The energy of one photon is very small compared to electric energy we normally experience as shown by Example 6.4.

EXAMPLE 6.4

A microwave source emits a pulse of radiation at 1 GHz with a total energy of 1 joule. Find (a) the energy photon and (b) the number of photons in the pulse.

SOLUTION

We find the energy per photon where 1 GHz = 10^{+9} s^{-1}

a.
$$W_p = hf$$
$$W_p = (6.63 \times 10^{-34} \text{ J} - \text{s})\,(10^9 \text{ s}^{-1}) \tag{6-3}$$
$$W_p = 6.63 \times 10^{-25} \text{ J}$$

b. The number of photons is

$$N = \frac{W}{W_p} = \frac{1 \text{ J}}{6.63 \times 10^{-25} \text{ J/photon}}$$

$$N = 1.5 \times 10^{24} \text{ photons}$$

Figure 6.2 shows the energy carried by a single photon at various wavelengths. This energy is expressed in electron volts where (1 eV = 1.602×10^{-19} J). This unit is conventionally employed and provides convenient magnitude for photon energy discussion.

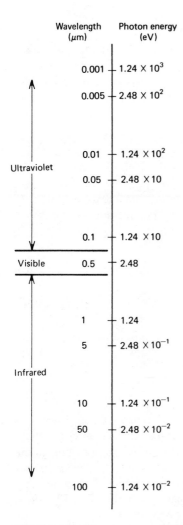

FIGURE 6.2   The energy carried by one photon varies inversely with the wavelength of the EM radiation.

ENERGY

When dealing with typical sources and detectors of light, it is impractical and unnecessary to consider this discrete nature of the radiation. Instead, we deal with so-called *macroscopic* properties which result from the collective behavior of a vast number of photons moving together. Here the general energy principles involve a description of the net energy content of the radiation as it propagates through a region of space. This description is then given in *joules* of energy in the propagating light. A simple statement of energy is insufficient, however, because of the motion of the energy and the spatial distribution of the energy.

POWER

Because EM radiation is energy in motion, a more complete description is also the joules per second or *watts* of power carried. Thus, one might describe a situation where a source emits 10 watts of light, meaning that 10 joules of energy in the form of light radiation are emitted every second. Even this description is incomplete, however, without specifying how the power is spatially distributed.

INTENSITY

A more complete picture of the radiation emerges if we also specify the spatial distribution of the power *transverse* to the *direction* of propagation. Thus, if the 10-watt source just discussed was concentrated in a beam with a cross-sectional area of 0.2 m², then we can specify the *intensity* as the watts per unit area, in this case as (10 W)/(0.2 m²) or 50 W/m². In general, the intensity is

$$I = \frac{P}{A} \qquad (6\text{-}4)$$

Where

I = intensity in W/m²
P = power in W
A = beam cross-sectional area in m²

DIVERGENCE

We have still not quite exhausted the necessary descriptors of the energy because of the tendency of light to travel in straight lines. Because the radiation travels in straight lines, it is possible for the intensity of the light to *change* even though the power remains *constant*. This is best seen in Figure 6.3. Here we have a 10-watt source with an area A_1 at the source. However, due to the nature of the source and the straight-line propagation, we see that some distance away the same 10 watts is

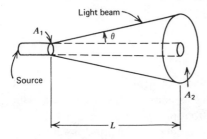

FIGURE 6.3   Sources of EM radiation exhibit divergence through the spreading of the beam with distance from the source.

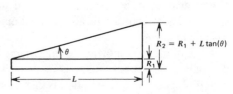

FIGURE 6.4   Diagram to aid in solving divergence problems such as Example 6.5.

distributed over a larger area A_2, and hence the *intensity* is *diminished*. Such spreading of radiation is called *divergence* and specified as the angle θ, made by the outermost edge of the beam to the central direction of propagation, as shown in Figure 6.3.

A complete description of the energy state of light propagating through a region of space demands knowledge of the power carried, the cross-sectional area over which the power is distributed, and the divergence.

EXAMPLE 6.5

Find the intensity of a 10-W source (a) at the source and (b) 1 meter away for the case shown in Figure 6.3 if the radius at A_1 is 0.05 m and the divergence is 2 degrees.

SOLUTION

The intensity at the source is simply the power over the area. Thus, for a circular cross section,

$$A_1 = \pi R_1{}^2 = (3.14)(0.05 \text{ m})^2$$
$$A_1 = 7.85 \times 10^{-3} \text{ m}^2$$

a. The intensity at the source is

$$I_1 = \frac{P}{A_1}$$

$$\text{(6-4)}$$

$$I_1 = \frac{10 \text{ W}}{7.85 \times 10^{-3} \text{ m}^2} = 1273 \text{ W/m}^2$$

b. We must find the beam area at 1 m to get A_2 and then the intensity at 1 meter. We can find the radius using Figure 6.4 as a guide. From elementary trigonometry we see that, for the right triangle formed

$$R_2 = R_1 + L \tan \theta \qquad \text{(6-5)}$$

then

$$R_2 = 0.05 \text{ m} + (1 \text{ m}) \tan (2°)$$
$$R_2 = 0.085 \text{ m}$$
$$A_2 = \pi R_2{}^2 = (3.14) (0.085 \text{ m})^2$$
$$A_2 = 0.0227 \text{ m}^2$$

$$I_2 = \frac{10 \text{ W}}{0.0227 \text{ m}^2}$$

$$I_2 = 440.53 \text{ W/m}^2 \text{ or } \mathbf{441 \text{ W/m}^2}$$

When dealing with sources originating from a very small point, but propagating in *all* directions, we have a condition of *maximum divergence*. Such *point sources* have an intensity that decreases as the *inverse square* of the distance from the point. This can be seen from a consideration of the divergence. Suppose a source delivers a power P of light as shown in Figure 6.5. The intensity at a distance R is found by dividing the total power by the surface area of a sphere of radius R from the source.

$$I = \frac{P}{A} \text{ in W/m}^2 \tag{6-4}$$

Now, the surface area of a sphere of radius R is $A = 4\pi R^2$, thus over the entire surface surrounding a point source

$$I = \frac{P}{4\pi R^2} \text{ in W/m}^2 \tag{6-6}$$

This equation shows that the intensity of a point source decreases as the inverse square of the distance from the point.

CHROMATICITY

Another factor of significance in the description of light includes the *spectral content* of the radiation. A source such as a laser beam, which delivers light of a *single* wavelength (or very nearly), is called a *monochromatic* source. A source such as an incandescent bulb may deliver a very *broad* spectrum of radiation, and is referred to as a *polychromatic* source.

COHERENCY

A less familiar characteristic of the radiation is its coherency. We have seen that light is described through electric and magnetic effects which oscillate in time and space. Whenever we consider oscillating phenomena, it is of interest to determine the phase relation of oscillations of different parts of the beam. We say that when all points along some cross section of a radiation beam are in phase, the beam has *spatial coherence*. When the radiation in a line along the beam has a fixed-phase

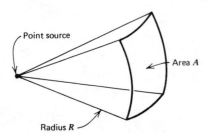

FIGURE 6.5 The intensity of light from a point source depends on the distance from the source R and the area considered A.

relation, we say the beam has *temporal coherence*. In general, conventional sources of light, such as incandescent and fluorescent light bulbs, produce beams with no coherence. Only with the development of the laser was a convenient source of coherent radiation made available.

6.2.3 Luminous Energy Principles

Because of the importance of visible light to the human environment, a special system of units is often used to describe both the energy state and the chromaticity of EM radiation. This system emerges principally from a human engineering application, in that the energy content is described in terms of its relation to the sensitive band of eyes to EM radiation. Light that falls within this band is called *luminous radiation*. The system of measurement, which is often referred to as *photometric*, is *relative* in that comparisons are made to a *standard source* of (visible) radiation.

STANDARD SOURCE

To define the photometric units, a standard source of EM radiation is used to represent a combination of power and chromaticity. The standard consists of a flask of molten platinum at 2046 K as schematically shown in Figure 6.6. Other sources are then evaluated by a comparison of their emission relative to that of this standard source.

LUMINOUS INTENSITY

The SI defines a measure of luminous intensity by a unit referred to as the *candela* (cd). The standard source is said to emit 1 candela from an area of one sixtieth of a square cm of its surface. This unit thus refers to the total radiation emitted in all

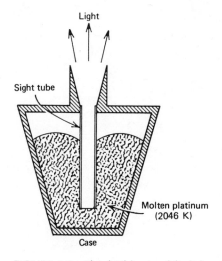

FIGURE 6.6   This highly simplified drawing shows the essential features of the standard source employed to define luminous units.

directions with all wavelengths from that section of surface. Often, sources are described by the *luminance* (L) which is another measure of luminous intensity but which refers to unit area. Thus, the standard source has a luminance of 60 cd/cm², where the common unit of measure is the cd/cm².

Note that these units include divergence together with intensity because the standard source not only emits a certain intensity of radiation, but also with an inverse square law divergence. Thus, if we have some source of radiation emitting principally UV radiation, it would have *very little* luminance because the nature of the radiation emitted in comparison with the standard source is small, and the object would appear dim or black to the human eye. Several additional luminous units are employed to further specify the characteristics of a luminous source.

LUMINOUS FLUX

The total quantity of EM radiation passing some area per second *relative* to the standard source is referred to as the luminous flux Φ. The unit of measurement is the lumen (lm) which is *defined* such that 1 candela emitted into one steradian solid angle is 1 lumen (1 lumen = 1 cd/sr). Here, a solid angle is defined as ratio of spherical surface area to radius squared. The SI unit of solid angle is the steradian (sr) which is defined so that an entire sphere has 4π sr. This is obvious from the fact that a sphere of radius R has a surface area of $4\pi R^2$ and thus,

$$\text{Sphere solid angle} = \frac{\text{Area}}{R^2} = \frac{4\pi R^2}{R^2} = 4\pi \text{ sr}$$

In Figure 6.7, this point is shown more clearly where an area of magnitude A is indicated on a spherical surface of radius R. The solid angle Ω formed by this area is given by

$$\Omega = \frac{A}{R^2} \tag{6-7}$$

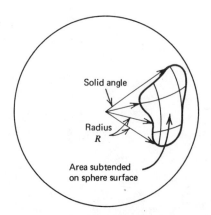

FIGURE 6.7   Solid angle is defined by the ratio of area subtended on a spherical surface divided by the radius squared.

Note that *any* other shape on the surface which has the same area would have the same solid angle. In this case, 1 sr is defined when the area in Equation (6-7) is equal to R^2.

Now luminous flux can be defined as

$$\Phi = I/\Omega \qquad (6\text{-}8)$$

ILLUMINATION

When describing the luminous radiation falling on a surface, we use a quantity defined by the luminous flux per unit area called the *illumination E*. The common representation of this quantity is lux (lx) which is 1 lumen per square meter (lm/m^2).

$$E = \frac{\Phi}{A} \qquad (6\text{-}9)$$

Where

Φ = luminous flux (lm)
A = area illuminated (m^2)
E = illumination (lm/m^2)

EFFICACY

This quantity relates the ratio of luminous output of a source to total output including the entire spectrum of EM radiation. The unit is the lumen per watt (lm/W).

EXAMPLE 6.6

A point source uniformly emits 90 cd. A detector of 2 cm² area is placed 1.5 m away from the source. Find (a) the luminous flux of the source, (b) the solid angle of the detector, and (c) the illumination of the detector.

SOLUTION

a. Because the source emits 90 cd in all directions, it emits 90 cd into 4π sr since a spherical surface has 4π sr. Hence,

$$\Phi = I/\Omega = (90 \text{ cd})/(4\pi \text{ sr}) \qquad (6\text{-}8)$$
$$\Phi = 7.16 \text{ cd/sr (lm)}$$

b. The solid angle of the detector is found from

$$\Omega = \frac{A}{R^2} = \frac{2 \times 10^{-4} \text{ m}^2}{(1.5 \text{ m})^2} \qquad (6\text{-}7)$$

$$\Omega = 8.89 \times 10^{-5} \text{ sr}$$

c. The illumination on the detector is found first by noting that 7.16 lm fall on the spherical surface at 1.5 m radius

$$E = \frac{\Phi}{4\pi R^2} = \frac{7.16 \text{ lm}}{(4\pi)(1.5)^2} \tag{6-9}$$

$$E = 0.25 \text{ lm/m}^2$$

We will not be using the luminous energy principles to any great extent in the work of this chapter, but familiarity is essential to an overall background in EM radiation studies.

RELATION TO ENVIRONMENT

Most people have little appreciation of the significance of luminous units to their environment. To better appreciate the magnitudes of these units, Table 6.1 lists some characteristics of both tungsten filament and fluorescent lamps and their luminous units. Note that all the fluorescent lamps are not only more efficient but also are a distributed source rather than highly localized as the filament type.

Another important factor in the study of lighting is the illumination provided by sources and conditions. Table 6.2 shows the illumination produced by several sources familiar to our environment.

EXAMPLE 6.7

a. Find the illumination on a table 2 m from a single 60-W tungsten lamp.
b. Is this light sufficient to read by?

SOLUTION

From Table 6.1, a 60-W tungsten lamp has a luminous flux of 835 lumens. Assuming this to be a point source, the illumination on any surface at a distance 2 m from a point source is

a.
$$E = \frac{\Phi}{4\pi R^2} = \frac{835 \text{ lm}}{(4)(3.14)(2 \text{ m})^2} \tag{6-9}$$

$$E = 16.6 \text{ lux} = 16.6 \text{ lm/m}^2$$

b. Table 6.2 shows that this level is insufficient for reading.

EXAMPLE 6.8

Find the luminous intensity in candela of a 200-W tungsten lamp.

SOLUTION

We assume the lamp is a point source which puts out 3650 lm into 4π sr. Since 1 candela per sr equals 1 lumen, it follows the source outputs:

TABLE 6.1   LAMP CHARACTERISTICS

Lamp Size and Type (watt input-electrical)	Luminous Flux (lumens)	Efficacy (lumens/watt-electrical)
10 W tungsten	78	7.8
40 W tungsten	465	11.7
60 W tungsten	835	13.9
200 W tungsten	3650	18.3
8 W fluorescent	490	35.0
40 W fluorescent	2320	58.0
100 W fluorescent	4400	44.0

$$I = \Phi\Omega \qquad\qquad (6\text{-}8)$$
$$I = (3650 \text{ cd/sr}) (4\,\pi \text{ sr})$$
$$I = 4.59 \times 10^4 \text{ cd}$$

6.3 PHOTODETECTORS

An important part of any application of light to an instrumentation problem is how to measure or detect radiation. In most process-control related applications, the radiation lies in the range from IR through visible and sometimes UV bands. The measurement transducers used generally are called *photodetectors* to distinguish them from other spectral ranges of radiation such as RF detectors in radio frequency (RF) applications.

In this section, we will study the principal types of photodetectors through a description of their operation and specifications.

6.3.1 Photodetector Characteristics

Several characteristics of photodetectors are particularly important in typical applications of these devices in instrumentation. In the discussions that follow, the various types of detectors are described in terms of these characteristics.

TABLE 6.2   ILLUMINATION OF COMMON SOURCES

Type of Lighting	Illumination (typical-lux) (lm/m^2)
Direct sunlight	10^5
Daylight room	10^3
Minimum for reading	10^2
Full moonlight	0.2
Starlight (clear night)	10^{-4}

SPECTRAL RESPONSE

Most detectors are able to function over some specified range of radiation wavelengths which defines the *spectral* response of the device. In most cases, the response is flat within some allowed deviation within this band of radiation.

TIME CONSTANT

Much of the instrumentation associated with optical or EM radiation techniques is associated with time-varying phenomena. For this reason, the response time, as specified through the time constant defined in Chapter 2, becomes an important part of the specifications.

DETECTIVITY

Various photodetectors vary remarkably so that no general contrast in detectivity or sensitivity can be made. An important part of the specification, however, is to note the type of transduction and sensitivity to the dynamic variable, which is usually given as intensity of the radiation at a particular wavelength.

6.3.2 Photoconductive Detectors

One of the most common photodetectors is based on the change in *conductivity* of a semiconductor material with radiation *intensity*. The change in conductivity appears as a change in *resistance* so that these devices also are called *photoresistive* cells. Because resistance is the parameter used as the transduced variable, we describe the device from the point of view of *resistance changes* versus *light intensity*.

PRINCIPLE

In Section 4.4.1, we noted that a semiconductor is a material in which an energy gap exists between conduction electrons and valence electrons. In a semiconductor photodetector, a photon is absorbed and thereby excites an electron from the valence to the conduction band. As many electrons are excited into the conduction band, the semiconductor resistance *decreases*, making the resistance an inverse function of radiation intensity. For the photon to provide such an excitation it must carry at least as much energy as the gap. From Equation (6-3) this indicates a maximum wavelength

$$E_p = \frac{hc}{\lambda_{max}} = \Delta W_g \tag{6-10}$$

$$\lambda_{max} = \frac{hc}{\Delta W_g}$$

Where

h = Planck's constant, $6.63 \times 10^{-34} J - s$
ΔW_g = semiconductor energy gap (J)
λ_{max} = maximum detectable radiation wavelength (m)

Note that any radiation with wavelength greater than that predicted by Equation (6-10) *cannot* cause any resistance change in the semiconductor.

EXAMPLE 6.9

Germanium has a band gap of 0.67 eV. Find the maximum wavelength for resistance change by photon absorption. Note $1.6 \times 10^{-19} J = 1$ eV.

SOLUTION

We find the maximum wavelength from

$$\lambda_{max} = \frac{hc}{\Delta W_g}$$

$$\lambda_{max} = \frac{(6.63 \times 10^{-34} J - s)(3 \times 10^8 \text{ m/s})}{(0.67 \text{ eV})(1.6 \times 10^{-19} J/\text{eV})} \tag{6-10}$$

$$\lambda_{max} = 1.86 \ \mu\text{m}$$

which lies in the *IR*.

It is important to note that the operation of a thermistor involves *thermal* energy exciting electrons into the conduction band. In order to prevent the photoconductor from showing similar thermal effects, it is necessary either to operate the devices at a controlled temperature or to make the gap too large for thermal effects to produce conduction electrons. Both approaches are employed in practice. The upper limit of the cell spectral response is determined by many other factors, such as reflectivity and transparency to certain wavelengths.

CELL STRUCTURE

The two most common photoconductive semiconductor materials are cadmium sulfide (CdS) with a band gap of 2.42 eV and cadmium selenide (CdSe) with a 1.74 eV gap. Because of these large gap energies, both materials have a very high resistivity at room temperature. This gives bulk samples a resistance much too large for practical applications. To overcome this, a special configuration is used, as shown in Figure 6.8, which minimizes resistance geometrically while providing maximum surface area for the detector. This result is based upon the equation presented earlier (Section 4.3.1).

$$R = \rho \ l/A \tag{4-7}$$

Where

R = resistance (Ω)
ρ = resistivity ($\Omega - $m)
l = length (m)
A = cross-sectional area (m²)

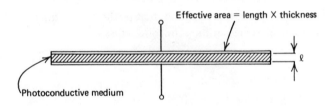

a) A long, narrow, thin sample gives optimum response and resistance

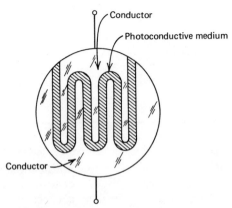

b) By folding the above pattern back and forth we concentrate the sensitive material in space

FIGURE 6.8   Photoconductive cell structure.

Note that the pattern of Figure 6.8*a* gives a minimum *l* and maximum A. By using a thin narrow strip and by winding this arrangement back and forth as in Figure 6.8*b*, we get a maximum surface area.

CELL CHARACTERISTICS

The characteristics of photoconductive detectors vary considerably different semiconductor materials are used as the active element. These characteristics are summarized for typical values in Table 6.3.

The nominal dark resistance and variation of resistance with intensity are usually provided in terms of a graph or table of resistance versus intensity at a particular wavelength within the spectral band. Typical values at dark resistance vary from hundreds of ohms to several MΩ for various types of photoconductors. The variation with radiation intensity is usually nonlinear with resistance decreasing as the radiation intensity increases.

TABLE 6.3   PHOTOCONDUCTOR CHARACTERISTICS

Photoconductor	Time Constant	Spectral Band
CdS	$\sim$ 100 ms	0.47 to 0.71 μm
CdSe	$\sim$ 10 ms	0.6 to 0.77 μm
PbS	$\sim$ 400 μs	1 to 3 μm
PbSe	$\sim$ 10 μs	1.5 to 4 μm

SIGNAL CONDITIONING

Like the thermistor, a photoconductive cell exhibits a resistance that decreases nonlinearly with the dynamic variable, in this case, radiation intensity. Generally, the change in resistance is very pronounced where a resistance may change by several hundred orders of magnitude from dark to normal room daylight.

If an absolute intensity measurement is desired, calibration data would be used in conjunction with any accurate resistance measurement method.

Sensitive control about some ambient radiation intensity is obtained using the cell in a bridge circuit adjusted for a null at the ambient level.

Various op-amp circuits using the photoconductor as a circuit element are used to convert the resistance change to a current or voltage change.

It is important to note that the cell is a variable resistor and therefore has some maximum power dissipation which cannot be exceeded. Most cells have a dissipation from 50 mW to 500 mW, depending on size and construction.

EXAMPLE 6.10

A CdS cell has a dark resistance of 100 kΩ and a resistance in a light beam of 30 kΩ. The cell time constant is 72 ms. Devise a system to trigger a 3-volt comparator within 10 ms of the beam interruption.

SOLUTION

There are many possible solutions to this problem. Let us first find the cell resistance at 10 ms using

$$R(t) = R_i + (R_f - R_i)\,[1 - e^{-t/\tau}]$$
$$R(10 \text{ ms}) = 30\text{ k} + 70\text{ k}\,(0.1296) = 39.077\text{ k}\Omega \tag{1-7}$$

so that we must have a $+3$-volt signal to the comparator when the cell resistance is 39.077 kΩ. The circuit of Figure 6.9 will accomplish this. Here, the cell is R_2 in the feedback of an inverting amplifier with a -1.0-volt constant input. The output is

$$V_{out} = -\left(\frac{R_2}{R_1}\right)(-1\text{ V}) = \frac{R_2}{R_1}$$

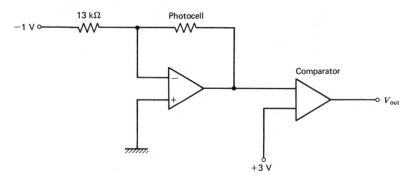

FIGURE 6.9   This circuit provides the requirements of Example 6.10.

when $R_2 = 39.077$ kΩ, we make $V_{out} = 3$ volt so that

$$R_1 = 39.077 \text{ k}\Omega$$
$$R_1 \simeq 13 \text{ k}\Omega$$

which assures that the comparator will trigger at 10 ms from beam interruption. To see that the comparator will *not* trigger with the beam present, set $R_1 = 30$ kΩ and the amplifier output is

$$V_{out} = -\frac{30 \text{ k}\Omega}{13 \text{ k}\Omega}(-1 \text{ V})$$

$$V_{out} = 2.3 \text{ V}$$

which is insufficient to trigger the comparator.

___

6.3.3 Photovoltaic Detectors

An important class of photodetectors generates a voltage which is proportional to incident EM radiation intensity. These devices are called *photovoltaic* cells because of their *voltage-generating* characteristics. They actually convert the EM energy into electrical energy. Applications are found as both EM radiation detectors and power sources converting solar radiation into electrical power. The emphasis of our consideration is on instrumentation-type applications.

PRINCIPLE

Operating principles of the photovoltaic cell are best described by Figure 6.10. We see that the cell is actually a giant diode, constructed using a *pn* junction between appropriately doped semiconductors. Photons striking the cell pass through the thin *p*-doped upper layer and are absorbed by electrons in the *n* layer, causing formation

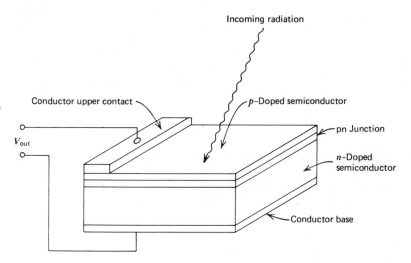

FIGURE 6.10   Structure of a photovoltaic cell.

of conduction electrons and holes. The depletion zone potential of the *pn* junction then separates these conduction electrons and holes causing a difference of potential to develop across the junction. The upper terminal is positive and the lower negative. It is also possible to build a cell with a thin *n*-doped layer on top so that all polarities are opposite. In general, the open-circuit voltage V developed in these cells varies logarithmically with the radiation intensity by

$$V_\bullet = V_0 \ln(I) \qquad\qquad (6\text{-}11)$$

Where

I = intensity in W/m^2
V_0 = calibration voltage, constant
V = output voltage, unloaded

All photovoltaic devices have low but finite internal resistance. When connected in a circuit with some load resistance, the cell voltage is reduced somewhat from the value predicted by Equation (6-11).

The constant voltage amplitude V_0 is a function of the cell material only and shows that the voltage produced is independent of the cell geometry. The current delivered from such a cell into a load depends on the intensity via Equation (6-11) and also increases with cell surface area. Cells are often arranged in series and parallel combinations to obtain desired voltage and current output levels.

The spectral range of the photovoltaic cell depends on the photon energy required to excite an electron-hole pair in the doped semiconductor and other properties of the materials.

EXAMPLE 6.11

A photovoltaic cell produces 0.33 V open-circuit when illuminated by 10 W/m² radiation intensity. A current of 2.2 mA is delivered into a 100 Ω load at that intensity. Calculate (a) the internal resistance and (b) the open-circuit voltage at 25 W/m².

SOLUTION

a. We find the internal resistance by noting that

$$I = \frac{V}{R_{in} + R_L}$$

where V is the open-circuit cell voltage, R_L the load resistance, I the current, and R_{in} the internal resistance. We find then

$$R_{in} = \frac{0.33 \text{ V} - (2.2 \text{ mA}) (100 \text{ }\Omega)}{2.2 \text{ mA}}$$

$$R_{in} = 50 \text{ }\Omega$$

b. The open-circuit voltage at 25 W/m² requires knowledge of V_0 in Equation (6-8). This can be found from

$$V_0 = \frac{V}{\ln(I)}$$

$$V_0 = \frac{0.33 \text{ V}}{\ln(10)} = 0.143 \text{ V}$$

so that at 25 W/m² we have

$$V = (0.143 \text{ V}) \ln (25)$$
$$V = 0.46 \text{ V}$$

CELL CHARACTERISTICS

The properties of photoconductive cells depend on the materials employed for the cell and the nature of the doping used to provide the n and p layers. Some cells are used only at low temperatures to prevent thermal effects from obscuring radiation detection. The silicon photovoltaic cell is probably the most common. Table 6.4 lists several types of cells and their typical specifications.

Thermal effects can be quite pronounced, producing changes of the order of mV/°C in output voltage of fixed intensity.

TABLE 6.4   TYPICAL PHOTOVOLTAIC CELL CHARACTERISTICS

Cell Material	Time Constant	Spectral Band
Silicon (Si)	$\sim$ 20 μs	0.44 μm to 1 μm
Selenium (Se)	$\sim$ 2 ms	0.3 μm to 0.62 μm
Germanium (Ge)	$\sim$ 50 μs	0.79 μm to 1.8 μm
Indium Arsenide (InAs)	$\sim$ 1 μs	1.5 μm to 3.6 μm (cooled)
Indium Antimonide (InSb)	$\sim$ 10 μs	2.3 μm to 7 μm (cooled)

SIGNAL CONDITIONING

Generally, signal conditioning depends on the application where the cell is used. Simple op-amp configuration provides a measure of either open-circuit voltage or current at any specified load impedance. Fast measurements require circuits which account for the internal capacity of the cell and the inherent time constant of the cell.

6.3.4 Photodiode Detectors

The *pn* junction of any diode is sensitive to EM radiation which may strike the junction. This sensitivity is usually in the form of an alteration of the I-V characteristic of the junction due to a change in current carriers. Special diodes which allow the junction to be exposed to incident EM radiation frequently are used for photodetectors. Generally, the junction is very small, requiring the use of lenses to focus the radiation on the junction. The most significant advantage of these detectors is their very fast response times. Most photodiodes have a time constant near 1 μs, but units are common with time constants less than 1 ns. These latter devices are employed in high-speed measurement or communication applications.

The most common diodes are silicon, which are used between 0.82 μm and 1.1 μm, and germanium, between 1.4 μm and 1.9 μm radiation wavelength. Signal conditioning usually involves standard diode circuits where incident radiation will cause a shift in the diode operating point. High-speed measurements require special, frequency-compensated circuits.

6.3.5 Photoemissive Detectors

This type of photodetector was developed many years ago but it still represents one of the most sensitive types. A wide variety of spectral ranges and sensitivities can be selected from the many types of photoemissive detectors available.

PRINCIPLES

To understand the basic operational mechanism of photoemissive devices let us consider the two-element vacuum phototube shown in Figure 6.11. Such photodetectors have been largely replaced by other detectors in modern measurements. In

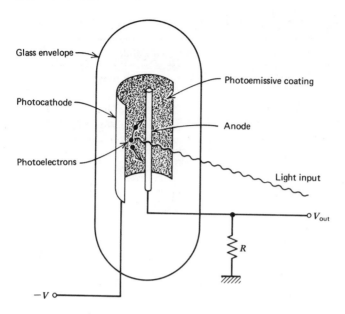

Glass envelope

Photoemissive coating

Photocathode

Anode

Photoelectrons

Light input

V_{out}

R

$-V$

FIGURE 6.11   Structure of the basic photoemissive diode.

Figure 6.11, we note that the cathode is maintained at some negative voltage with respect to the wire anode which is grounded through resistor R. The inner surface of the cathode has been coated with a *photoemissive* agent. This material is a metal for which electrons are easily detached from the metal surface. Easily means it does not take much energy to cause an electron to leave the material. In particular, then, a photon can strike the surface and impart sufficient energy to an electron to eject it from this coating. The electron will then be driven from the cathode to anode and thence through resistor R. Thus, we have a current that depends on the intensity of light striking the cathode.

PHOTOMULTIPLIER TUBE

The simple diode described above is the basis of one of the most sensitive photo-detectors available, as shown in Figure 6.12a. As above, a cathode is maintained at a large negative voltage and coated with a photoemissive material. In this case, however, we have many following electrodes, called *dynodes*, maintained at successively more positive voltages. The final electrode is the *anode* which is grounded through a resistor R. A photoelectron from the cathode strikes the first dynode with sufficient energy to eject several electrons. All of these electrons are accelerated to the second dynode where *each* strikes the surface with sufficient energy to again eject several electrons. This process is repeated for each dynode until the electrons that reach the anode are greatly multiplied in number, where they constitute a current through R. Thus, the photomultiplier (unlike some other transducers) has a *gain* associated with its detection. One single photon striking the cathode may result in a million electrons at the anode! It is this effect that gives the photomultiplier its

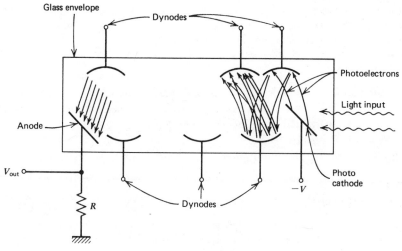

Glass envelope

Dynodes

Anode

Photoelectrons

Light input

V_{out}

R

Dynodes

$-V$

Photo
cathode

a) Basic structure of the photomultiplier tube

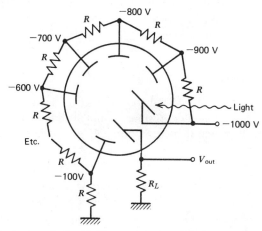

-800 V

R R

-700 V

-900 V

R

-600 V

R

Light

-1000 V

R

Etc.

R

V_{out}

-100V

R_L

R

b) Typical voltage divider used for dynode potentials

FIGURE 6.12   The photomultiplier tube.

excellent *sensitivity*. Many other electrode designs arrangements are used having the same principle of operation shown in Figure 6.12*a*.

SPECIFICATIONS

The specifications of photomultiplier tubes depend upon several features:

1.   The *number* of dynodes and material from which they are constructed determines the amplification or current gain. Gains of 10^5 to 10^7 (relating direct photoelectrons from the cathode to electrons at the anode) are typical.

2.   The spectral response is determined by two factors. The first is the spectral response of the photoemissive material coated on the cathode. The second is the transparency of the glass envelope or window through which the EM radiation must pass. Using various materials it is possible to build different types of photomultipliers which, taken together, span wavelengths from 0.12 μm to 0.95 μm.

The combination of cathode coating and window material is described by a standard system to indicate the spectral response. Thus, a designation of S and a number identifies a particular band. For example, S-3 designates flat response from about 0.35 μm to 0.7 μm.

The time constant for photomultiplier tubes ranges from 20 μs down to 0.1 μs, typically.

SIGNAL CONDITIONING

The signal conditioning is usually a high-voltage negative supply directly connected to the photocathode. A resistive voltage divider (Figure 6.12b) provides the respective dynode voltages. Usually, the anode is grounded through a resistor, and a voltage drop across this resistor is measured. The cathode typically requires -1000 to -2000 V, and each dynode is then divided evenly from that. A 10-dynode tube with a -1000-V cathode thus has -900 V on the first dynode, -800 V on the second, and so on, as shown in Figure 6.12b.

6.3.6 Other Detectors

A variety of *other* detectors of EM radiation is used in specialized measurement applications. The following brief discussions are intended to make the reader aware of some of these together with their basic principle of operation.

PYROELECTRIC DETECTOR

This detector has a unique design, giving it a very wide and flat spectral response from visible through far infrared. The device uses a *pyroelectric crystal* which produces a voltage when heated above ambient temperature. This effect is very similar to the piezoelectric effect (Section 5.4.3) where stress applied to a crystal produced a voltage. If EM radiation is focused on a pyroelectric crystal (which has been treated to absorb such radiation), the energy carried by the radiation heats the crystal. This heating produces a voltage in direct proportion to the intensity of the radiation. Response times of better than 5 ns have been reported for pyroelectric detectors.

PHOTON DRAG DETECTOR

This detector is based on a phenomenon that occurs when infrared radiation passes through a sample of germanium. Germanium is partially transparent to IR radiation. The transmission of such radiation causes a potential difference to appear across the semiconductor material which is proportional to the intensity of the radiation. The photon drag IR detector also has a very fast response time, measured in the ns.

6.4 PYROMETRY

One of the most significant applications of optoelectronic transducers is in the noncontact measurement of temperature. The early term *pyrometry* is now extended to include any of several methods of temperature measurement which rely on EM radiation. These methods depend on a direct relation between an object's temperature and the EM radiation emitted. In this section, we consider the mechanism by which such radiation and temperature are related and how it is used for temperature measurement.

6.4.1 Thermal Radiation

All objects having a finite absolute temperature emit EM radiation. The nature and extent of such radiation depend on the temperature of the object. The understanding and description of this phenomenon occupied the interest and attention of physicists for many years, but we can briefly note the results of these researches through the following argument. It is well known that EM radiation is generated by the acceleration of electrical charges. We also have seen that the addition of thermal energy to an object results in vibratory motion of the molecules of the object. A simple marriage of these concepts, coupled with the fact that molecules consist of electrical charges, leads to the conclusion that an object with finite thermal energy emits EM radiation because of *charge* motion.

Since an object is *emitting* EM radiation and such radiation is a form of energy, then the object must be *losing* energy, but if this were true, then its temperature would decrease as the energy radiates away. (In fact, for an isolated small sample, this does occur.) In general, however, a state of equilibrium is reached where the object gains as much energy as it radiates, and so remains at a fixed temperature. This energy gain may be from thermal contact with another object or the absorption of EM radiation from surrounding objects. It is most important to note this interplay between EM radiation *emission* and heat *absorption* by an object in achieving *thermal equilibrium*.

BLACKBODY RADIATION

To develop a quantitative description of thermal radiation, consider first an *idealized* object. Such an object is one that absorbs *all* radiation impinging on it, regardless of wavelength, and is called an *ideal absorber*. This object also emits radiation without regard to any special pecularities to particular wavelengths and is called an *ideal emitter*.

Assume this ideal object is now placed in thermal equilibrium so that its temperature is controlled. In Figure 6.13, the EM radiation emitted from such an ideal object is plotted to show the intensity and spectral content of the radiation for several temperatures. The abscissa is the radiation wavelength, and the ordinate is the energy emitted per second (power dissipated) per unit area at a particular wavelength. Note that the area under the curve indicates the total energy per second (power dissipated) per unit area emitted by the object. Several curves are shown on the plot for different temperatures. We see that at low temperatures, the radiation

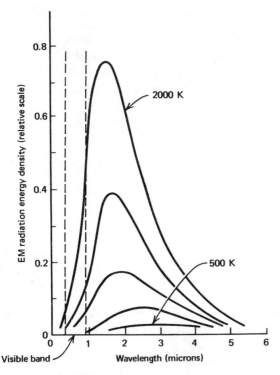

FIGURE 6.13   Idealization of the EM radiation emitted by a perfect black body as a function of temperature.

emitted is predominantly in the long wavelength (far-infrared to microwave) region. As the temperature is increased, the maximum emitted radiation is in the shorter wavelengths and finally, at very *high* temperatures, the maximum emitted radiation is near the *visible* band. It is because of this shift in emission peak with temperature that an object begins to glow as its temperature is increased. We see then that for the blackbody, the temperature and emitted radiation are in one-to-one correspondence in the following respects:

1.   *Total Radiation.* A study of blackbody radiation shows that the total emitted radiation energy per second for all wavelengths increases with the fourth power of the temperature, or

$$E \propto T^4 \tag{6-12}$$

Where

E = radiation emission in J/s per unit area or W/m²
T = temperature (K) of object

2.   *Monochromatic Radiation.* It is also clear from Figure 6.13 that the radiation energy emitted at any particular wavelength increases as a function of temperature. Thus, the J/s per

area at some given wavelength increases with temperature. This is manifested by the object getting brighter at the (same) wavelength as its temperature increases.

BLACKBODY APPROXIMATION

Most materials emit and absorb radiation at *preferred* wavelengths, giving rise to *color*, for example. Thus, these objects cannot display a radiation energy versus wavelength curve like that of an ideal blackbody. Correction factors are applied to relate the radiation curves of *real* objects to that of an ideal blackbody. For calibration purposes, a blackbody is constructed as shown in Figure 6.14. Here, the radiation emitted from a small hole in a metal enclosure is close to an ideal blackbody.

6.4.2 Broadband Pyrometers

One type of temperature measurement system based on emitted EM radiation uses the exponential relation between total emitted radiation energy and given temperature. This shows that the total EM energy emitted for all wavelengths, expressed as joules per second per unit area, varies as the fourth power of the temperature. A system that responded to this energy could thus measure the temperature of the emitting object. In practice, it is virtually impossible to build a detection system to respond to radiation of all wavelengths. However, a study of the curves in Figure 6.14 shows that most of the energy is carried in the IR and visible bands of radiation. Collection of radiation energy in these bands provides a good approximation of the *total* energy radiated.

TOTAL RADIATION PYROMETER

One type of broadband pyrometer is designed to collect radiation extending from the visible through the infrared wavelengths and is referred to as a *total radiation pyrometer*. One form of this device is shown in Figure 6.15. Here, the radiation from an object is collected by the spherical mirror S and focused on a broadband detector D. The signal from this detector then is a representation of the incoming radiation intensity and thus, the object's temperature. In these devices, the detector

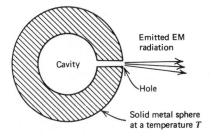

FIGURE 6.14   The EM radiation emitted by a small hole in a metal sphere with a cavity as shown simulates a black body when the temperature of the sphere is *T*.

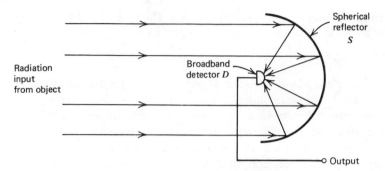

FIGURE 6.15   A total radiation pyrometer attempts to determine an object's temperature by input of radiation of all wavelengths emitted by the object.

is often a series of micro thermocouples attached to a blackened platinum disc. The radiation is absorbed by the disc which heats up and thereby results in an emf developed by the thermocouples. The advantage of such a detector is that it responds to visible and IR radiation with little regard to wavelength.

IR PYROMETER

Another popular version of the broadband pyrometer is one that is mostly sensitive to IR wavelengths. This device often uses a lens formed from silicon or germanium to focus the IR radiation on a suitable detector. IR pyrometers are often hand-held, "pistol-shaped" devices that read the temperature of an object toward which they are pointed.

CHARACTERISTICS

Broadband pyrometers often have a readout directly in temperature, either analog or digital. Generally, the switchable range is 0°C to 1000°C with an accuracy of ± 5°C to ± 0.5°C, depending on cost. Accurate measurements require the input of emissivity information by variation in a reading scale factor. Such a correction factor accounts for the fact that an object is not an ideal blackbody and does not conform exactly to the radiation curves.

APPLICATIONS

As the technology of IR pyrometers has advanced, these devices have experienced a vast growth in industrial applications. Some applications are:

1.   *Metal Production Facilities.* In the numerous industries associated with the production and working of metals, temperatures in excess of 500 °C are common. Contact temperature measurements usually involve a very limited lifetime of the measurement element. With broadband pyrometers, however, a noncontact measurement can be made, and the result converted into a process-control loop signal.

2.   *Glass Industries.* Another area where high temperatures must be controlled is in the production, working, and annealing of glasses. Here again, the broadband pyrometers find

ready application in process-control loop situations. In carefully designed control systems, glass furnace temperatures have been regulated to within ±0.1 K.

3. *Semiconductor Processes.* The extensive use of semiconductor materials in electronics has resulted in a need for carefully regulated, high-temperature processes producing pure crystals. In these applications, the pyrometer measurements are used to regulate induction heating equipment, crystal pull rates, and other related parameters.

As the accuracy of IR pyrometers is improved in the temperature ranges below 500 K, many applications that have historically used contact measurements will use IR pyrometers.

6.4.3 Narrowband Pyrometers

Another class of pyrometer depends on the variation in *monochromatic* radiation energy emission with temperature. These devices often are called *optical pyrometers* because they generally involve wavelengths only in the *visible* part of the spectrum. We know that the intensity at any particular wavelength is proportional to temperature. If the intensity of one object is *matched* to another, the temperatures are the same. In the optical pyrometer, the intensity of a heated platinum filament is varied until it matches an object whose temperature is to be determined. Because the temperatures are now the same and filament temperature is calibrated versus a heat setting, the temperature of the object is determined.

Figure 6.16 shows a typical system for implementation of an optical pyrometer. Here, the system is focused on the object whose temperature is to be determined, where the filter picks out only the desired wavelength, which is usually in the red. The viewer also sees the platinum filament superimposed on an image of the object. At low heating the filament appears dark against the background object as in Figure 6.17*a*. As the filament is heated, it eventually appears as a bright filament against the background object as in Figure 6.17*c*. Somewhere in between is the point when the brightness of the filament and the measured object match. At this setting, the filament disappears with respect to the background object, and the object temperature is read from the filament heating dial.

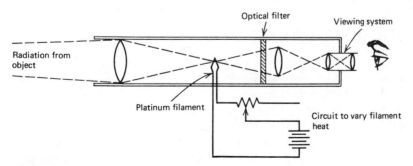

FIGURE 6.16   An optical pyrometer matches the intensity of the object to a heated, calibrated filament, usually as a wavelength in the red.

a) Filament heat too low

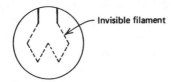

b) Filament heat adjusted correctly

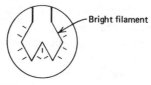

c) Filament heat too high

FIGURE 6.17 Examples of the appearance of the filament during optical pyrometer adjustment.

The range of optical pyrometer devices is determined in the low end at the point where an object becomes visible in the red (~500 K) and is virtually limited by the melting point of platinum at the upper end (~3000 K). Accuracy is typically ±5 K to ±10 K and is a function of operator error in matching intensities and emissivity corrections for the object. These devices are not easily adapted to control processes because they require acute optical comparisons, usually by a human operator. Applications are predominantly in spot measurements where constant monitoring or control of temperature is *not* required.

6.5 OPTICAL SOURCES

One limitation in the application of EM radiation devices to process control has been the lack of convenient characteristics of available optical sources. Often, complicated collimating lens systems are required, heat dissipation may be excessive, wavelength characteristics may be undesirable, or a host of other problems may arise. The development of sources relying on Light Amplification by Stimulated Emission of Radiation (LASER) has provided EM radiation sources having good characteristics for application to process-control measurements. In this section, we will consider the general characteristics of both conventional and laser light sources and their applications to measurement problems. Our discussion is confined to sources in the visible or IR wavelength bands, although it should be noted that many applications exist in other regions of the EM radiation spectrum.

6.5.1 Conventional Light Sources

Before development of the laser, two primary types of light sources were employed. Both of these are fundamentally *distributed* because radiation emerges from a physically distributed source. They also are both *divergent*, incoherent, and often not particularly monochromatic.

INCANDESCENT SOURCES

A common light source is based on the principle of thermal radiation discussed in Section 6.4. Thus, if a fine current carrying wire is heated to a very high temperature by I^2R losses, it emits considerable EM radiation in the visible band. A standard lamp is an example of this type of source as are flashlight lamps, automobile headlights, and so on. Because the light is distributed in a very broad wavelength spectrum (Figure 6.13), it clearly is *not* monochromatic. Such light actually results from molecular vibrations induced by heat, and light from one section of the wire is not associated with the light from another section. From this argument we see that the light is incoherent. The divergent nature of the light is inherent in the observation that no direction of emission is preferential. In fact, the employment of lenses or mirrors to render the light collimated is familiar to anyone who uses a flashlight. Note also that a large fraction of the emitted radiation lies in the IR spectrum, which shows up as a radiant heat loss rather than effective lighting. In fact, to a great extent, the elevated temperature of the glass bulb of an incandescent lamp is caused through absorption by the glass of the IR radiation emitted by the filament!

From this we see that an incandescent source is *polychromatic, divergent, incoherent,* and *inefficient* for visible light production. Yet, this source has been a workhorse for lighting for many years. As such, it is most deserving, but for use as a measurement transducer the above limitations are severely restrictive.

ATOMIC SOURCES

The light sources that provide the red "neon" signs used in advertisements and the familiar "fluorescent" lighting are examples of another type of light source. Such light sources are atomic in that they depend on rearrangements of electrons within atoms of the material from which the light originates. Figure 6.18 shows a schematic representation of an atom with the nucleus and associated electrons. If one of these electrons is excited from its normal energy level to a different position as indicated by an $a \to b$ transition, energy must be provided to the atom. This electron returns to its normal level in a very short time ($\sim 10^{-8}$ s for most atoms). In doing so, it gives up energy in the form of emitted EM radiation, as indicated by the

$b' \rightarrow a'$ transition in Figure 6.18. This process is often represented by an *energy diagram* as in Figure 6.19. Here the lines represent all possible positions or energies of electron excited states. The excitation $a \rightarrow b$ of the previous example is indicated by the upward arrow. The deexcitation and resulting radiation is shown as $b' \rightarrow a'$. The other arrows show that a multitude of possible excitations and deexcitations may occur. Note that the wavelength of the emitted radiation is *inversely* proportional to the energy of the transition (Equation 6-3). Thus, $b' \rightarrow a'$ will have a shorter wavelength than $c' \rightarrow d'$.

In *atomic* sources, a mechanism is provided to cause excitation of electron states, so that the EM radiation emitted by the resulting deexcitation appears as visible light.

In a neon light, the excitation is provided by collision between electrons and ions in the gas, where the electrons are provided by an electric current through the gas. In the case of neon gas, the deexcitation results in light emission whose wavelength is predominantly in the red-orange part of the visible spectrum. Thus, the light is nearly monochromatic, although light of other wavelengths representing different deexcitation modes is still present.

In a fluorescent light, a two-step process occurs. The initial deexcitation produces light predominantly in the *ultraviolet* (UV) part of the spectrum which is absorbed on an inner coating of the bulb. Electrons of atoms in the coating material are excited by UV radiation and then deexcite by many level transitions, producing radiation of a broad band of wavelengths in the *visible* region. Thus, the radiation

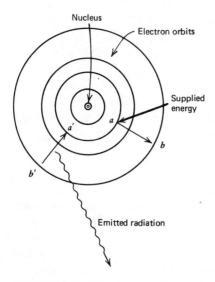

FIGURE 6.18   This shows a schematic representation of electron transitions in an atom which can cause the emission of EM radiation.

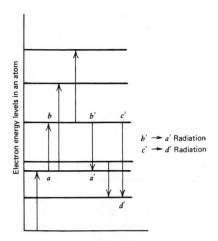

FIGURE 6.19   An energy level diagram schematically shows the electron orbit energies and possible transitions.

emitted is *polychromatic*. Note that the radiation in these sources is also divergent and incoherent.

FLUORESCENCE

Certain materials exhibit a peculiar characteristic with regard to the deexcitation transition time of electrons. This characteristic is that a transition may take much longer on the average than the normal 10^{-8} s. In some cases, the average transition time may be even hours or days. Such levels are called *long-lived states* and show up in materials which fluoresce or "glow in the dark" following exposure to an intense light source. What actually happens is that the material is excited by exposure to the light source and fills electrons into some long-lived excited states. Because the transition time may be several minutes, the object continues to emit light when taken into a darkroom until the excited levels are finally depleted. Such long-lived state materials actually form a basis for development of a laser.

6.5.2 Laser Principles

STIMULATION EMISSION

The basic operation of the laser depends on a principle formulated by Albert Einstein regarding the emission of radiation by excited atoms. He found that if several atoms in a material are excited to the *same* level and one of the atoms emits its radiation before the others, then the passage of this radiation by such excited atoms can also *stimulate* them to deexcite. It is significant that when stimulated to deexcite, the emitted radiation will be *inphase* and *in the same direction* as the stimulating radiation! This effect is shown in Figure 6.20 where atom **(a)** emits radiation spontaneously. When this radiation passes by atoms indicated by **(b)**, **(c)**, and so on, they are also stimulated to emit in the *same* direction and *in phase* (coherently). Such radiation is also *monochromatic* because only a single transition energy is involved. Such *stimulated* emission is the first requirement in the realization of a laser.

LASER STRUCTURE

To see how the concept of stimulated emission is employed in a laser, consider Figure 6.21. Here we have a host material that also contains atoms having long-lived states described earlier. If some of these atoms spontaneously deexcite, their radiation stimulates other atoms in the radiation path to deexcite, giving rise to pulses of radiation indicated by P_1, P_2, and so on. Now consider one of these pulses directed perpendicularly to mirrors M_1 and M_2 as shown in P_L. This pulse reflects between the mirrors at the speed of light, stimulating atoms in its paths to emit. The majority of excited atoms is quickly deexcited in this fashion. If mirror M_2 is only 60% reflecting, some of this pulse in each reflection will be passed. The overall result is that following excitation, a pulse of light emerges from M_2 which is *monochromatic, coherent,* and has very little divergence. This system can also be made to operate continuously by providing a continuous excitation of the atoms to replenish those deexcited by stimulated emission.

PROPERTIES OF LASER LIGHT

The light that comes from the laser is characterized by the following properties:

1. *Monochromatic.* Laser light comes predominantly from a particular energy level transi-

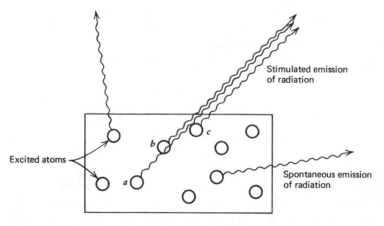

FIGURE 6.20 Stimulated emission of radiation gives rise to monochromatic coherent radiation pulses moving in various directions.

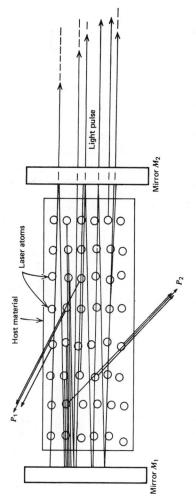

FIGURE 6.21 In the most elementary sense the laser simply collects the stimulated radiation emitted perpendicular to reflecting surfaces.

tion and is therefore almost monochromatic. (Thermal vibration of the atoms and the presence of impurities cause some other wavelengths to be present.)

2. *Coherent.* Laser light is coherent as it emerges from the laser output mirror and remains so for a certain distance from the laser; this is called the *coherence length*. (Slight variations in coherency induced by thermal vibrations and other effects cause the beam to eventually lose coherency.)

3. *Divergence.* Because the laser light emerges prependicular to the output mirror, the beam has very little divergence. Typical divergency may be 0.001 radians.

4. *Power. Continuous-operation* lasers may have power outputs of 0.5 mW to 100 W or more. *Pulse-type lasers* have power levels up to terrawatts, but for only very short time pulses—microsecond or even manoseconds in duration.

Table 6.5 summarizes some characteristics of typical industrial lasers. The workhorses of measurement applications are the He-Ne continuous-operation lasers. These lasers are relatively cheap and operate in the visible (red) wavelength regions.

EXAMPLE 6.12

A He-Ne laser with an exit diameter of 0.2 cm, power of 7.5 mW, and divergence of 1.7×10^{-3} rad is to be used with a detector 150 m away. If the detector has an area of 5 cm^2, find the power of the laser light to which the detector must respond.

SOLUTION

To solve this problem, we must ultimately find the intensity of the laser beam

TABLE 6.5 LASER CHARACTERISTICS

Material	Wavelength (μm)	Power	Applications
Helium-Neon (gas)	0.6328 (red)	0.5 to 100 mW (cw)	General purpose, ranging, alignment communication, etc.
Argon (gas)	0.4880 (green)	0.1 to 5 W (cw) 10–100 W (pulsed)	Heating, small part welding, communication
Carbon dioxide (gas)	10.6 (IR)	0 to 1 kW (cw) 0 to 100 kW (pulse)	Cutting, welding, communication, vaporization, drilling
Ruby (solid)	0.6943 (red)	0 to 1 GW (pulse)	Cutting, welding, vaporization, drilling, ranging
Neodymium (solid)	1.06 (IR)	0 to 1 GW (pulse)	Cutting, welding, vaporization, drilling, communication

150 m away. We first find the beam area at 150 m using the known divergence. From Figure 6.4 we find (see Example 6-5) the radius at 150 m, by

$$R_2 = R_1 + L \tan \theta$$
$$R_2 = 10^{-3} + (150 \text{ m}) \tan (1.7 \times 10^{-3} \text{ rad}) \qquad (6\text{-}5)$$
$$R_2 = 0.256 \text{ m}$$

Thus, the area is

$$A_2 = \pi R_2{}^2 = (3.14)(0.256 \text{ m})^2$$
$$A_2 = 0.206 \text{ m}^2$$

The intensity at 150 m is

$$I_2 = P/A$$
$$I = \frac{7.5 \times 10^{-3} \text{ W}}{0.206 \text{ m}^2} \qquad (6\text{-}4)$$
$$I = 0.036 \text{ W/m}^2$$

Then we find the power intercepted by the detector as

$$P_{\text{Det}} = IA_{\text{Det}} = (0.036 \text{ W/m}^2)(5 \times 10^{-4} \text{ m}^2)$$
$$P_{\text{Det}} = 1.82 \times 10^{-5} \text{ W!}$$

Thus, the detector must respond to a power of 18.2 μW of power. Although small, such low power is measured by many detectors. The reader can show (by similar calculation) that a source such as a flashlight with a divergence of 2° or 34.9×10^{-3} rad would have resulted in a much lower power at the detector of $\mathbf{4.35 \times 10^{-8}}$ W!

6.6 APPLICATIONS

Several applications of optical transduction techniques in process control will be discussed. The intention is to demonstrate only the typical nature of such application and not design details. Pyrometry for temperature measurement has already been discussed and is not considered in these examples.

6.6.1 Label Inspection

In many manufacturing processes, a large number of items are produced in batch runs where an automatic process attaches labels to the items. Inevitably, some items are either missing labels or the labels are incorrectly attached. The system of Figure

6.22 examines the presence and alignment of labels on boxes moving on a conveyer belt system. If the label is missing or improperly aligned, the photodetector signals are incorrect in terms of light reflected from the sources and a solenoid pushout rejects the item from the conveyer. The detectors in this case could be CdS cells and the sources either focused incandescent lamps or a small He-Ne laser. Source/detector system A detects the presence of a box and initiates measurements by source/detector systems B and C. If the signals received by detectors B and C are identical and at a preset level, the label is correct, and the box moves on to the accept conveyer. In any other instance, a misalignment or missing label is indicated and the box is rejected onto the reject conveyer.

EXAMPLE 6.13

Devise signal conditioning circuitry for the application of Figure 6.22 using CdS cells as the detectors. If *both* cells have resistance of $1000 \pm 100 \ \Omega$ or less, the label is considered *correct*. No label produces $2000 \ \Omega$ or more.

SOLUTION

One of the many possible methods for implementing this solution is shown in Figure 6.23. Here, if the label is misaligned or missing, one or both comparator outputs is high, thus driving the summing amplifier to close the *reject* relay. If the box is present, then power is applied to the reject solenoid and the box is ejected. The resistors are chosen so that if the cell resistance exceeds $1100 \ \Omega$, the comparator outputs go high. The relay is chosen to close if either (or both) comparator signal is present.

6.6.2 Turbidity

It is also possible to measure the turbidity of liquids in a process in-line, that is, without the necessity of taking periodic samples by a method similar to that in Figure 6.24. In this case, a laser beam is split and passed through two samples to matched photodetectors. One sample is a carefully selected *standard* of allowed (acceptable) turbidity. The other is an in-line sample of the process liquid itself. If the in-line sample attenuates the light more than the standard, the signal conditioning system triggers an alarm or takes other appropriate action to reduce turbidity.

6.6.3 Ranging

The development of the laser and fast (small τ) photodetectors has introduced a number of methods for measuring distances and the rate of travel of objects by noncontact means. Distances can be measured by measuring the time of flight of light pulses scattered off a distant object. Because the speed of light is constant, we use a simple equation to find distance providing the time of flight T is known. Thus,

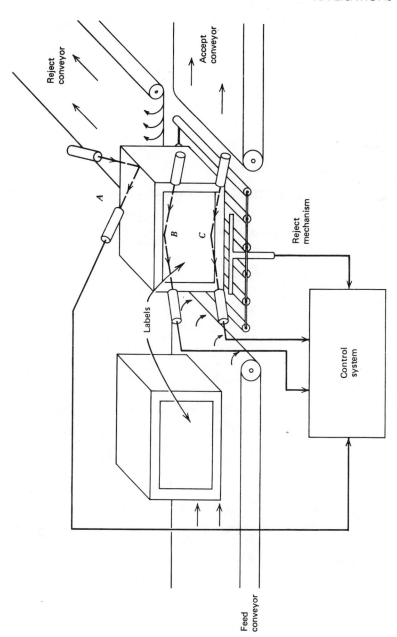

FIGURE 6.22 Label inspection using optical techniques.

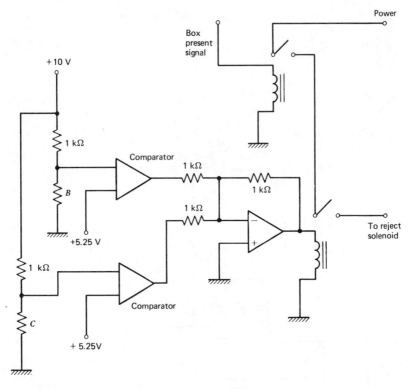

FIGURE 6.23 One possible circuit to implement the requirements of Example 6.12.

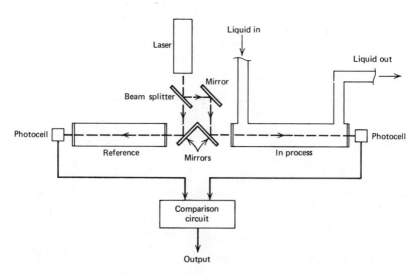

FIGURE 6.24 Turbidity measurements can be made in line with the optical system shown.

if a pulse of light is directed at a distant object and the reflection is detected a time T later, then the distance is

$$D = cT/2 \qquad\qquad (6\text{-}13)$$

Where

D = distance to the object (m)
c = speed of light (m/s)
T = time for light round trip (s)

EXAMPLE 6.14

An object is approximately 300 m away. Find the approximate time difference to be measured to calculate the distance using a light pulse reflected from the object.

SOLUTION

From

$$D = \frac{cT}{2} \qquad\qquad (6\text{-}13)$$

We have

$$T = \frac{2D}{c} = \frac{(2)(3 \times 10^2 \text{ m})}{3 \times 10^8 \text{ m/s}}$$

$$T = 2 \ \mu s$$

This ranging method can be employed for measurement of shorter distances limited by time measurement capability, and detectability of the reflected signal for longer distances. Surveying instruments for measuring distance have been developed by this method. Velocity or rate of motion can be measured by an electronic computing system which records the changing reflected pulse travel time and computes velocity. These and the interferometric methods such as those used in Doppler radar are beyond the scope of this text.

SUMMARY

EM radiation measurement allows noncontact measurement techniques for many variables, such as temperature, level, and others. This chapter presented the essential elements of EM radiation and its application to process-control measurement. In summary, this included the following:

1. EM radiation is defined as a form of energy characterized by constant propagation speed. The wavelength and frequency are related by

$$\lambda = cf \qquad (6\text{-}1)$$

2.   As a form of energy, EM radiation is best described by the intensity in watts per unit area, divergence (spreading) in radians, and spectral content.

3.   When associated with human eye detection, light is described by a set of luminous units which are all relative to a standard, visible source.

4.   There are four basic photodetectors: photoconductive, photovoltaic, photoemissive, and photodiode. Each has its special characteristics relative to spectral sensitivity, detectable power, and response time.

5.   Pyrometry relates to measurement of temperature by measuring the intensity of EM radiation from an object as a function of its temperature.

6.   Total radiation and IR pyrometers may be used in process-control applications for measurement of temperature from about 300 K to almost no limit. Accuracy varies from ± 10 K to ± 0.5 K or better in some cases.

7.   Conventional light sources are usually divergent and incoherent. Incandescent types are polychromatic, but atomic sources may be almost monochromatic.

8.   A laser is based on a concept of stimulated emission wherein one photon can stimulate excited atoms to emit many more photons.

9.   The light from a laser is nearly nondivergent, coherent, and monochromatic.

10.   Applications of optical techniques are particularly useful where contact measurement is difficult.

PROBLEMS

6.1   Find the frequency of 3-cm wavelength EM radiation. What band of phenomena does this radiation represent?

6.2   If a source of green light has a frequency of $6.5 \times 10^{+14}$ Hz, what is its wavelength in μm and angstrom units?

6.3   A light beam is passed through 100 m of liquid $n = 1.7$ index of refraction. How long does it take light to traverse the 100 m with and without the liquid?

6.4   A light beam from a flashlight has a power of 100 mW, an exit diameter of 4 cm, and a divergence of $1.2°$. Calculate the intensity at 60 m, and the size of the beam.

6.5   If the lumens/watt of the flashlight in Problem 6.4 is 0.62, find the illumination at 1 m and 60 m.

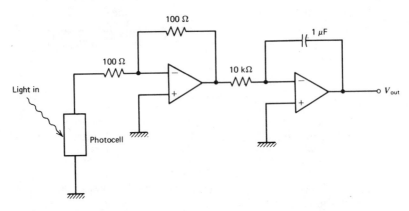

FIGURE 6.25   Circuit for Problem 6.9.

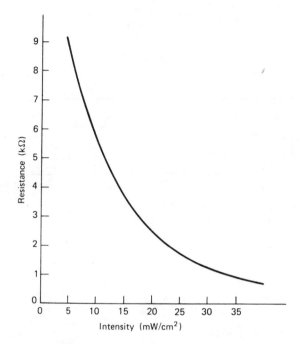

FIGURE 6.26   Figure for Problem 6.13.

6.6   A detector will just respond to an illumination of 0.14 lux. Find the maximum distance from a 10 W tungsten bulb for detectability. Assume a point source.

6.7   A CdS cell has a time constant of 73 ms and a dark resistance of 150 kΩ. At some time it is struck by a light pulse for which its final resistance is 45 kΩ. A reaction must be initiated after only 10 ms. Find the resistance of the cell at that time.

6.8   A photomultiplier has a current gain of 3×10^6. A weak light beam produces 50 electrons/s at the photocathode. What anode to ground resistance must be used to get a 3.0 μV voltage from this light pulse? (1e $= 1.6 \times 10^{-19}$ coulomb)

6.9   A silicon photovoltaic cell is employed in the circuit shown in Figure 6.25. The cell has an internal resistance of 65 Ω. A light pulse of 20 ms duration and a 2 W/m^2 intensity strikes the detector. Sketch the output voltage versus time and find the maximum voltage produced. The cell calibration voltage is $V_0 = 0.6$ V.

6.10   A 10 W Argon laser has a beam diameter of 1.5 cm. If focused to a 1-mm diameter, find the intensity of the radiation beam at the focus.

6.11   In a turbidity system such as that of Figure 6.24, the tanks are 2 m in length. A laser with a 2.2 mrad divergence, 2.1 mW power, and a 1 mm exit radius is employed. If the sample nominally detracts from the beam power by 12% per m, find the intensity of the beam at the detector. The laser is 1.5 m from the beam splitter. (*Note:* The beam splitter halves the power and does *not* affect divergence.)

6.12   A special timing circuit can resolve 2.4 ns time differences. Find the closest distance that could be measured using laser ranging.

6.13   For the turbidity system of Figure 6.24, two matched photoconductive cells with curves of R versus I (given by Figure 6.26) are employed. Design a signal conditioning system that outputs the deviation of the flowing system turbidity in volts and triggers an alarm if the intensity is reduced by 10% from the nominal of 15 mW/cm^2.

7

FINAL
CONTROL

INSTRUCTIONAL OBJECTIVES

In this chapter, the general techniques used to implement the final control element function are presented. After you read it, you should be able to:

1. Define the three parts of final control operation.
2. Give two examples of electrical signal conversion.
3. Make a diagram and describe the operating principles of the flapper/nozzle pneumatic system.
4. Describe the operating principle of ac, dc, and stepping motors.
5. Explain how a pneumatic positioning actuator functions in both the direct and reverse modes.
6. Contrast quick-opening, linear, and equal percentage control valves in terms of the flow versus stem position.
7. Explain how control valve sizing techniques allow selection of the proper size of control valve.

7.1 INTRODUCTION

In a typical process-control application, the measurement and evaluation of some controlled variable are carried out by a low-energy analog or digital representation of the variable. The control signal that carries feedback information back to the process for necessary corrective action is expressed also by the same low level of representation. In general, the controlled process itself may involve a high-energy condition, such as the flow of thousands of cubic meters of liquid or several hundred thousand newton hydraulic forces, as in a steel rolling mill. The function of the final control element is to translate low-energy control signals into a level of action commensurate with the process under control. This can be considered an amplification of the control signal, although in many cases, the signal is also converted into an entirely *different* form.

In this chapter, the general techniques used to implement the final control element function are presented together with specific examples in several areas of process control.

An ideal transducer used to measure some dynamic variable in a process-control application has negligible effect on the process itself. Typically, it is a passive, low-energy transduction of a dynamic variable, which may be a high-energy parameter. It follows that transducer selection is based mainly on required measurement specifications and necessary protections (of the transducer) from harmful effects of the process environment. In transducer selection, the process-control technologist need not have intimate knowledge of the mechanisms of the process itself but rather its environmental effects.

These arguments do not apply, however, when considering the *final control element*. By necessity, the final control element has a profound effect on the process, and must therefore be selected after detailed considerations of the process operational mechanisms. Such a selection, therefore, cannot be the responsibility of the process-control technologist alone. In this view, the process-control technologist should have sufficient background on the final control element and its associated signal conditioning to know how such devices interface with preceding process controllers and transducers. The technologist should be able to communicate and work closely with process engineers on these subjects. The objectives of this chapter were selected to fulfill such responsibility.

7.2 FINAL CONTROL OPERATION

Final control element operations involve the steps necessary to convert the control signal (generated by a process controller) into proportional action on the process itself. Thus, to use a typical 4–20 mA control signal to vary a large flow rate, from say 10.0 m³/min to 50.0 m³/min, certainly requires some intermediate operations. The specific intermediate operations vary considerably depending on the process-control design, but certain generalizations can be made regarding the steps leading from the control signal to the final control element itself. For a typical process-control application the conversion of a process-controller signal to a control function can be represented by the steps shown in Figure 7.1. Here, the input control signal may take many forms, including an electric current, digital signal, or pneumatic pressure.

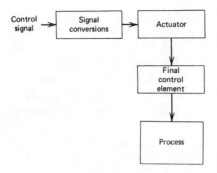

FIGURE 7.1   Elements of the final control operation.

7.2.1 Signal conversions

This step refers to modifications that must be made to the control signal to properly interface with the next stage of control, that is, the actuator. Thus, if a valve control element is to be operated by an electric motor actuator, then a 4–20 mA d-c control signal must be modified to operate the motor. If a d-c motor is used, modification might be current to voltage conversion and amplification.

Standard types of signal modification are discussed in Section 7.3. The devices that perform such signal conversion are often called *transducers* because they convert control signals from one form to another, such as current to pressure, current to voltage, and so on.

7.2.2 Actuators

(See Figure 7.1.) The results of signal conversions provide an amplified and/or converted signal which is designed to operate (actuate) a mechanism to change a controlling variable in the process. The direct effect is usually implemented by something in the process such as a valve or heater which must be operated by some device. The *actuator* is a translation of the (converted) control signal into action *on* the control element. Thus, if a valve is to be operated, then the actuator is a device that converts the control signal into physical action of opening or closing the valve. Several examples of actuators in common process-control use are discussed in Section 7.4.

7.2.3 Control Element

Finally, (Figure 7.1) we get to the final control element itself. This device has direct influence on the process dynamic variable and is designed as an integral part of the process itself. Thus, if flow is to be controlled, then the control element, a valve, must be built directly into the flow system. Similarly, if temperature is to be controlled, then some mechanism or control element which has a direct influence on temperature must be involved in the process. This could be a heater/cooler combination that is electrically actuated by relays or a pneumatic valve to control influx of reactants.

In Figure 7.2, a control system is shown to control the degree of baking of, say, crackers, as determined by the *cracker color*. The optical measurement system produces a 4–20 mA conditioned signal which is an analog representation of cracker color (and, therefore, proper baking). The controller compares the measurement to a setpoint and outputs a 4–20 mA signal which regulates the conveyer belt feed motor speed to adjust baking time in the oven. The final control operation is then represented by a *signal conversion* which transforms the 4–20 mA signal into 50–100 to volt signal as required for motor speed control. The motor itself is the *actuator*, whereas the conveyer belt assembly is the *control element*.

Because applications of process-control techniques in industry are as varied as the industry itself, it is impractical to consider more than a few final control techniques. By studying some examples, the reader should be prepared to analyze and understand many other techniques that arise in industry.

7.3 SIGNAL CONVERSIONS

The principal objective of signal conversion is to convert the low-energy control signal to a high-energy signal to drive the actuator. Controller output signals are typically in one of three forms: (1) electrical current, usually 4 to 20 mA, (2) pneumatic pressure, usually 3 to 15 psi, and (3) digital signals, usually TTL level voltages in serial or parallel format. There are many different schemes for conversion of these signals to other forms depending on the desired final form and

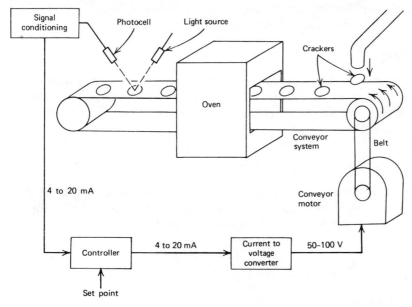

FIGURE 7.2   A process-control system showing the final control operations.

on evolving technology for producing such conversions. In the following sections, a number of the more common conversion schemes are presented. You should always be receptive to the advances of technology and the new methods of signal conditioning and conversion which will come with this advancement.

7.3.1 Analog Electrical Signals

Many of the methods of analog signal conditioning discussed in Chapter 2 are used in conversions necessary for final control. The following paragraphs summarize some of the more common approaches.

RELAYS

A common conversion is to use the controller signal to activate a relay when simple on/off or two-position control is sufficient. In some cases, the low-current signal is insufficient to drive a heavy industrial relay, and an amplifier is used to boost the control signal to a level sufficient to do the job.

AMPLIFIERS

High-power ac or dc amplifiers can often provide the necessary conversion of the low-energy control signal to a high-energy form. Such amplifiers may serve for motor control, heat control, light-level control, and a host of other industrial needs.

EXAMPLE 7.1

A magnetic amplifier requires a 5–10 volt input signal from 4–20 mA control signal. Design a signal conversion system to provide this relationship.

SOLUTION

We first must convert the current to a voltage and then provide the required gain and bias. We can get a voltage using a resistor in the current line of, say, 100 Ω. Then the 4–20 mA becomes 0.4–2.0 volts. Now the amplifier system must provide an output given by

$$V_{out} = KV_{in} + V_B$$

where K is the gain and V_B is an appropriate bias voltage. We know that 0.4 volts input must provide 5 volts output, and 2 volts input must provide 10 volts output. This allows us to find K and V_B, using simultaneous equations as

$$5 = 0.4\,K + V_B$$
$$10 = 2\,K + V_B$$

subtracting we get

$$5 = 1.6\,K$$
$$K = \mathbf{3.125}$$

which we use in either equation above to find

$$V_B = \mathbf{3.75}.$$

Thus, the result is

$$V_{out} = 3.125\,V_{in} + 3.75$$

The circuit of Figure 7.3 shows how this can be implemented using an op amp configuration.

MOTOR CONTROL

Many motor control circuits are designed as packaged units which accept a low-level dc signal directly to control motor speed. If such a system is not available, it is possible to build circuits using amplifiers and SCRs or TRIACs to perform this control. The details of such a control system are beyond the intent of this text, however. The basic elements of electrical motors and some words about their control are discussed later in this chapter.

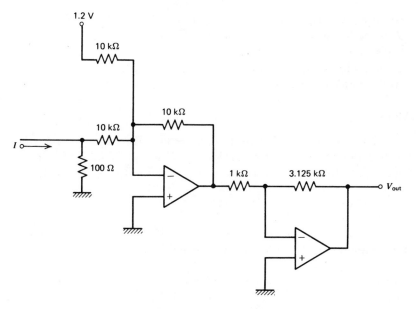

FIGURE 7.3   An op amp circuit to provide the signal conditioning requirements of Example 7.1.

7.3.2 Digital Electrical Signals

Conversions of digital signals to forms required by final control operations usually are carried out using systems already discussed in Chapter 3. We mention again, however, the basic elements of the output interface between computer and final control.

ON/OFF CONTROL

There are many cases in process control where the control algorithm is accomplished by simple commands to outside equipment to change speed, turn on (or off), move up, and so on. In such cases, the computer can simply load an output line with a 1 or 0 as is appropriate. Then, it is a simple matter to use this signal to close a relay or activate some other outside circuit.

DAC

When the digital output must provide a smooth control, as it does in valve positioning, the computer must provide an input to a DAC which then determines an appropriate analog output. When a computer must provide outputs to many final control elements, a data output module or system such as that described in Chapter 3 can be employed. These integrated modules contain channel addressing, DAC, and other required elements of a self-contained output interface system.

EXAMPLE 7.2

A 4-bit digital word is intended to control the setting of a 2 Ω d-c resistive heater. Heat output varies as a 0-24 volt input to the heater. Using a 10-volt DAC followed by an amplifier and a unity gain high-current amplifier, calculate
 (a) the settings from minimum to maximum heat dissipation
 (b) how the power varies with LSB changes

SOLUTION

 (a) The 4-bit word has a total of 16 states; thus, the DAC outputs voltage from 0 V for a 0000_2 input to 9.375 V for a 1111_2 input. If we use a gain of 2.56, then the heater input will be 0-24 volts in 1.5 V steps. The heat dissipation is found from

$$P = \frac{V^2}{R} \tag{7-1}$$

Then, for the minimum $P = 0$ because $V = 0$ and for the maximum

$$P_{max} = \frac{(24 \text{ V})^2}{2 \ \Omega}$$

$$P_{max} = 288 \text{ W}$$

 (b) The variation in heating with voltage is *not* linear because the power varies as the square of the voltage. We can find the increase in power for an infinitesimal change in voltage by differentiation of Equation 7-1

$$\frac{dP}{dV} = \frac{2\,V}{R}$$

For our finite digital step we have,

$$\text{Where:} \quad \Delta P = \frac{V^2}{R} - \frac{(V - \Delta V)^2}{R}$$

is the change in heating per step change in voltage which corresponds to a bit change. Since $\Delta V = 1.5$ volts, we have

$$\Delta P = \frac{1.5\,(2V - 1.5)}{R} \text{ Watts/LSB}$$

This means that the first bit change produces a power change of 1.125 W and the last bit a power change of 34.875

DIRECT ACTION

As the use of digital and computer techniques in process control becomes more widespread, new methods of final control are developed which can be actuated directly by the computer. Thus a stepping motor, to be discussed later, interfaces very easily to the digital signals which a computer outputs. In another development, special integrated circuits are made which reside within the final control element and allow the digital signal to be connected directly.

7.3.3 Pneumatic Signals

The general field of pneumatics covers a broad spectrum of applications of fluid pressure to industrial needs. One of the most common is to provide a force by the fluid pressure acting on a piston or diaphragm. Later in this chapter, we will deal with this application in process control. In this section, however, we are interested in pneumatics as a means of propagating information, that is, as a signal carrier, and how that signal can be converted to other forms.

PRINCIPLES

In a pneumatic system, information is carried by the pressure of gas in a pipe. If we have a pipe of any length and raise the pressure of gas in one end, this increase in pressure will propagate down the pipe until the pressure throughout is raised to the new value. The pressure signal travels down the pipe at a speed in the range of the speed of sound in the gas (say air), which is about 330 m/s (1082 ft/s). Thus, if a transducer varies gas pressure at one end of a 330-meter pipe (about 360 yards) in response to some controlled variable, then that same pressure occurs at the other end of the pipe after a delay of approximately 1 second. For many process-control installations, this delay time is of no consequence, although it is very slow compared to an electrical signal. This type of signal propagation was used for many years in process control before electrical/electronic technology advanced to a level of reliability and safety to enable use with confidence. Pneumatics is still employed in many installations either because of danger of electrical equipment or as a carryover from previous years, where conversion to electrical methods would not be cost effective. In general, pneumatic signals are carried with dry air as the gas where signal information has been adjusted to lie within the range of 3 to 15 psi. In metric systems, the range of 20 to 100 kPa is used. There are three types of signal conversion of primary interest and these are covered below.

AMPLIFICATION

A pneumatic amplifier, also called a *booster* or *relay*, raises the pressure and/or air flow volume by some linearly proportional amount from the input signal. Thus, if the booster has a pressure gain of 10, the output would be 30 to 150 psi for an input of 3 to 15 psi. This is accomplished via a regulator which is activated by the control signal. A schematic diagram of one type of pressure booster is shown in Figure 7.4. Note that, as the signal pressure varies, the diaphragm motion will move the plug in the body block of the booster. If motion is down, the gas leak is reduced and pressure in the output line is increased. The device shown is reverse acting because a high-signal pressure will cause output pressure to decrease. Many other designs are used.

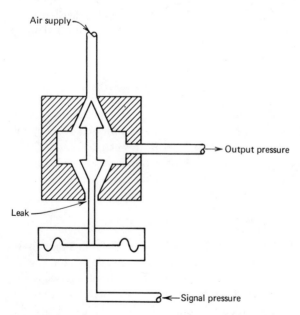

FIGURE 7.4 A pneumatic amplifier or booster converts the signal pressure to a higher pressure or the same with greater air volume.

NOZZLE/FLAPPER SYSTEM

A very important signal conversion is from pressure to mechanical motion and vice versa. This can be provided by a nozzle/flapper system (sometimes called a nozzle/baffle system). A diagram of this device is shown in Figure 7.5a. A regulated supply of pressure, usually over 20 psig, provides a source of air through the restriction. The nozzle is open at the end where the gap exists between the nozzle and flapper and air escapes in this region. If the flapper moves down and closes off the nozzle opening so that no air leaks, the signal pressure will rise to the supply pressure. As the flapper moves away the signal pressure will drop due to the leaking gas. Finally, when the flapper is far away, the pressure will stabilize at some value determined by the maximum leak through the nozzle. Figure 7.5b shows the

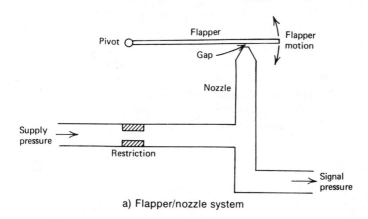

a) Flapper/nozzle system

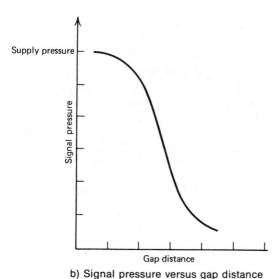

b) Signal pressure versus gap distance

FIGURE 7.5   Principles of a flapper/nozzle system.

relationship between signal pressure and gap distance. Note the great sensitivity in the central region. A nozzle/flapper is designed to operate in the central region where the slope of the line is greatest. In this region, the response will be such that a very small motion of the flapper can change the pressure by an order of magnitude. More discussion of this system is given in Chapter 9 under the discussion on pneumatic controllers.

CURRENT TO PRESSURE CONVERTERS

The current to pressure converter, or simply *I/P converter*, is a very important element in process control. Often, when we want to use the low-level electric current signal to do work, it is much easier to let the work be done by a pneumatic signal. The I/P converter gives us a linear way of translating the 4 to 20 mA current into a 3 to 15 psig signal. There are many designs for these converters but the basic principle almost always involves the use of a nozzle/flapper system. Figure 7.6 illustrates a simple way to construct such a converter. Notice that the current through a coil produces a force which will tend to pull the flapper down and close off the gap. This means that a high current produces a high pressure so that the device is direct acting. Adjustment of the springs and perhaps the position relative to the pivot to which they are attached allows the unit to be calibrated so that 4 mA corresponds to 3 psig and 20 mA corresponds to 15 psig.

7.4 ACTUATORS

If a valve is used to control fluid flow, some mechanism must physically open or close the valve. If a heater is to warm a system, some device must turn the heater ON or OFF or vary its excitation. These are examples of the requirement of an

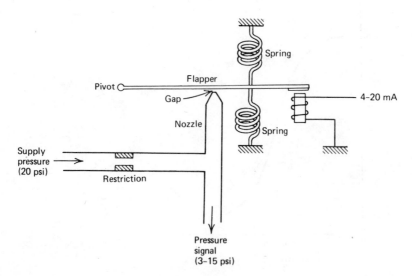

FIGURE 7.6 Principles of a current to pressure converter.

actuator in the process-control loop. Notice the distinction of this device from both the input control signal and the control element itself (valve, heater, and so on, as shown in Figure 7.1). Actuators take on many diverse forms to suit the particular requirements of process-control loops. We will consider several types of electrical and pneumatic actuators.

7.4.1 Electrical Actuators

In the following paragraphs a short description of several common types of electrical actuators is given. The intention is to present only the essential features of the devices and not an in-depth study of operational principles and characteristics. In any specific application one would be expected to consult detailed product specifications and books associated with each type of actuator.

SOLENOID

A solenoid is an elementary device which converts an electrical signal into a mechanical motion, usually rectilinear, that is, in a straight line. As shown in Figure 7.7, the solenoid consists of a coil and plunger. The plunger may be free standing or spring loaded. The coil will have some voltage or current rating and may be a dc or ac type. Solenoid specifications include the electrical rating and the plunger pull or push force when excited by the specified voltage. This force may be expressed in newtons or kilograms in the SI system and in pounds or ounces in the English system. Some solenoids are rated only for intermittent duty due to thermal constraints. In this case, the maximum duty cycle (percentage on to total time) will be specified. Solenoids are used when a large sudden force must be applied to perform some job. In Figure 7.8, a solenoid is used to change the gears of a two-position transmission. Note that an SCR is used to activate the solenoid coil.

ELECTRICAL MOTORS

Electrical motors are devices which accept electrical input and produce a continuous rotation as a result. Motor styles and sizes vary as demands for rotational speed

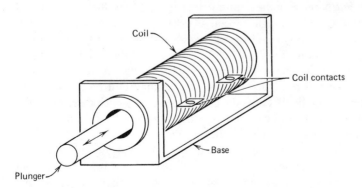

FIGURE 7.7   A solenoid converts an electrical signal to a physical displacement.

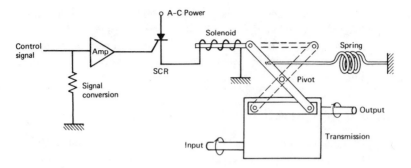

FIGURE 7.8   A solenoid is used to change gears.

(revolutions per minute or rpm), starting torque, rotational torque, and other specifications vary. There are numerous cases where electrical motors are employed as actuators in process control. Probably the most common control situation is where motor speed is driving some part of a process and that speed must be controlled to control some variable in the process; the drive of a conveyor system, for example. There are many types of electrical motors, each with its special set of characteristics. We will simply discuss the three most common varieties, the dc motor, ac motor, and stepping motor.

DC MOTOR

In its simplest form, a dc motor uses a permanent magnet (PM) to produce a static magnetic field across two pole pieces. Between the poles is connected a coil of wire which is free to rotate (the armature) and which is connected to a source of dc current through a switch mounted on the shaft (a commutator). This system is shown schematically in Figure 7.9a. For the condition shown, the current in the coil will produce a magnetic field with a north/south orientation like that shown in Figure 7.9b. The repulsion of the PM south and the coil south (and the norths) will cause a torque which will rotate the coil as shown. If the commutator were *not split*, the coil would simply rotate until the PM and coil north and south poles were lined up and then stop, *but* because of the commutator, the coil finds that when rotated around the current direction through the coil reverses so that the condition shown in Figure 7.9c occurs. Thus, the rotational torque is again present, and the coil continues to rotate. From this simple model you can see that the coil will continue to rotate. The speed will depend on the current. Actually, the armature current is not determined by the coil resistance because of a counter emf produced by the rotating wire in a magnetic field. Thus, the effective voltage, which determines the current from the wire resistance and ohms law, is the difference between the applied voltage and the counter emf produced by the rotation.

Many dc motors use an electromagnet instead of a PM to provide the static field. The coil used to produce this field is called the field coil. The current for this field coil can be provided by placing the coil in series with the armature or in parallel (shunt). In some cases the field is composed of two windings, one of each type. This

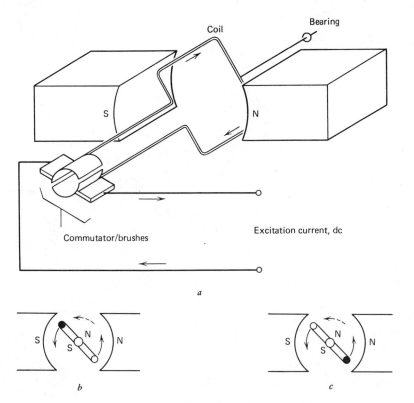

FIGURE 7.9   Permanent magnet dc motor.

is a compound dc motor. The schematic symbols of each type of motor are shown in Figure 7.10. Characteristics of dc motors with a field coil are:

1.   **Series Field.** This motor has large starting torque but is difficult to speed control. Good in applications of starting heavy, nonmobile loads and where speed control is not very important, such as quick-opening valves.

2.   **Shunt Field.** This motor has a smaller starting torque but very good speed control characteristics by varying armature excitation current. Good in applications where speed is to be controlled, such as conveyor systems.

3.   **Compound Field.** This motor attempts to obtain the best features of both of the two previous types. Generally, starting torque and speed control capability fall predictably between the two pure cases.

AC MOTORS

There are many types of ac motors. We will give the basic principles of several types. A synchronous ac motor's speed of rotation is determined by the frequency of the ac voltage which drives it. Its primary application is in timing because of the high stability of the power-line frequency. Operation of this type of motor can be seen from a simple example shown in Figure 7.11. Here, the rotor is a PM, and the

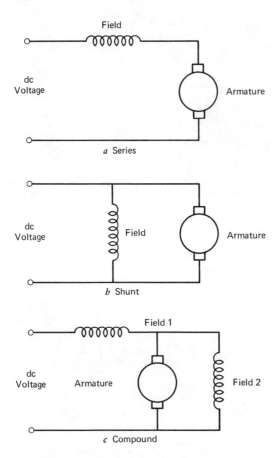

FIGURE 7.10   Three dc motor configurations.

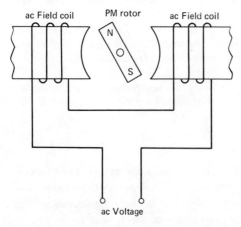

FIGURE 7.11   Simple ac motor with a pm rotor.

field is provided by coils driven from the ac line. Due to the inertia of the PM, the starting torque is not very high but once rotation is started the PM will rotate in phase with the field reversals caused by the oscillations of the ac line voltage. It is clear then that the rate of rotation is determined by the ac line frequency. An induction motor replaces the PM with a very heavy wire coil into which is induced a current from the changing field of the ac excited field coils. Figure 7.12 illustrates this motor. As before, once rotation is started the rotor will continue rotation in phase with the line frequency-induced changes of field coil excitation. The difficulty with these motors is that they are not self-starting and special modifications are necessary to get them to begin rotation. Clearly then, the starting torque is very low. One method of providing self-starting is to drive the motor with two or more phases of ac excitation. In general, however, ac motors do not have a high starting torque or convenient methods of speed control.

STEPPING MOTOR

The stepping motor has increased in importance in recent years because of the ease with which this rotating machine can be interfaced with digital circuits. A stepping motor is a rotating machine which actually completes a full rotation by sequencing through a series of discrete rotational steps. Each step position is an equilibrium position in that, without further excitation, the rotor position will stay at the latest step. Thus, continuous rotation is achieved by the input of a train of pulses, each of which causes an advance of one step. Note that it is not really continuous rotation, but discrete, stepwise rotation. The rotational rate is determined by the number of steps per revolution and the rate at which the pulses are applied. A driver circuit is necessary to convert the pulse train into proper driving signals for the motor.

EXAMPLE 7.3

A stepper motor has 10° per step and must rotate at 250 rpm. What input pulse rate, in pulses per second, is required?

SOLUTION

A full revolution has 360° so that with 10° per step it will take 36 steps to complete one revolution. Thus,

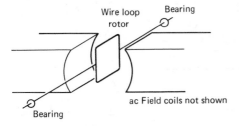

Wire loop rotor   Bearing

ac Field coils not shown

Bearing

FIGURE 7.12   The induction motor depends on a rotor field induced by the ac field coils (not shown).

$$\left(250 \ \frac{\text{rev}}{\text{min}}\right)\left(36 \ \frac{\text{pulses}}{\text{rev}}\right) = 9000 \ \text{pulses/min}$$

therefore,

$$(9000 \ \text{pulses/min})(1 \ \text{min}/60 \ \text{s}) = \textbf{150 pulses/sec}$$

The operation of a stepping motor can be understood from the simple model shown in Figure 7.13, which has 90° per step. In this motor, the rotor is an PM which is driven by a particular set of electromagnets. In the position shown, the system is in equilibrium and no motion occurs. The switches are typically solid state devices, such as transistors, SCRs, or TRIACs. The switch sequencer will direct the switches through a sequence of positions as the pulses are received. The next pulse

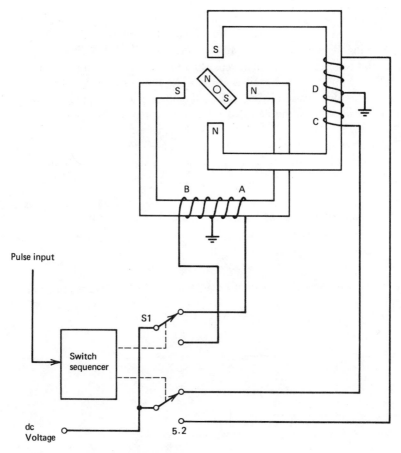

FIGURE 7.13 An elementary stepping motor.

in Figure 7.13 will change S2 from C to D resulting in the poles of that electromagnet reversing fields. Now, because the pole north/south orientation is different, the rotor is repelled and attracted so that it moves to the new position of equilibrium shown in Figure 7.14b. With the next pulse, S1 is changed to B causing the same kind of pole reversal and rotation of the PM to a new position, as shown in Figure 7.14c. Finally, the next pulse causes S2 to switch to C again and the PM rotor again steps to a new equilibrium position, as in Figure 7.14d. The next pulse will send the system back to the original state and the rotor to the original position. This sequence is then repeated as the pulse train comes in, resulting in a stepwise continuous rotation of the rotor PM. Although this example illustrates the principle of operation, the most common stepper motor uses no PM but instead a rotor of magnetic material (not a magnet) with a certain number of teeth. This rotor is driven by a phased arrangement of coils with a different number of poles so that the rotor can never be in perfect alignment with the stator. Figure 7.15 illustrates this for a rotor with eight "teeth" and a stator with twelve "poles." Note that one set of four teeth are aligned but the other four are not. If excitation is placed on the next set of poles (B) and taken off the first set (A), then the rotor will step once to come into alignment with the B set of poles.

The direction of rotation of stepper motors can be changed by just changing the order in which different poles are activated and deactivated.

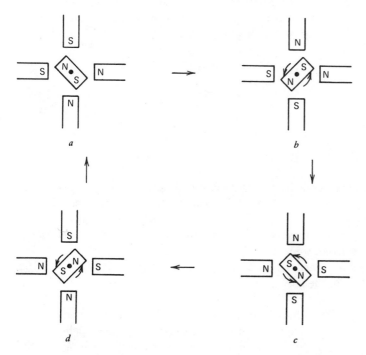

FIGURE 7.14 The four positions of the elementary stepper.

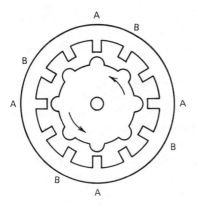

FIGURE 7.15 Cross section of a stepping motor with eight rotor teeth and 12 stator poles. Note that the rotor lines up with the A poles. With the next step the rotor will line up with the B poles.

7.4.2 Pneumatic Actuators

The actuator often translates a control signal into a large force or torque as required to manipulate some control element. The *pneumatic actuator* is most useful for such translation. The principle is based on the concept of pressure as force per unit area. If we imagine that a net pressure difference is applied to a diaphragm of surface area A, then a net force acts on the diaphragm given by

$$F = (p_1 - p_2)A \qquad (7\text{-}2)$$

Where

$p_1 - p_2$ = pressure difference (Pa)
A = diaphragm area (m^2)
F = force (N)

If we need to double the available force for a given pressure, it is merely necessary to double the diaphragm area. Very large forces can be developed by standard signal pressure ranges of 3–15 psi (20–100 kPa). Many types of pneumatic actuators are available, but perhaps the most common are those associated with *control valves*. We will consider these in some detail to convey the general principles.

The action of a *direct* pneumatic actuator is shown in Figure 7.16. Figure 7.16a shows the condition in the *low* signal pressure state where the spring S maintains the diaphragm and connected control shaft in a position as shown. The pressure on the opposite (spring) side of the diaphragm is maintained at atmospheric pressure by the open hole H. Increasing the control pressure (gauge pressure) applies a force on the diaphragm forcing the diaphragm and connected shaft down against the spring force. Figure 7.16b shows this in the case of maximum control pressure and maximum travel of the shaft. Note that the pressure and force are linearly related, as

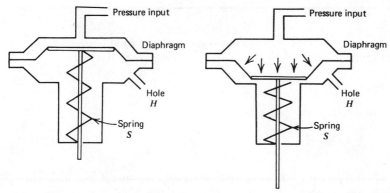

a) Direct actuator in the low pressure state

b) Direct actuator in the high pressure state

FIGURE 7.16   A direct pneumatic actuator for converting a pressure signal into mechanical motion of a shaft.

shown in Equation (7-2), and that the compression of a spring is linearly related to forces as discussed in Chapter 5. Then, we see that the shaft position is linearly related to the applied control pressure

$$\Delta x = \frac{A}{k} \Delta p \qquad (7\text{-}3)$$

Where

Δx = shaft travel (m)
Δp = applied gauge pressure, (Pa), pascals
A = diaphragm area (m²)
k = spring constant (N/m)

A *reverse* actuator, shown in Figure 7.17, moves the shaft in the opposite sense from the direct actuator, but obeys the same operating principle. Thus, the shaft is pulled in by the application of a control pressure.

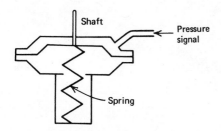

FIGURE 7.17   A reverse pneumatic actuator for converting a pressure signal to mechanical shaft motion.

EXAMPLE 7.4

Suppose a force of 400 N must be applied to open a valve. Find the diaphragm area if a control gauge pressure of 70 kPa ($\sim$10 psi) must provide this force.

SOLUTION

Here we must calculate the area from

$$F = A(p_1 - p_2) \tag{7-2}$$

where our applied pressure is $p_1 - p_2$ because a gauge pressure is specified. Then

$$A = \frac{F}{p} = \frac{400 \text{ N}}{7 \times 10^4 \text{ Pa}}$$

$$A = 5.71 \times 10^{-3} \text{ m}^2$$

or about 8.5 cm in diameter.

7.4.3 Hydraulic Actuators

In some cases where very large forces are required, the area of a required pneumatic diaphragm for the standard control signals may be too large for practical considerations. In such cases, a hydraulic actuator may be employed. The basic principle here is shown in Figure 7.18. The basic idea is the same as for pneumatic except that an incompressible fluid is used to provide the pressure, which can be made very large by adjustment of the area of the forcing piston A_1. The hydraulic pressure is given by

$$p_H = F_1/A_1 \tag{7-4}$$

Where

p_H = hydraulic pressure (Pa)
F_1 = applied piston force (N)
A_1 = forcing piston area (m^2)

The resulting force on the working piston is

$$F_W = p_H A_2 \tag{7-5}$$

Where

F_W = force of working piston (N)
A_2 = working piston area (m^2)

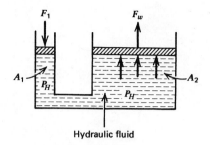

FIGURE 7.18 A hydraulic actuator converts a small force F_1 into an amplified force F_w.

Thus, the working force is given in terms of the applied force by

$$F_W = \frac{A_2}{A_1} F_1 \qquad\qquad (7\text{-}6)$$

EXAMPLE 7.5

Find the working force resulting from 200 N applied to a 1 cm radius forcing piston (a) if the working piston has a radius of 6 cm and (b) the hydraulic pressure.

SOLUTION

(a) We can find the working force from

$$F_W = \frac{A_2}{A_1} F_1 \qquad\qquad (7\text{-}6)$$

or

$$F_W = \left(\frac{R_2}{R_1}\right)^2 F = \left(\frac{6\text{ cm}}{1\text{ cm}}\right)^2 (200\text{ N})$$

$$F_W = 7200\text{ N !}$$

(b) Thus, the 200 N force provides 7200 N of force. The hydraulic pressure is

$$p_H = F_W/A_2 = \frac{7200\text{ N}}{(\pi)(6\times 10^{-2}\text{ m})^2}$$

$$p_H = 6.4 \times 10^5\text{ Pa}$$

This pressure is approximately 92 lb/in.^{2}

7.5 CONTROL ELEMENTS

The actual *control element* (which is a part of the process itself) can be many different devices. It is not the intention of this text to present many of these, but a general survey of standard devices would be valuable in presenting a complete picture of process control.

Several examples of control elements are described below in terms of different control problems.

7.5.1 Mechanical

Control elements that perform some *mechanical* operation in a process (by virtue of operations) are called mechanical control elements. Examples of these types are:

SOLID MATERIAL HOPPER VALVES

Consider the grain supply bin of Figure 7.19. Here, the control system is to maintain the flow of grain from the storage bin to provide a constant flow rate on the conveyer. This flow depends on the height of grain in the bin and hence the *hopper valve* must open or close to compensate for the variation. In this case an actuator operates a vane-type valve to control the grain flow rate. The actuator could be a motor to adjust shaft position, a hydraulic cylinder, or others covered above.

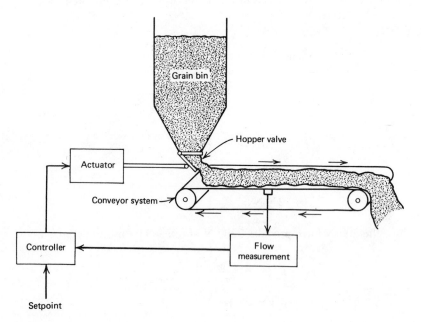

FIGURE 7.19   An example of a mechanical control element in the form of the hopper valve.

PAPER THICKNESS

In Figure 7.20, the essential features of a system for controlling paper thickness are shown. Here, the paper is in a wet fiber suspension and is passed between rollers. By varying the roller separation, paper thickness is regulated. The mechanical control element shown here is the movable roller. The actuator could be electrical, pneumatic, or hydraulic and adjusts roller separation based on a thickness measurement.

7.5.2 Electrical

There are numerous cases where a direct electrical effect is impressed in some process-control situation. The following examples illustrate some typical cases of electrical control elements.

MOTOR SPEED CONTROL

The speed of large electrical motors depends on many factors, including supply voltage level, load, and others. A process-control loop regulates this speed through direct change of operating voltage or current, as shown in Figure 7.21 for a d-c motor. Here, voltage measurements of engine speed from a tachometer are used in a process-control loop to determine the power applied to the motor brushes. In some cases motor speed control is an intermediate operation in a process-control application. Thus, in the operation of a kiln for solid chemical reaction, the rotation (feed) rate may be varied by motor speed control based on, for example, reaction temperature, as shown in Figure 7.22.

TEMPERATURE CONTROL

Temperature often is controlled by using electrical heaters in some application of industrial control. Thus, if heat can be supplied through heaters electrically in an endothermic reaction, then the process-control signal can be used to ON/OFF cycle a heater or set the heater within a continuous span of operating voltages, as in Figure 7.23. In this example, a reaction vessel is maintained at some constant temperature using an electrical heater. The process-control loop provides this by smoothly varying excitation to the heater.

7.5.3 Fluid Valves

The chemical and petroleum industries have many applications requiring control of fluid processes. Many other industries also depend in part on operations that involve fluids and the regulation of fluid parameters. The word *fluid* here represents either gases, liquids, or vapors. Many principles of control can be equally applied to any of these states of matter with only slight corrections. Many fluid operations require regulation of such quantities as density and composition, but by far the most

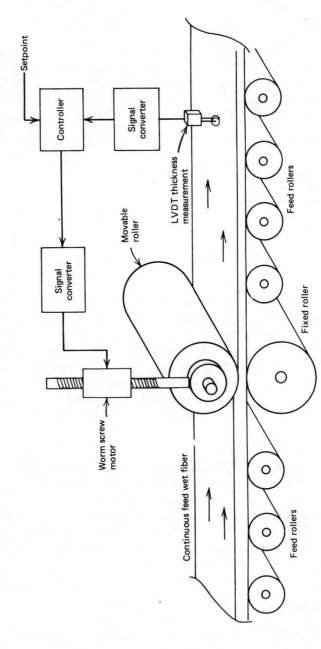

FIGURE 7.20 A continuous operation paper thickness controlling system using the mechanical final control elements.

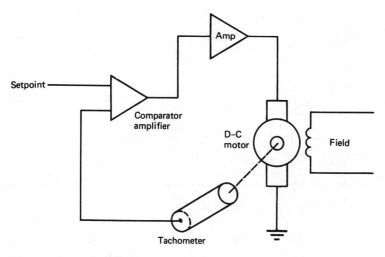

FIGURE 7.21   Electrical final control as found in the control of a d-c motor speed.

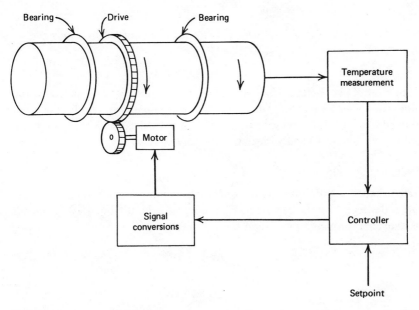

FIGURE 7.22   An electrical control system with an electrical final control element that controls the rotational rate of a reaction kiln.

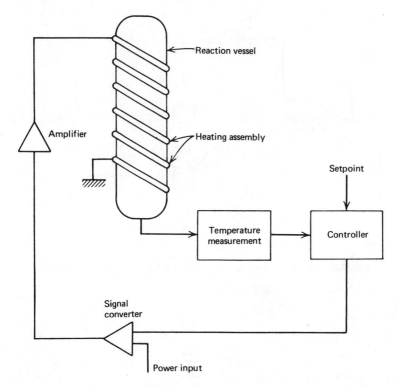

FIGURE 7.23   Control of heat to a reaction vessel can be provided by purely electrical means.

important control parameter is flow rate. A regulation of flow rate emerges as the regulatory parameter for reaction rate, temperature, composition, or a host of other fluid properties. For this reason we will consider in some detail that process-control element specifically associated with flow—the *control valve*.

CONTROL VALVE PRINCIPLES

Flow rate in process control is usually expressed as volume per unit time. If a mass flow rate is desired, it can be calculated from the particular fluid density. If a given fluid is delivered through a pipe, then the volume flow rate is

$$Q = Av \qquad (7\text{-}7)$$

Where

Q = flow rate (m³/s)
A = pipe area (m²)
v = flow velocity (m/s)

EXAMPLE 7.6

Alcohol is pumped through a pipe of 10 cm diameter at 2 m/s flow velocity. Find the volume flow rate.

SOLUTION

A pipe of 10 cm diameter has a cross-sectional area of

$$A = \frac{\pi D^2}{4} = \frac{(\pi)(10^{-1} \text{ m})^2}{4}$$

$$A = 7.85 \times 10^{-3} \text{ m}^2$$

Thus, the flow rate is

$$Q = Av = (7.85 \times 10^{-3} \text{ m}^2)(2 \text{ m/s})$$
$$Q = 0.0157 \text{ m}^3/\text{s}$$

(7-7)

A control valve regulates the flow rate in a fluid delivery system. In general, a close relation exists between the pressure along a pipe and the flow rate so that if the pressure is changed, then the flow rate is also changed. A control valve changes flow rate by changing the pressure in a flow system because it introduces a constriction in the delivery system. In Figure 7.24, we see that the placement of a constriction in a pipe introduces a pressure difference across the pipe. We can show that the flow rate through the constriction is given by

$$Q = K\sqrt{\Delta p}$$

(7-8)

Where

K = proportionality constant ($m^3/s/Pa^{1/2}$)
$\Delta p = p_2 - p_1$ = pressure difference (Pa)

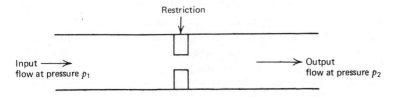

FIGURE 7.24   Flow rate through a restriction in a line is a function of the pressure drop across the restriction.

EXAMPLE 7.7

A pressure difference of 1.1 psi occurs across a constriction in a 5 cm diameter pipe. The constriction constant is 0.009 m³/s per kPa¹/². Find (a) the flow rate in m³/s and (b) the flow velocity in m/s.

SOLUTION

First we note that 1.1 psi is

$$\Delta p = (1.1 \text{ psi})(6.895 \text{ kPa/psi})$$
$$\Delta p = 7.5845 \text{ kPa}$$

(a) The flow rate is

$$Q = K\sqrt{\Delta p} = (0.009)(7.5845)^{1/2}$$
$$Q = 0.025 \text{ m}^3/\text{s} \tag{7-8}$$

(b) The flow velocity is found from

$$Q = Av$$

$$v = \frac{Q}{A} = 4\left[\frac{0.025 \text{ m}^3/\text{s}}{\pi(5 \times 10^{-2})^2}\right] \tag{7-7}$$

$$v = 12.7 \text{ m/s}$$

The constant K depends on the size of the valve, the geometrical structure of the delivery system, and, to some extent, on the material flowing through the valve. Now the actual pressure of the entire fluid delivery (and sink) system in which the valve is used (and, hence, the flow rate) is not a predictable function of the valve opening only. But because the valve opening does change flow rate, it thereby provides a mechanism of flow control.

CONTROL VALVE TYPES

The different types of control valves are classified by a relationship between the valve stem position and the flow rate through the valve. This *control valve characteristic* is assigned with the assumptions that the stem position indicates the extent of the valve opening and that the pressure difference is determined by the valve alone. Correction factors allow one to account for pressure differences introduced by the whole system. Figure 7.25 shows a typical control valve using a pneumatic actuator attached to drive the stem and, hence, open and close the valve. There are three basic types of control valves whose relationship between stem position (as percentage of full range) and flow rate (as a percentage of maximum) is shown in Figure 7.26.

1. *Quick Opening.* This type of valve is used predominantly for full ON/full OFF control

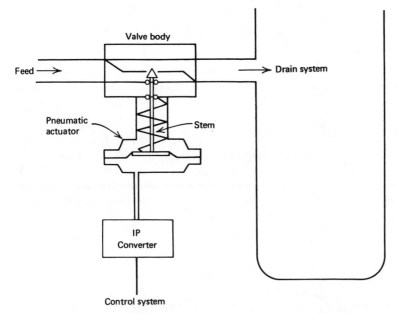

FIGURE 7.25 The essential features of a control valve are shown. Many variations in the construction of the valve body exist.

applications. The valve characteristic of Figure 7.26 shows that a relatively small motion of valve stem results in maximum possible flow rate through the valve. Such a valve, for example, may allow 90% of maximum flow rate with only a 30% travel of the stem.

2. *Linear.* This type of valve, as shown in Figure 7.26, has a flow rate that varies linearly with the stem position. It represents the ideal situation where the valve alone determines the pressure drop. The relationship is expressed as

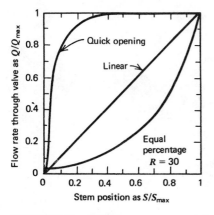

FIGURE 7.26 Different responses of the three main types of control valves to stem position.

$$\frac{Q}{Q_{max}} = \frac{S}{S_{max}} \tag{7-9}$$

Where

Q = flow rate (m³/s)
Q_{max} = maximum flow rate (m³/s)
S = stem position (m)
S_{max} = maximum stem position (m)

3. *Equal-Percentage.* A very important type of valve employed in flow control has a characteristic such that a given percentage change in stem position produces an equivalent change
in flow, that is, an equal percentage. Generally, this type of valve does not shut off the flow
completely in its limit of stem travel. Thus, Q_{min} represents the minimum flow when the
stem is at one limit of its travel. At the other extreme, the valve allows a flow Q_{max} as its
maximum, open valve, flow rate. For this type, we define *rangeability* R as the ratio

$$R = \frac{Q_{max}}{Q_{min}} \tag{7-10}$$

The curve in Figure 7.26 shows a typical equal-percentage curve which depends on the
rangeability for its exact form. The curve shows that increase in flow rate for a given change
in valve opening depends on the extent to which the valve is already open. This curve is
typically exponential in form and is represented by

$$Q = Q_{min}R^{S/S_{max}} \tag{7-11}$$

where all terms have been defined above.

EXAMPLE 7.8

An equal percentage valve has a maximum flow of 50 m³/s and a minimum of
2 m³/s. If the full travel is 3 cm, find the flow at 1 cm opening.

SOLUTION

The rangeability is

$$R = Q_{max}/Q_{min}$$
$$R = (50 \text{ m}^3/\text{s})/(2 \text{ m}^3/\text{s}) = 25 \tag{7-10}$$

Then the flow at 1-cm opening is

$$Q = Q_{min}R^{S/S_{max}}$$
$$Q = (2 \text{ m}^3/\text{s})(25)^{1 \text{ cm}/3 \text{ cm}} \tag{7-11}$$
$$Q = 5.85 \text{ m}^3/\text{s}$$

CONTROL VALVE SIZING

Another important factor associated with all control valves involves corrections to Equation (7-8) due to the nonideal characteristics of the materials that flow. A standard nomenclature is used to account for these corrections depending on the liquid, gas, or steam nature of the fluid. These correction factors allow selection of the proper size of valve to accommodate the rate of flow which the system must support. The correction factor most commonly used at present is measured as the number of U.S. gallons of water per minute that flow through a fully open valve with a pressure differential of 1 pound per square inch. The correction factor is called the *valve flow coefficient* and is designated as C_v. Using this factor, a liquid flow rate in U.S. gallons per minute is

$$Q = C_v \sqrt{\frac{\Delta p}{S_G}} \qquad (7\text{-}12)$$

Where

Δp = pressure across the valve (psi)
S_G = specific gravity of liquid

Typical valves of C_v for different size valves are shown in Table 7.1. Similar equations are used for gases and vapors to determine the proper valve size in specific applications.

TABLE 7.1
CONTROL
VALVE FLOW
COEFFICIENTS

Valve size (inches)	C_v
$\frac{1}{4}$	0.3
$\frac{1}{2}$	3
1	14
$1\frac{1}{2}$	35
2	55
3	108
4	174
6	400
8	725

EXAMPLE 7.9

Find (a) the proper C_v for a valve that must pump 150 gallons of ethyl alcohol per minute with a specific gravity of 0.8 at maximum pressure of 50 psi and (b) the required valve size.

SOLUTION

a. We find the correct sizing factor from

$$Q = C_v \sqrt{\frac{\Delta p}{S_G}} \qquad (7\text{-}12)$$

then

$$C_v = Q \sqrt{\frac{S_G}{\Delta p}}$$

$$C_v = \left(150 \ \frac{\text{gal}}{\text{min}}\right) \sqrt{\frac{0.8}{50 \ \text{lb/in.}^2}}$$

$$C_v = 18.97$$

b. A valve $1\frac{1}{2}$ in. in diameter (3.8 cm) would be selected from Table 7.1.

FLUID CONTROL EXAMPLE

The chemical and process-control industry uses fluid control systems extensively. Examples of such applications are many and varied. Consider, for example, control of distillation column composition by regulation of a fixed point column temperature. Such regulation is achieved by controlling the feed rate as shown in Figure 7.27. Here a thermocouple measures temperature which is transmitted to the con-

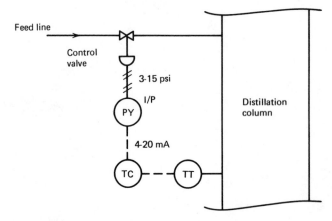

FIGURE 7.27 Feed control to a distillation column based on temperature.

troller as a 4–20 mA control signal. The controller outputs a 4–20 mA signal proportional to proper control valve position. This is converted to a 3–15 psi (20–100 kPa) pneumatic signal by an IP converter which, in turn, operates a pneumatic actuator connected to the control valve. The valve size is determined by the characteristics of the gas or vapor that is flowing. The size of the required actuator is determined from the valve size.

SUMMARY

The operation of the final control element has three separate functions; the ultimate goal is to translate a low-level control signal into a large-scale process. The following specific details were considered:

1. The final control function can be implemented by *signal conditioning*, an *actuator*, and a *final control element*.
2. Signal conditioning involves changing a control signal into that form and power necessary to energize the actuator. Simple electronic amplification, digital to analog conversion, electrical to pneumatic conversion, and pneumatic to hydraulic conversion are all typical signal conditioning operations.
3. The current to pressure converter is frequently employed in process-control systems. This device is based on a flapper/nozzle (nozzle/baffle) system which converts linear displacement into a pressure change.
4. Actuators are an intermediate step between the converted control signal and the final control element. Common electrical actuators are solenoids, digital stepping motors, ac and dc motors.
5. Pneumatic and hydraulic actuators are often used in process control because they allow very large forces to be produced from modest pressure systems. A pneumatic actuator converts a pressure signal to a shaft extension according to

$$\Delta x = \frac{A}{k} \Delta p \qquad (7\text{-}3)$$

where the force that causes this extension is given by

$$F = A(p_1 - p_2) \qquad (7\text{-}2)$$

6. Actual final control elements are as varied as the applications of process control in industry. Examples include motor driven conveyer belts, paper thickness roller assemblies, and heating systems.
7. The most general type of final control element is a *control valve*. This device is designed for use in process-control applications involving liquid, gas, or vapor flow rate control. Three types are commonly used: *quick opening*, *linear*, and *equal percentage*.

PROBLEMS

7.1   A 4–20 mA control signal is loaded by a 100 Ω resistor and must produce a 20–40 volt motor drive output. Find an equation relating input current to required output voltage.

7.2   Implement the equation of the previous problem if a power amplifier is available which outputs 0–100 volts with a gain of 10.

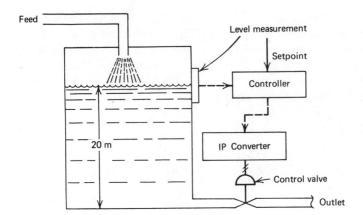

FIGURE 7.28   Figure for Problem 7.12.

7.3   A motor to be driven by a digital signal has a speed variation of 200 rev/min per volt with minimum rpm at 5 volts and the maximum at 10 volts. Find the minimum speed word, maximum speed word, and the speed change per LSB change. Use a 5-bit, 15-volt, ref DAC.

7.4   A stepping motor has 130 steps per revolution. Find the digital input rate that produces 10.5 rev per second.

7.5   What force is generated by 90 kPa acting on a 30 cm^2 area diaphragm?

7.6   A hydraulic system uses pistons of diameter 2 cm and 40 cm. What force on the small piston will raise a 500 kg mass?

7.7   What pneumatic pressure is required on the small piston of the previous problem to produce the necessary force?

7.8   The SCR in Figure 7.6*b* requires a 4-volt trigger. Design a system by which the gears are shifted when a CDS photocell resistance drops below 2.5 kΩ.

7.9   Design a system by which a control signal of 10–50 mA is converted into a force of 200–1000 N. Use a pneumatic actuator and specify the required diaphragm area if the pressure range is to be 3–15 psi. An IP converter is available which converts 0–5 volts into 3–15 psi.

7.10   Find the proper valve size in inches and centimeters for pumping a liquid flow rate of 600 gal/min with a maximum pressure difference of 55 psi. The liquid specific gravity is 1.3.

7.11   An equal percentage control valve has a rangeability of 32. If the maximum flow is 100 m^3/min, find the flow at 2/3 and 4/5 open settings.

7.12   The level of water in a tank is to be controlled at 20 m and output is nominally 65 m^3/hr through a control valve as shown in Figure 7.28. Under nominal conditions determine the required valve size in inches and centimeters.

7.13   If the valve actuator of Problem 7.12 has a rangeability of 30, a maximum stem travel of 5 cm, and is to be half open under the nominal condition, find the minimum flow, maximum flow and stem opening for a 100 m^3/hr flow.

7.14   A stepping motor has 7.5° per step. Find the rpm produced by a pulse rate of 2000 pps on the input.

7.15   A quick-opening valve, with a characteristic given by Figure 7.26, moves from closed to full open from 5 turns of a shaft. The shaft is driven through a 10 to 1 reducer from a stepping motor of 3.6° per step. If the maximum input pulse rate for the stepping motor is 250 steps per second, find the fastest time for the valve to move from closed to 90% open.

7.16   A feed hopper requires 30 lbs of force to open. Find the pneumatic actuator area to provide this force from 9 psi.

8
CONTROLLER
PRINCIPLES

INSTRUCTIONAL OBJECTIVES

This chapter presents the *operational* modes of a *process-control loop*. A mode is determined by the nature of controller responses to a dynamic variable measurement and setpoint comparison. After you have read it, you should be able to:

1. Define process load, process lag, and self-regulation.
2. Describe two-position and floating-control mode.
3. Define the proportional controller mode.
4. Give an example and description of an integral-control mode.
5. Describe the derivative-control mode.
6. Contrast proportional-integral and proportional-derivative control modes.
7. Describe three mode controllers.
8. Provide a description of the controller output for a fixed error input of any of the above controller modes.

8.1 INTRODUCTION

The last element of the process-control loop to be studied is the *controller*. This element inputs the result of a measurement of the controlled variable and determines an appropriate output to the final control element. Essentially, the controller is some form of computer, either analog or digital, pneumatic or electronic, which, using input measurements, solves certain equations to calculate the proper output. The equations necessary to obtain control exist in only a few forms, independent both of the process itself and whether the controller function is provided by an analog or digital computer. We call these equations the *modes* of controller operation. The nature of the process itself and the particular variable controlled determine which mode or modes of control are to be used and the value of certain constants in the mode equations. In this chapter we will study the various modes of controller operation. Later chapters will examine how the modes are implemented by analog or digital means and by pneumatic and electronic means.

8.2 PROCESS CHARACTERISTICS

The selection of what controller modes to use in a process is a function of the characteristics of the process. It is not our intention to discuss how the modes are selected but rather to define the meaning of each mode. At the same time, it is helpful in understanding the modes if certain pertinent characteristics of the process are considered. In this section we will define a few properties of processes which are important for selecting the proper modes to be used in control.

8.2.1 Process Equation

The purpose of a process-control loop is to regulate some *dynamic variable* in a process. This *controlled* variable, a process parameter, may depend on many other parameters (in the process) and thus suffer changes from many different sources. We have selected one of these others to be our *controlling parameter*. This means that if a measurement of the controlled variable shows a deviation from the setpoint, then the controlling parameter is changed, which in turn changes the variable.

As an example, consider the control of liquid temperature in a tank, as shown in Figure 8.1. Here the *controlled variable* is the liquid temperature T_L. This temperature depends on many parameters in the process, to wit, the input flow rate via pipe A, the output flow rate via pipe B, the ambient temperature T_A, the steam temperature T_S, inlet temperature T_0, and the steam flow rate Q_S. In this case, the steam flow rate is the *controlling parameter* chosen to provide control over the variable (liquid temperature). If one of the other parameters changes, a change in temperature results. In order to bring the temperature back to the setpoint value, we only change the steam flow rate, that is, heat input to the process. This could be described by a *process equation* where liquid temperature T_L is a function as

$$T_L = F(Q_A, Q_B, Q_S, T_A, T_S, T_0) \tag{8-1}$$

Where

Q_A, Q_B = flow rates in pipes A and B
Q_S = steam flow rate
T_A = ambient temperature
T_0 = inlet fluid temperature
T_S = steam temperature

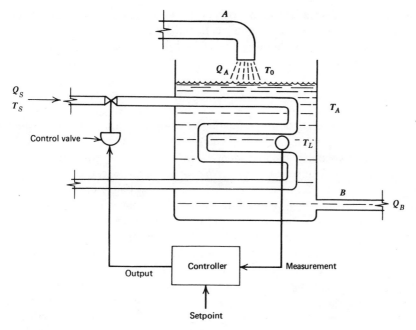

FIGURE 8.1   Control of temperature by process control.

Note that to provide control via Q_S, we need not know the functional relationship exactly nor require linearity of the function. The purpose of the control loop is to adjust Q_S and thereby regulate T_L regardless of how the other parameters in Equation (8-1) vary with each other. In many cases the relationship of Equation (8-1) is not even analytically known.

8.2.2 Process Load

Usually, nominal values are assumed for the various process parameters from which a value of the controlling parameter is chosen. This provides the desired *setpoint* value of the *controlled variable*. As such, *process load* refers to the set of all parameters, *excluding* the controlled variable. When all parameters have their nominal values, we speak of the *nominal load* on the system. The required controlling variable value under these conditions is the *nominal value* of that parameter. If the setpoint is changed, the control parameter is altered to cause the variable to adopt this new operating point. The load is still nominal, however, because the other parameters are assumed unchanged. Suppose one of the parameters changes from nominal, causing a corresponding shift in the controlled variable. We then say that a *process load* change has occurred. The controlling variable is adjusted to compensate for this load change and its effect on the dynamic variable to bring it back to the setpoint. In the example of Figure 8.1, a process load change is due to any change in any of the five parameters affecting liquid temperature. The extent of the load change on the controlled variable is formally determined by process equations such as Equation (8-1). In practice, we are only concerned that variation in the *controlling parameter* brings the controlled variable back to the setpoint. We are not particularly interested in the cause, nature, or extent of the load change.

8.2.3 Process Lag

As previously noted, process-control operations are essentially a time-variation problem. At some point in time, a process load change causes a change in the controlled variable. The process-control loop responds to assure, some finite time later, that the variable returns to the setpoint value. Part of this time is consumed by the process itself and is called the *process lag*. Thus, referring to Figure 8.1, assume the inlet flow is suddenly doubled. Such a large process load change radically changes (reduces) the liquid temperature. The control loop responds by opening the steam inlet valve to allow more steam and heat input to bring the liquid temperature back to the setpoint. The loop itself reacts faster than the process. In fact, the physical opening of the control valve is the slowest part of the loop. Once steam is flowing at the new rate, however, the body of liquid must be heated by the steam before the setpoint value is reached again. This time delay or *process lag* in heating is a function of the process and NOT the control system. Clearly, there is no advantage in designing control systems *many* times faster than the process lag itself.

8.2.4 Self-Regulation

A significant process characteristic is its tendency to find a specific value of the dynamic variable for nominal load with no control operations. The control operations may be significantly affected by such *self-regulation*. The process of Figure 8.1 has self-regulation as shown by the following argument.

(1) Suppose we fix the steam valve at 50% and open the control loop so that no changes in valve position are possible. (2) The liquid heats up until the energy carried away by the liquid equals that input energy from the steam flow. (3) If the load changes, a new temperature is adopted (because the system temperature is *not* controlled). (4) The process is *self-regulating*, however, because the temperature will not "run away," but stabilizes at some value under given conditions.

An example of a process *without* self-regulation is a tank from which liquid is pumped out at a fixed rate. Assume that the influx just matches the outlet rate. Then the liquid in the tank is fixed at some nominal level. If the influx increases slightly, however, the level rises until the tank overflows! No self-regulation of the level is provided.

8.3 CONTROL SYSTEM PARAMETERS

We have just described the basic characteristics of the process that are related to control. Let us now examine the general properties of the controller shown in Figure 8.2.

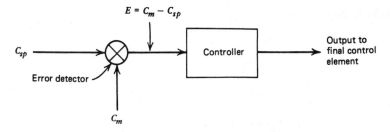

$$E = C_m - C_{sp}$$

C_{sp} →

Error detector

C_m

Controller

Output to
final control
element

FIGURE 8.2   The error detector and controller block diagram.

To review: (1) inputs to the controller are a measured indication of both the *controlled variable* and a *setpoint* representing the desired value of the variable, expressed in the same fashion as the measurement; (2) the controller output is a signal representing action to be taken when the measured value of the controlled variable deviates from the setpoint.

8.3.1 Error

The deviation or error of the controlled variable from the setpoint is given by

$$E = C_m - C_{sp} \qquad (8\text{-}2)$$

Where

E = error
C_m = measured value of variable
C_{sp} = setpoint of variable

The above equation expresses error in an absolute sense, usually in units of the measured analog of the control signal. Thus, if the setpoint in a 4–20 mA range corresponds to 9.9 mA and the measured value is 10.7 mA, we would have an error of +0.8 mA. Obviously, this current error has little direct meaning unless related to the dynamic variable itself. We could work back through the loop and prove that it corresponds to flow rate of 1.1 m³/min, for example. This would show the significance of the error relative to the actual process-control loop.

In actual controller operation, however, we use *neither* representation of error. In fact, it is the error compared to its full-scale range that becomes most useful and meaningful. Thus, in discussing controller modes of operation, we use error expressed as a percentage of full-scale variable range. This ratio is formed by first writing the error as the difference of the measured value and setpoint value, both expressed as percentage of full scale, or

$$E_p = \left[\frac{C_m - C_{min}}{C_{max} - C_{min}} - \frac{C_{sp} - C_{min}}{C_{max} - C_{min}} \right] \cdot 100$$

which reduces algebraically to

$$E_p = \left[\frac{C_m - C_{sp}}{C_{max} - C_{min}} \right] \cdot 100 \tag{8-3}$$

Where

E_p = error as percent full scale from setpoint
C_m = measured value of variable
C_{sp} = setpoint of variable
C_{max} = maximum value of variable
C_{min} = minimum value of variable

Now, if the variable is expressed in a standard range, that is, 4–20 mA, then current *can be used* directly in Equation (8-3). In this view, if the range is 4–20 mA and the setpoint is set at 10.5 mA, then a measurement of 13.7 mA means an error of

$$E_p = \left[\frac{13.7 \text{ mA} - 10.5 \text{ mA}}{20 \text{ mA} - 4 \text{ mA}} \right] \cdot 100$$

$$E_p = +20\%$$

EXAMPLE 8.1

The temperature in Figure 8.1 has a range of 300 K to 440 K and a setpoint of 384 K. Find the percent full-scale error when the temperature is 379 K.

SOLUTION

The percent full scale is

$$E_p = \left[\frac{C_m - C_{sp}}{C_{max} - C_{min}} \right] \cdot 100$$

$$E_p = \left[\frac{379 - 384}{440 - 300} \right] \cdot 100 \tag{8-3}$$

$$E_p = -3.6\%$$

Note that a *positive* error indicates a measurement *above* the setpoint, whereas a *negative* error indicates a measurement *below* the setpoint.

8.3.2 Variable Range

Generally, the dynamic variable under control has a range of values within which control is to be maintained. This range can be expressed as the minimum and maximum value of the dynamic variable or the nominal value plus and minus the spread about this nominal. If a standard 4–20 mA signal transmission is employed, then 4 mA represents the minimum value of the variable and 20 mA the maximum.

8.3.3 Control Parameter Range

Another range is associated with the *controller output*. Here we assume the final control element has some minimum and maximum effect on the process. The controller output range is the translation of output to the range of possible values of the final control element. This range also is expressed as the 4–20 mA standard signal again with the minimum and maximum effects in terms of the minimum and maximum current.

Often, the output is expressed as a percentage where 0% is the minimum controller output and 100% the maximum (obviously). Thus, in the example of Figure 8.1, the valve in the fully open position corresponds to a 100% controller signal output. Note that very often the minimum does *not* correspond to zero effect. For example, it may be that the steam flow should never be less than that flow with the valve half open. In this case a 0% minimum controller corresponds to the flow rate with a half-open valve.

The controller output has a percent of full scale when the output varies between specified limits is given by

$$P = \left[\frac{S_p - S_{min}}{S_{max} - S_{min}} \right] \cdot 100 \tag{8-4}$$

Where

P = controller output as percent of full scale
S_p = value of the output
S_{max} = maximum value of controlling parameter
S_{min} = minimum value of controlling parameter

EXAMPLE 8.2

A controller outputs a 4–20 mA signal to control motor speed from 140–600 rpm with a linear dependence. Calculate (a) current corresponding to 310 rpm and (b) the value of (a) expressed as the percent of control output.

SOLUTION

(a) We find the slope m, and intersect S_0, of the linear relation between current I and speed S, where

$$S_p = mI + S_0$$

by knowing S_p and I at the two given positions, we write two equations:

$$140 = 4m + S_0$$
$$600 = 20m + S_0$$

solving these simultaneous equations, we get $m = 28.75$ rpm/mA and $S_0 = 25$ rpm. Thus, at 310 rpm we have $310 = 28.75I + 25$ which gives $I = 9.91$ mA.

(b) Expressed as a percentage of the 4–20 mA range, this controller output is

$$P = \frac{S_p - S_{min}}{S_{max} - S_{min}} \cdot 100$$

$$P = \left[\frac{9.91 - 4}{20 - 4}\right] \cdot 100 \qquad (8\text{-}4)$$

$$P = 36.9\%$$

8.3.4 Control Lag

The control system itself also has a lag associated with its operation which must be compared to the process lag (Section 8.2.2). When a dynamic variable experiences a sudden change, the process-control loop reacts by outputting a command to the final control element to adopt a new value to compensate for the change detected. *Control lag* refers to the time for the process-control loop to make necessary adjustments to the final control element. Thus, in Figure 8.1, if a sudden change in liquid temperature occurs, it requires some finite time for the control system to physically actuate the steam control valve.

8.3.5 Dead Time

Another time variable associated with process control is both a function of the process-control system and the process itself. This is the elapsed time between the instant a deviation (error) occurs and when the corrective action first occurs. An example of *dead time* occurs in the control of a chemical reaction by varying reactant flow rate through a very long pipe. When a deviation is detected, a control system quickly changes a valve setting to adjust flow rate. But if the pipe is quite long, there is a period of time during which no effect is felt in the reaction vessel. This is the time required for the new flow rate to move down the length of the pipe. Such *dead times* can have a very profound effect on the performance of control operations on a process.

8.3.6 Cycling

We frequently refer to the behavior of the dynamic variable error under various modes of control. One of the most important modes is an *oscillation* of the error about zero. This means the variable is *cycling* above and below the setpoint value.

Such cycling may continue indefinitely in which case we have *steady-state cycling*. Here we are interested in both the *peak amplitude* of the *error* and the *period* of the *oscillation*.

If the cycling amplitude decays to zero, however, we have a *cyclic transient error*. Now we are interested in the *initial error*, the *period* of the *cyclic oscillation*, and *decay time* for the error to reach zero.

8.3.7 Controller Modes

The controller was defined (Section 8.3) by the statement that a controller generates a control signal to the final element, based on a measured deviation of the dynamic variable from the setpoint. It is natural to ask how the controller responds to the deviation. In a thermostatically controlled temperature system used in the home, the controller response is simple. If the temperature drops below the thermostat setpoint, a bimetallic relay turns on a heater.

But, consider the case of the system shown in Figure 8.1. Here, no simple ON-OFF decision can be made because the setting of the steam valve can be smoothly varied from one extreme to another. Thus, if a deviation from liquid temperature setpoint occurs, what should the controller do? Should it open the valve a little or a lot? Should it open the valve fast or slow?

These questions are answered by specifying the *mode* of the controller operation. One distinction is clear from the above examples. The domestic thermostat involves a mode which is *discontinuous*, where the controller command initiates a discontinuous change in the control parameter. The process of Figure 8.1 is *continuous* because smooth variation of the control parameter is possible. Section 8.4 covers various controller modes in detail.

The choice of operating mode for any given process-control system is a complicated decision. It involves not only process characteristics but cost analysis, product rate, and other industrial factors. At the outset, the process-control technologist should have a good understanding of the operational mechanism of each mode and their advantages and disadvantages. The operation of each mode is defined below and examples are given with some general statements of application details. In each case, the output of the controller is described by a factor P. This is the *percent of controller output* relative to its total range as defined in Equation (8-4).

For example, if a controller outputs a 4–20 mA current signal to the final control element and has a $P = 25\%$, then the corresponding current is

$$I = I_{min} + P(I_{max} - I_{min})$$
$$I = 4 \text{ mA} + (0.25)(20 - 4) \text{ mA}$$
$$I = 8 \text{ mA}$$

If this current is used to drive a valve actuator for which 4 mA is closed and 20 mA full open, then the valve is 25% open. If the valve is an equal percentage type with a rangeability of 30, then the flow rate from Section 7.5.3 is

$$Q = Q_{min}R^{S/S_{max}}$$
$$Q = Q_{min}(30)^{0.25}$$
$$Q = 2.34\,Q_{min}$$

(7-11)

The above example shows that if the percentage output of the controller is known, then the actual value of the controlled variable can be determined.

The input of the controller is described by the error E_p defined in Equation (8-3) as the percentage error of measured variable from the setpoint relative to full scale. In general, the controller operation is expressed as a relation:

$$P = F(E_p)$$

(8-5)

where $F(E_p)$ represents the relation by which the appropriate controller output is determined. In some cases, a graph of P versus E_p also is employed to aid in a definition of the control mode.

8.4 DISCONTINUOUS CONTROLLER MODES

This section discusses the various controller modes which show discontinuous changes in controller output as controlled variable error occurs. It is important that you understand these modes, both because of their frequent use in process control, and because they form the basis of the continuous modes to be discussed in the next section.

8.4.1 Two-Position Mode

The most elementary controller mode is the ON-OFF or two-position mode. This is an example of a *discontinuous* mode. It is the simplest but also the cheapest and often suffices when its disadvantages are tolerated. Although an analytic equation cannot be written, we can, in general, write

$$P = \begin{cases} 100\% & E_p > 0 \\ 0\% & E_p < 0 \end{cases}$$

(8-6)

This relation shows that when the measured value exceeds the setpoint, *full* controller output results. When it is less than the setpoint, the controller output is zero. A space heater is a common example. If the temperature drops below a setpoint, the heater is turned ON. If above the setpoint, it turns OFF.

NEUTRAL ZONE

In virtually any practical implementation of the two-position controller, there is an overlap as E_p increases through zero or decreases through zero. In this span, *no change* in controller output occurs. This is best shown in Figure 8.3 which plots P versus E_p for a two-position controller. Here we see that until an increasing error

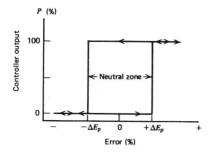

FIGURE 8.3   Two-position controller action with a neutral zone.

changes by ΔE_p *above zero*, the controller output will not change state. In decreasing, it must fall ΔE_p *below zero* before the controller changes to the 0% rating. The range $2\Delta E_p$, which is referred to as the *neutral zone* or *differential gap*, is often purposely designed above a certain minimum quantity to prevent excessive cycling. The existence of such a neutral zone is an example of desirable hysteresis in a system.

EXAMPLE 8.3

A liquid level control system linearly converts a displacement of 20–30 meters into a 4–20 mA control signal. A relay serves as the two-position controller to open or close an inlet valve. The relay closes at 12 mA and opens at 10 mA. (a) Find the relation between displacement level and current, and (b) the neutral zone or displacement gap in meters.

SOLUTION

(a) The relation between level and a current is a linear equation such as

$$H = KI + H_0$$

we find K and H_0 by writing two equations

$$20 \text{ m} = K(4 \text{ mA}) + H_0$$
$$30 \text{ m} = K(20 \text{ mA}) + H_0$$

solving these simultaneous equations yields $K = 0.625$ m/mA and $H_0 = 17.5$ m, at the intersection of the linear relations.

(b) Now the relay closes at 12 mA which is a high level H_H of

$$H_H = (0.625 \text{ m/mA}) (12 \text{ mA}) + 17.5 \text{ m}$$
$$H_H = 25 \text{ m}$$

The low level H_L occurs at 10 mA which is

$$H_L = (0.625 \text{ m/mA}) (10 \text{ mA}) + 17.5 \text{ m}$$
$$H_L = 23.75 \text{ m}$$

Thus, the neutral zone is $H_H - H_L = (25 - 23.75)$ m or **1.25** m.

APPLICATIONS

Generally, the two-position control mode is best adapted to large-scale systems with relatively *slow* process rates. Thus, in the example of either a room heating or airconditioning system, the capacity of the system is very large in terms of air volume, and the overall effect of the heater or cooler is relatively slow. Sudden large-scale changes are not common to such systems. The process under two-position control must allow continued oscillation in the dynamic variable because, by its very nature, this mode of control always produces such oscillation. For large systems, these oscillations will be of long duration, which is partly a function of the neutral-zone size. To illustrate this, consider the following example.

EXAMPLE 8.4

As a water tank loses heat the temperature drops by 2 K per minute. When a heater is on, the system gains temperature at 4 K per minute. A two-position controller has a 0.5-minute control lag and a neutral zone of ±4% of the setpoint about a setpoint of 323 K. Plot the heater temperature versus time. Find the oscillation period.

SOLUTION

Let us assume we start at the setpoint value; then the temperature will drop linearly at

$$T_1(t) = T(t_s) - 2(t - t_s) \qquad (8\text{-}7)$$

The heater will start at a temperature of 310 K (4% below setpoint) after which the temperature will rise according to

$$T_2(t) = T(t_s) + 4(t - t_s) \qquad (8\text{-}8)$$

When the temperature reaches 336 K, the heater goes off, and the system temperature drops by 2 K/min until 310 K is reached, and so on. The system response is then plotted as in Figure 8.4, using Equations (8-7) and (8-8). Notice the period is 21.5 minutes. There is also a 1 K undershoot and a +2 K overshoot due to the lag.

In general, some overshoot and undershoot of the dynamic variable will occur as in Example 8.4 above. This is due to the finite time required for the control element to impress its full effect on the process. Thus, the finite warmup time and cooloff time of the heater (included in Example 8.4) caused some over- and under-

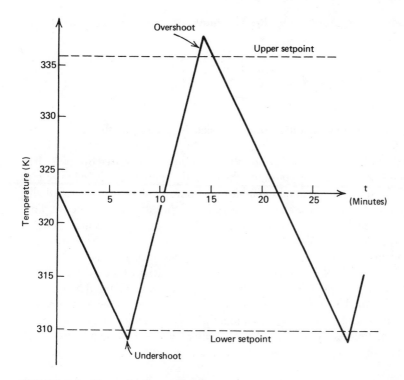

FIGURE 8.4   Figure for Example 8.4.

shoot of temperature. In some cases, if the final control element lag is large, substantial errors can result, and the neutral zone must be reduced to reduce these errors. In general, the cycling, as noted in the previous example, is a function of the neutral zone. If the neutral zone of this example is reduced to $\pm 2\%$, the reader can verify that, although the control is now maintained tighter, the cycling period is 10.7 min. The solution of these values is left as a problem for the reader.

8.4.2 Multiposition Mode

A logical extension of the previous two-position control mode is to provide *several intermediate* rather than only two settings of the controller output. This discontinuous control mode is used in an attempt to reduce the cycling behavior and also overshoot and undershoot inherent in the two-position mode. In fact, however, it is usually more expedient to use some other mode when the two-position is not satisfactory. This mode is represented by

$$P = P_i \quad E_p > |E_i| \ i = 1, 2, \ldots n \tag{8-9}$$

The meaning here is that as the error exceeds certain set limits $\pm E_i$, the controller output is adjusted to preset values P_i. The most common example is the three-position controller where

$$P = \begin{cases} 100 & E_p > E_1 \\ 50 & -E_1 < E_p < E_1 \\ 0 & E_p < -E_1 \end{cases} \qquad (8\text{-}10)$$

This means that as long as the error is between E_2 and E_1 of the setpoint, the controller stays at some nominal setting indicated by a controller output 50%. If the error exceeds the setpoint by E_1 or more, then the output is increased by 100%. If it is less than the setpoint by $-E_1$ or more, the controller output is reduced to zero.

Figure 8.5 illustrates this mode graphically. Some small neutral zone usually exists about the change points but not by design, and thus they are not shown. This type of control mode usually requires a more complicated final control element because it must have more than two settings. Figure 8.6 shows a graph of dynamic variable and final control element setting versus time for a hypothetical case of three-position control. Note the change in control element setting as the variable changes about the two trip points. On this graph the finite time required for the final control element to change from one position to another is also shown. Notice the overshoot and undershoot of the error around the upper and lower setpoints. This is due to both the process lag time and the controller lag time indicated by the finite time required for the control element to reach a new setting.

8.4.3 Floating Control Mode

In the two previous modes of controller action, the output was uniquely determined by the magnitude of the error input. If the error exceeded some preset limit, the output was changed to a new setting as quickly as possible. In floating control, the specific output of the controller is *not* uniquely determined by the error. If the error

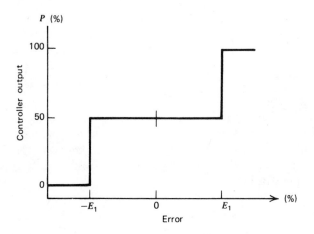

FIGURE 8.5 Three-position controller action.

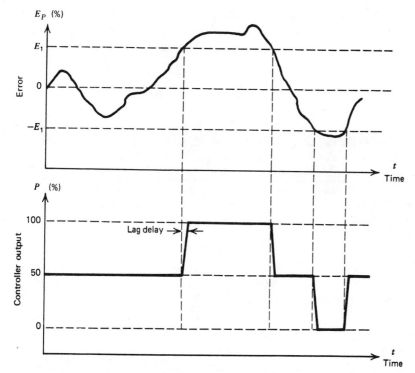

FIGURE 8.6   Relationship between error and three-position controller action including the effects of lag.

is zero, the output will not change but remains (floats) at whatever setting it was when the error went to zero. When the error moves off zero, the controller output again begins to change. Actually, as with the two-position mode, there is typically a neutral zone about zero error where no change in controller position occurs.

SINGLE SPEED

In the single-speed floating control mode, the output of the control element changes at a fixed rate about the setpoint when the error exceeds the neutral zone. An equation for this action is

$$\frac{dP}{dt} = \pm K_F \qquad |E_p| > \Delta E_p \qquad (8\text{-}11)$$

Where

$\frac{dP}{dt}$ = rate of change of controller output with time

K_F = rate constant (%/s)

ΔE_p = half the neutral zone

If Equation (8-11) is integrated for the actual controller output, we get

$$P = \pm K_F t + P(0) \qquad |E_p| > \Delta E_p \qquad (8\text{-}12)$$

Where

$P(0)$ = controller output at $t = 0$

which shows that the present position depends on the time history of errors that have previously occurred. Because such a history usually is not known, the actual value of P floats at an *undetermined* value. Note that if the deviation persists, then Equation (8-11) shows that the controller either saturates at 100% or 0% and remains there until an error drives it toward the opposite extreme. A graph of single-speed floating control is shown in Figure 8.7a.

EXAMPLE 8.5

Suppose a process error lies within the neutral zone with $P = 25\%$. At $t = 0$, the error falls *below* the neutral zone. If $K = +2\%$ per second, find the time when the output saturates.

SOLUTION

The relation between controller output and time is

$$P = K_F t + P(0) \qquad (8\text{-}12)$$

when $P = 100$

$$100\% = (2\%/s)(t) + 25\%$$

which, when solved for t, yields

$$t = 37.5 \text{ s}$$

In Figure 8.7b a graph shows controller output versus time and error versus time for a hypothetical case illustrating typical operation. In this example, we assume the controller is reverse acting, which means the controller output decreases when the error exceeds the neutral zone. This corresponds to a negative K_F in Equation (8-11). Most controllers can be adjusted to act in either the reverse or direct mode. Here the controller starts at some output $P(0)$. At time t_1, the error exceeds the neutral zone. The controller output now decreases at a constant rate until t_2 when the error *again* falls below the neutral zone limit. At t_3, the error falls below the lower limit of the neutral zone causing controller output to change until the error again moves within the allowable band.

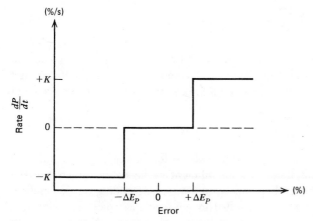

a) Single-speed floating controller action. The ordinate is the rate of change of controller output with time

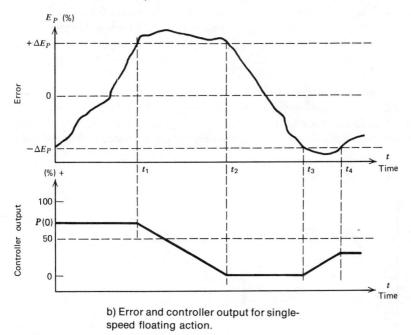

b) Error and controller output for single-speed floating action.

FIGURE 8.7   Single-speed floating controller.

MULTIPLE SPEED

In the floating multiple-speed control mode not one but several possible speeds (rates) are changed by controller output. Usually, the rate increases as the deviation exceeds certain limits. Thus, if we have certain speed change points E_{pi} depending on the error, then each has its corresponding output rate change K_i. We can then say

$$\frac{dP}{dt} = \pm K_{Fi} \qquad |E_p| > E_{Pi} \qquad (8\text{-}13)$$

If the error exceeds E_{Pi}, then the speed is K_{Fi}. If the error rises to exceed E_{P2}, the speed is increased to K_{F2}, and so on. Actually, this mode is a discontinuous attempt to realize an integral control mode (Section 8.4.4). A graph of this mode is shown in Figure 8.8.

APPLICATIONS

Primary applications of the floating control mode are for the single-speed controllers with a neutral zone. This mode has an inherent cyclic nature much like the two-position, although this cycling can be minimized depending on the application. Generally, the method is well-suited to self-regulation processes with very small lag or dead time, which implies small-capacity processes. When used with large-capacity systems, the inevitable cycling must be accounted for.

An example of single-speed floating control is that of liquid flow rate through a control valve. Such a system is shown in Figure 8.9. The load is determined by the inlet and outlet pressures P_{in} and P_{out}, whereas the flow is determined in part by the pressure P within the DP cell and control valve. This is an example of a system with self-regulation. Here, we assume some valve opening has been found commensurate with the desired flow rate. If the load changes (either P_{in} or P_{out}), then an error occurs. If larger than the neutral zone, the valve begins to open or close at a constant rate until an opening is found which supports the proper flow rate at the new load conditions. Clearly, the rate is very important here because very fast process lags cause the valve to continue opening (or closing) beyond that optimum

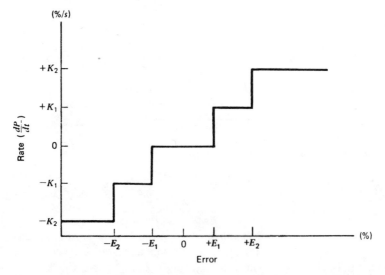

FIGURE 8.8 Multiple-speed floating control mode action.

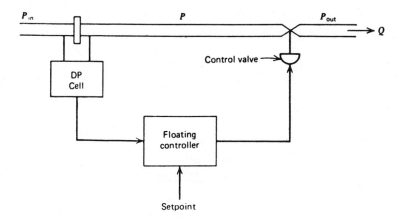

FIGURE 8.9   Single-speed floating control action applied to a flow control system.

self-regulated position. This is shown in Figure 8.10, where the response to a sudden deviation is shown for various floating rates.

8.5 CONTINUOUS CONTROLLER MODES

The most common controller action used in process control is one or a combination of continuous controller modes. In these modes, the output of the controller changes smoothly in response to the error or rate of change of error. These modes are an extension of the discontinuous types discussed in the previous section.

8.5.1 Proportional Control Mode

The two-position mode had the controller output of either 100% or 0% depending on the error being greater or lesser than the neutral zone. In multiple-step modes more divisions of controller outputs versus error are developed. The natural extension of this concept is the *proportional mode,* where a smooth, linear relationship

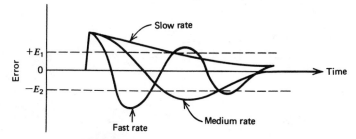

FIGURE 8.10   The rate of controller output change has a strong effect on error recovery in a floating controller.

exists between the controller output and the error. Thus, over some range of errors about the setpoint, each value of error has a unique value of controller output in one-to-one correspondence. The range of error to cover the 0% to 100% controller output is called the *proportional band* because the one-to-one correspondence exists only for errors in this range. This mode can be expressed by

$$P = K_P E_p + P_0 \qquad (8\text{-}14)$$

Where

    K_P = proportional constant between error and controller output (%/%)
    P_0 = controller output with no error (%)

The proportional band is given in percent by $100/K_P$. The controller output for errors exceeding the proportional band is saturated at either 100% or 0% depending on the sign of the error. In Figure 8.11, the graph of this control mode is shown. Note that the proportionality constant determines the proportional band. The value of P_0 is often chosen as 50% to give equal controller output swing as errors occur above and below the setpoint.

OFFSET

An important characteristic of the proportional control mode is that it produces a permanent *residual error* in the operating point of the controlled variable when a change in load occurs. This error is referred to as *offset*. It can be minimized by a

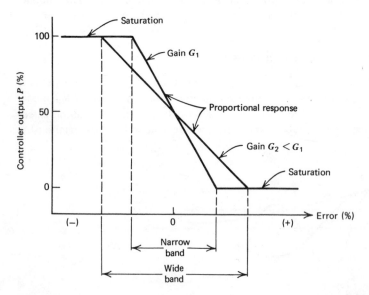

FIGURE 8.11   The proportional band of a proportional controller depends upon the gain in an inverse fashion.

larger constant K_P, which also reduces the proportional band. To see how offset occurs, consider a system under nominal load with the controller at 50% and the error zero as shown in Figure 8.12. If a transient error occurs, the system responds by changing controller output in correspondence with the transient to effect a return to zero error. Suppose, however, a load change occurs which requires a permanent change in controller output to produce the zero error state. Because a one-to-one correspondence exists between controller output and error, it is clear that a new, zero error controller output can *never* be achieved. Instead, the system produces a small permanent offset in reaching a compromise position of controller output under new loads.

EXAMPLE 8.6

Consider the proportional mode level control system of Figure 8.13. Valve A is linear with a flow scale factor of 10 m³/hr per percent controller output. The controller output is nominally 50% with a constant of $K_P = 10\%/\%$. A load change occurs when flow through valve B changes from 500 m³/hr to 600 m³/hr. Calculate the new controller output and offset error.

SOLUTION

Certainly, valve A must move to a new position of 600 m³/hr flow or the tank will empty! This can be accomplished by a 60% new controller output because

$$Q_A = \left(10 \ \frac{\text{m}^3/\text{hr}}{\%}\right)(60\%) = 600 \text{ m}^3/\text{hr}$$

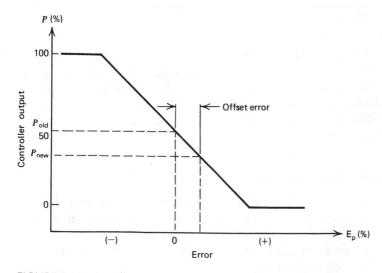

FIGURE 8.12   An offset error must occur if a proportional controller requires a new nominal controller output following a load change.

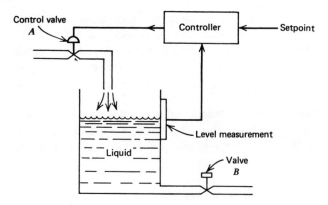

FIGURE 8.13   Level control system for Example 8.6.

as required. Because this is a proportional·controller, we have

$$P = K_P E_p + P_0 \qquad (8\text{-}14)$$

with the nominal condition $P_0 = 50\%$. Thus,

$$E_p = \frac{P - P_0}{K_P} = \frac{60 - 50}{10} \%$$

$$E_p = 1\%$$

so that a 1% offset error occurred due to the load change.

APPLICATION

The offset error limits use of the proportional mode to only a few cases, particularly those where a manual reset of the operating point is possible to eliminate offset. Proportional control generally is used in processes where large load changes are unlikely or with moderate to small process lag times. Thus, if the process lag time is small, the proportional band can be made very small (large K_P) which reduces offset error.

8.5.2 Integral Control Mode

This mode represents a natural extension of the principle of floating control in the limit of infinitesimal changes in the rate of controller output with infinitesimal changes in error. Instead of single speed or even multiple speeds, we have a continuous change in speeds depending upon error. This mode is often referred to as *reset action*. Analytically, we can write

$$\frac{dP}{dt} = K_I E_p \qquad (8\text{-}15)$$

Where

$$\frac{dP}{dt} = \text{rate of controller output change } (\%/s)$$

K_I = constant relating the rate to the error $(\%/s/\%)$

Units of K_I are in percentage controller output per second per percent error.

In some cases the inverse of K_I, called the integral time $T_I = 1/K_I$, expressed in seconds, is used to describe the integral mode.

If we integrate Equation (8-15), we can find the actual controller output at any time as

$$P(t) = K_I \int_0^t E_p(t)dt + P(0) \tag{8-16}$$

Where

$P(0)$ = the controller output at $t = 0$

This equation shows that the present controller output, $P(t)$, depends on the history of errors from when observation started at $t = 0$. We see from Equation (8-15), for example, that if the error doubles, the rate of controller output change doubles also. The constant K_I expresses the scaling between error and controller output. Thus, a large value of K_I means that a *small* error produces a *large* rate of change of P and vice versa. Figure 8.14a graphically illustrates the relationship between the P rate of change and error for two different values of K_I. Figure 8.14b shows how, for a fixed error, the different K_I values produce different values of P as a function of time as predicted by Equation (8-15). Thus, we see that the *faster* rate provided by K_I causes a *much greater* controller output at a particular time after the error is generated.

EXAMPLE 8.7

An integral controller is used for level control with a setpoint of 12 meters within a range of 10–15 m. The controller output is 22% initially. The constant $K_I = 0.15\%$ controller output per second per percentage error. If the level jumps to 13.5 m, calculate the controller output after 2 seconds for a constant E_p.

SOLUTION

We find E_p from

$$E_p = \frac{C_m - C_{sp}}{C_{max} - C_{min}} \times 100$$

$$E_p = \frac{13.5 - 12}{15 - 10} \times 100 \tag{8-3}$$

$$E_p = 30\%$$

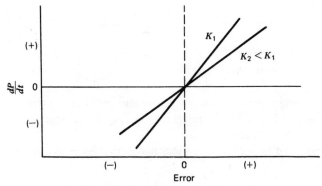

a The rate of output change depends upon gain and error

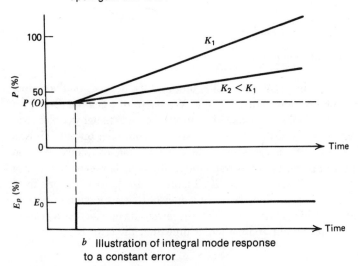

b Illustration of integral mode response to a constant error

FIGURE 8.14   Integral controller mode action.

The rate of controller output change is then

$$\frac{dP}{dt} = K_I E_p = (0.15s^{-1})(30\%)$$

$$\frac{dP}{dt} = 4.5\%/s$$

(8-15)

The controller output for constant error will be

$$P = K_I \int_0^t E_p dt + P(0)$$

$$P = K_I E_p t + P(0)$$

(8-16)

After 2 seconds we have

$$P = (0.15)(30\%)(2) + 22$$
$$P = \textbf{31\%}$$

Note that the integral controller constant K_I may be expressed in percentage change per *minute* per percentage error, whenever a typical process-control loop has characteristic response times in minutes rather than seconds. Thus, an integral mode controller with reset action set at 5.7 minutes means that K_I for our equations would be

$$K_I = \frac{1}{(5.7 \text{ min})(60 \text{ s/min})}$$

$$K_I = 2.92 \times 10^{-3} \text{ s}^{-1}$$

APPLICATIONS

Use of the integral mode is shown by the flow control system in Figure 8.9, except that we now assume that the controller operates in the *integral* mode. Operation can be understood using Figure 8.15. Here a load change induced error occurs at some time t_1. The proper valve position under the new load to maintain the constant flow rate is shown as a dashed line in the P graph of Figure 8.15. In the

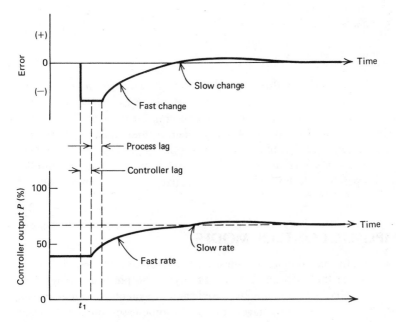

FIGURE 8.15 Illustration of integral mode controller output and error showing the effect of process and control lag.

integral mode, the value initially begins to change very rapidly as predicted by Equation (8-15). As the valve opens, the error decreases and slows the valve opening rate as shown. The ultimate effect is that the system drives the error to zero at a slowing controller rate. The effect of process and control system lag is shown as simple delays in the controller output change and in the error reduction when the controller action occurs. If the process lags are too large, the error can oscillate about zero or even be cyclic. Typically, the integral mode is not used alone but can be for systems with small process lags and correspondingly small capacities.

8.5.3 Derivative Control Mode

The last *pure* mode of controller operation provides that the controller output depends on the rate of change of *error*. This mode is also known as *rate* or *anticipatory* control. The mode *cannot* be used alone because when the error is *zero or constant,* the controller has either no output or the nominal output for zero error. The analytic expression is

$$P = K_D \frac{dE_p}{dt} + P_0 \tag{8-17}$$

Where

K_D = derivative gain constant ($\% - s/\%$).

$\dfrac{dE_p}{dt}$ = rate of change of error ($\%/s$).

P_0 = output with no error rate

The derivative gain constant also is called the rate of derivative time and commonly is expressed in minutes. The characteristics of this device can be noted from the graph of Figure 8.16 which shows controller output for the rate of change of error. This shows that, for a given rate of change of error, there is a unique value of controller output. The time plot of error and controller response further shows the behavior of this mode as shown in Figure 8.17. Notice that the extent of controller output depends upon the rate at which this error is changed and *not* on the value of the error.

8.6 COMPOSITE CONTROL MODES

It is very common (in the complex of industrial processes) to find control requirements that do *not* fit the application norms of any of the previously considered controller modes. It has been found both possible and expedient to *combine* several basic modes, thereby gaining advantages of each. In some cases, an added advantage is that the modes tend to eliminate some limitations they individually possess.

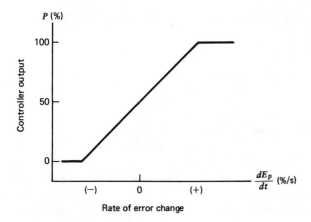

FIGURE 8.16   Derivative mode of controller action where an output of 50% has been assumed for the zero derivative state.

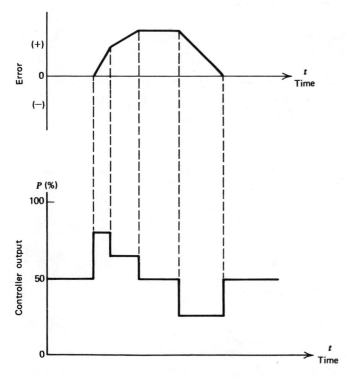

FIGURE 8.17   Derivative mode of controller action for a sample error signal.

We will consider only those combinations that are commonly used and discuss the merits of each.

8.6.1 Proportional-Integral Control (PI)

This is a control mode that results from a combination of the proportional mode and the integral mode. The analytic expression for this control process is found from an additive combination of Equations (8-14) and (8-15)

$$P = K_P E_p + K_P K_I \int E_p dt + P(0) \tag{8-18}$$

Where all the terms have been defined previously.

The main advantage of this composite control mode is that the one-to-one correspondence of the proportional mode is available and the integral mode eliminates the inherent offset. Notice that the proportional gain, by design, also changes the net integration mode gain, but that the integration gain, through K_I, can be independently adjusted. Recall that the proportional mode offset occurred when a load change required a new nominal controller output and could not be provided except by a fixed error from the setpoint. In the present mode, the integral function provides the required new controller output, thereby allowing the error to be zero after a load change. The integral feature effectively provides the **reset** of operating point when a load change occurs. This can be seen by the graphs of Figure 8.18. At

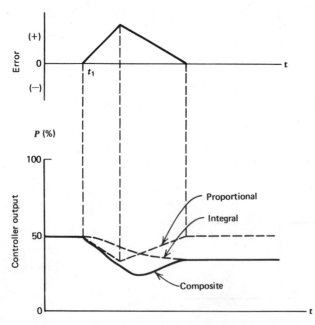

FIGURE 8.18 Proportional integral action showing the reset action of the integral contribution (assumes K_P is negative).

time t_1 a load change occurs which produces the error shown. Accommodation of the new load condition requires a new controller output. We see that the controller output is provided through a sum of proportional plus integral action, which finally leaves the error at zero. Notice that the proportional part is obviously just an image of the error.

APPLICATION

As noted, this composite proportional-integral mode eliminates the offset problem of proportional controllers. It follows that the mode can be used in systems with large load changes. Because of the integration time, however, the process must have relatively slow changes in load to prevent oscillations induced by the integral over-shoot. Another disadvantage of this system is that during startup of a batch process, the integral action causes a considerable overshoot of the error before settling to the operation point. This is shown in Figure 8.19, where we see the proportional band

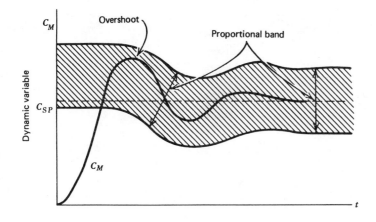

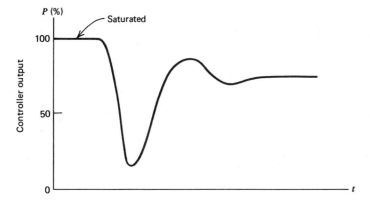

FIGURE 8.19   Overshoot and cycling when the proportional-integral mode is used in the startup of batch processes.

as a shaded region. Note that the effect of the integral action can be viewed as a *shifting* of the whole proportional band. The integral drives the system to a saturated state which cannot be corrected until considerable overshoot takes place.

EXAMPLE 8.8

Given the error of Figure 8.20*a*, plot a graph of a proportional-integral controller output as a function of time. $K_P = 5$, $K_I = 1.0$ s^{-1}, and $P(0) = 20\%$.

SOLUTION

We find the solution by an application of

$$P = K_P E_p + K_P K_I \int E_p dt + P(0) \tag{8-18}$$

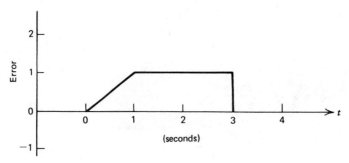

a Error signal

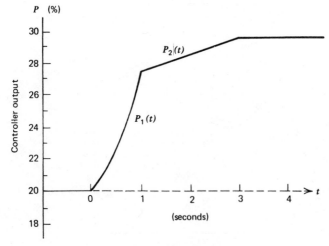

b Controiler output

FIGURE 8.20   Figure for Example 8.8.

to the spans from 0–1 s and 1–3 s. In the first span, we have $E_p = at$ with $a = 1\%/s$. Then,

$$P_1(t) = K_P at + \frac{1}{2} K_I K_P a t^2 + P(0)$$

or

$$P_1(t) = 5t + \frac{5}{2} t^2 + 20$$

which is plotted in Figure 8.20b from 0–1 s. Now, from 1–3 s, $E_p = 1$ and we have

$$P_2(t) = K_P + K_P K_I (t - 1) + P(0) + 2.5$$

or

$$P_2(t) = 5 + (t - 1) + 20 + 2.5$$

The resulting controller output is shown in Figure 8.20 from 1–3 s. After 3 s the error is zero and the controller output stays fixed. Note that the integral action has allowed a shift in controller output to provide zero error.

8.6.2 Proportional-Derivative Control Mode (PD)

A second combination of control modes finds many industrial applications. It involves the serial (cascaded) use of the proportional and derivative modes. The analytic expression for this mode is found from a combination of Equations (8-16) and (8-17)

$$P = K_P E_p + K_P K_D \frac{dE_p}{dt} + P_0 \qquad (8\text{-}19)$$

where the terms are all defined in terms given by previous equations.

It is clear that this system cannot eliminate the offset of proportional controllers. It can, however, handle fast process load changes as long as the *load change offset error* is acceptable. An example of the operation of this mode for a hypothetical load change is shown in Figure 8.21. Note the effect of derivative action in moving the controller output in relation to the error rate change.

EXAMPLE 8.9

Suppose the error, Figure 8.22a, is applied to a derivative-proportional controller with $K_P = 5$, $K_D = 0.5$ s, and $P_0 = 20\%$. Draw a graph of the controller output that results.

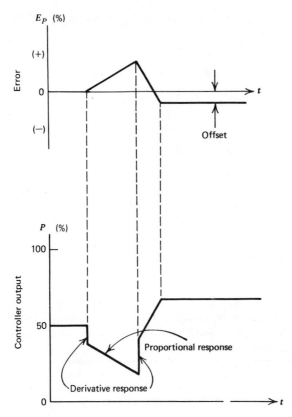

FIGURE 8.21 Proportional-derivative action showing the offset error from the proportional mode (negative K_P).

SOLUTION

In this case we evaluate

$$P = K_P E_p + K_D K_P \frac{dE_p}{dt} + P_0 \qquad (8\text{-}19)$$

over the two spans of the error. In the time of 0–1 s where $E_p = at$, we have

$$P_1 = K_P at + K_D K_P a + P_0$$

or since a = 1%/s

$$P_1 = 5t + 2.5 + 20$$

Note the instantaneous change of 2.5% produced by this error. In the span from 1–3 s we have

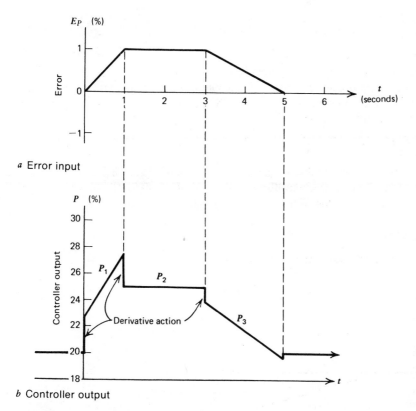

a Error input

b Controller output

FIGURE 8.22   Figure for Example 8.9.

$$P_2 = 5 + 20$$

The span from 3–5 s has an error of $E_p = -0.5t + 2.5$ so that we get for 3–5 s

$$P_3 = -2.5t + 12.5 - 1.25 + 20$$

or

$$P_3 = -2.5t + 31.25$$

This controlled output is plotted in Figure 8.22*b*.

8.6.3 Three-Mode Controller (PID)

One of the most powerful but complex controller mode operations combines the proportional, integral, and derivative modes. This system can be used for virtually *any* process condition. The analytic expression is

$$P = K_P E_p + K_P K_I \int E_p dt + K_P K_D \frac{dE_p}{dt} + P(0) \qquad (8\text{-}20)$$

where all terms have been defined earlier.

This mode eliminates the offset of the proportional mode and depresses the tendencies toward oscillations. In Figure 8.23, the response of the three-mode system to an error are shown.

EXAMPLE 8.10

Let us combine everything and see how the error of Figure 8.22a produces an output in the three-mode controller with $K_P = 5$, $K_I = 0.7$ s^{-1}, $K_D = 0.5$ s, and $P(0) = 20\%$. Draw a plot of the controller output.

SOLUTION

In Figure 8.22a the error can be expressed as follows

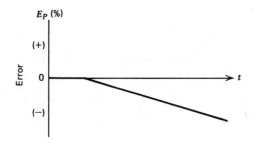

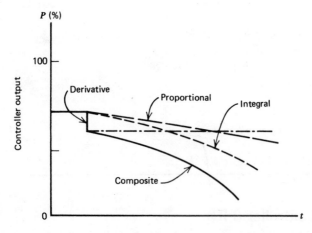

FIGURE 8.23   The three-mode controller action exhibits proportional, integral, and derivative action.

$$0\text{-}1 \text{ s} \quad E_p = t\%$$
$$1\text{-}3 \text{ s} \quad E_p = 1\%$$

$$3\text{-}5 \text{ s} \quad E_p = -\frac{1}{2}t + 2.5\%$$

We must apply each of these spans to the three-mode equation for controller output

$$P = K_P E_p + K_P K_I \int E_p dt + K_P K_D \frac{dE_p}{dt} + P(0) \qquad (8\text{-}20)$$

or

$$P = 5E_p + 3.5 \int E_p dt + 2.5 \frac{dE_p}{dt} + 20$$

From 0-1 s, we have

$$P_1 = 5t + 3.5 \int_0^t t \, dt + 2.5 + 20$$

or

$$P_1 = 5t + 1.75t^2 + 22.5$$

This is plotted in Figure 8.24 in the span of 0-1 s. Now from 1-3 s, we have

$$P_2 = 5 + 1.75 + 3.5 \int_1^t (1) dt + 20$$

or

$$P_2 = 3.5(t - 1) + 26.75$$

This controller variation is shown in Figure 8.24 from 1-3 s. Finally, from 3-5 s, we have

$$P_3 = 5\left(-\frac{1}{2}t + 2.5\right) + 3.5 \int_3^t \left(-\frac{1}{2}t + 2.5\right) dt - \frac{2.5}{2} + 28.75$$

or

$$P_3 = -0.875t^2 + 6.25t + 21.625$$

This is plotted in Figure 8.24 from 3-5 s.

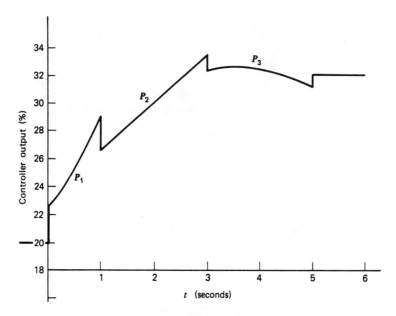

FIGURE 8.24   Figure for Example 8.10.

8.6.4 Special Terminology

There are a number of special terms used in process control for discussing the controller modes. The following summary defines some of these and shows how these terms relate to the equations presented in this chapter.

1. *Proportional Band (PB).* Although this term was defined earlier, let us note again that this is the percentage error which results in a 100% change in controller output. Thus, the PB is found from the proportional gain by the relation, PB = $100/K_P$.

2. *Repeats per minute.* This term is another expression of the integral gain for PI and PID controller modes. The term derives from the observation that the integral gain, K_I, has the effect of causing the controller output to change every unit time by the proportional mode amount. You also can see this by taking the derivative of the integral term in the controller equation. This gives a change in controller output, ΔP, *of*

$$\Delta P = K_I K_P E_p \Delta t$$

Since $K_P E_p$ is just the proportional contribution, in a unit time interval, $\Delta t = 1$, K_I just repeats the proportional term. For example, if $E_P = 0.5\%$ and $K_P = 10\%$, then $K_P E_p = 5\%$. Now, if $K_I = 10\%/(\%\text{-min})$, then every minute the output would increase by 5% times $10\%/(\%\text{-min})$ or 50% or "10 repeats per minute." It repeats the proportional amount ten times per minute.

3. *Rate Gain.* This is just another way of saying the derivative gain, K_D. Since K_D has the units of %-s/% (or %-min/%), one often expresses the gain as time directly. Thus, a *rate gain* of 0.05 min or a *derivative time* of 0.05 min both mean $K_D = 0.05$ %-min/%.

4. *Direct/Reverse Action.* This specifies whether the controller output should increase

(direct) or decrease (reverse) for a positive error. The action is specified by the sign of the proportional gain; $K_P > 0$ is direct and $K_P < 0$ is reverse.

SUMMARY

This chapter covers the general characteristics of controller operating modes without consideration of implementation of these functions. Numerous terms which are important to an understanding of controller operations are defined. The following highlighted items are:

1.   In considering controller operating modes it is important to know the *process load*, which is the nominal value of all process parameters, and the *process lag*, which represents a delay in reaction of the controller variable to a change of load variable.

2.   Some processes exhibit *self-regulation*, that is, the characteristic that a dynamic variable adopts some nominal value commensurate with the load, with no control action.

3.   The controller operation is defined through a relationship between percentage *error* or *deviation* relative to full scale

$$E_p = \left[\frac{C_m - C_{sp}}{C_{max} - C_{min}} \right] \cdot 100 \tag{8-3}$$

and the controller output as a percentage of the controlling parameter

$$P = \left[\frac{S_p - S_{min}}{S_{max} - S_{min}} \right] \cdot 100 \tag{8-4}$$

4.   *Control lag* and *dead time* refer to a delay in controller response when a deviation occurs and a period of no response of the process to a change in the controlling variable.

5.   Discontinuous controller modes refer to instances where the controller output does not change smoothly for input error. Examples are two-position, multiposition, and floating.

6.   Continuous controller modes are modes where the controller output is a smooth function of the error input or rate of change. Examples are proportional, integral, and derivative modes.

7.   Composite controller modes combine the continuous modes. Examples are the proportional-integral (PI), proportional-derivative (PD) and the proportional-integral-derivative (PID) (or three-mode).

PROBLEMS

8.1   A velocity control system has a range of 220–460 mm/s. If the setpoint is 327 mm/s and the measured value is 294 mm/s, calculate the percentage error.

8.2   A controlling parameter is a motor speed that varies from 800 rpm to 1750 rpm. If the speed is controlled by a 25–50 volt d-c signal, calculate (a) the speed produced by an input of 38 volts, and (b) speed calculated as a percentage of the full range speed.

8.3   A 5 m diameter cylindrical tank is emptied by a constant outflow of 1.0 m³/s. A two-position controller is used to open and close a fill valve with an open flow rate of 2.0 m³/s. For level control, the neutral zone is 1 m and the setpoint is 12 m; (a) calculate the cycling period, and (b) graph the level versus time.

8.4   For Example 8.4, verify that a ±2% neutral zone produces the results of limits of oscillation and period given after the example.

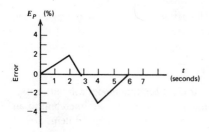

FIGURE 8.25   Figure for Problem 8.10.

8.5   The tank in Problem 8.3 above uses a three-position controller with rates of 2.0 m³/s at levels less than 11 m, 0.5 m³/s for levels between 11 m and 12 m, and 0 m³/s for levels greater than 12 m. Find the cycling period. If the lag is 5 s plot the level starting at 13 m.

8.6   For a proportional controller, the controlled variable is a process temperature with a range of 50–130°C and a setpoint of 73.5°C for which the controller output is 50%. Find the proportional offset that results from a change to a 55% controller output to maintain the setpoint if the proportional gain is (a) 0.1, (b) 0.7, (c) 2.0, (d) 5.0.

8.7   For the applications of Problem 8.6, find the percentage controller output with the 73.5°C setpoint and a proportional gain of 2 produced by (a) 61°C, (b) 122°C, (c) (82 + 5t)°C?

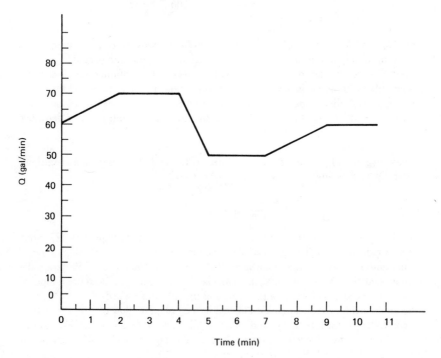

FIGURE 8.26   Figure for Problem 8.16.

8.8   An integral controller has a reset action of 2.2 minutes. Find the integral controller constant in s^{-1}. Find the output of this controller to a constant error of 2.2%.

8.9   How would a derivative controller with $K_D = 44s$ respond to an error changing as $E_p = 2.2 \sin (0.04t)$?

8.10   A proportional controller has a gain $K_P = 2$. Plot the response of the error of Figure 8.25 if $P(0) = 50\%$.

8.11   A proportional-integral controller has constants of $K_P = 2$, $K_I = 2.2s^{-1}$, $P(0) = 40\%$. Plot the response to the error of Figure 8.25.

8.12   A proportional-derivative controller has constants of $K_P = 2$, $K_D = 2s$, $P_0 = 40\%$. Plot the response to the error of Figure 8.25.

8.13   A three-mode controller has constants of $K_P = 2$, $K_D = 2s$, $K_I = 2.2s^{-1}$, and $P(0) = 40\%$. If the error is as given in Figure 8.25, plot the controller output.

8.14   A PI controller has settings of reverse action, PB = 20, 12 repeats per minute. Find (a) proportional gain, (b) integral gain, (c) the time that the controller output will reach 0% after an error of $+1.5\%$ occurs and remains. The controller output when the error occurred was 72%.

8.15   Suppose rate action was added to the controller of Problem 8.14 with a rate gain of 0.1 minutes. Specify the derivative gain and determine the time at which the controller output reaches 0% with this added mode.

8.16   A PI controller is used to control flow with a range of 20 gal/min to 100 gal/min. The setpoint is 60 gal/min. The controller output drives a valve with 3 to 15 psi for the 0 to 100% output. The controller settings are Direct, $K_P = 0.9\%/\%$, $K_I = 0.4\%/(\%\text{-min})$. Plot the pneumatic pressure at the valve for the flow of Figure 8.26. The pressure was 10.8 psi when the time was zero.

9

ANALOG CONTROLLERS

INSTRUCTIONAL OBJECTIVES

The objectives of this chapter provide a more comprehensive understanding of the principles of mode implementation. The objectives below have been chosen from this point of view. After a comprehensive study of this chapter, the reader should be able to:

1. Recognize the essential elements of an analog controller.
2. Diagram and describe the implementation of two-position, proportional, and integral control modes, using op amps.
3. Diagram and describe the implementation of proportional-integral, proportional-derivative, and three-mode controllers using op amps; diagram and describe the operation of a three-mode pneumatic controller.
4. Design the basic elements of a process-control loop using electronic analog techniques.

9.1 INTRODUCTION

In the previous chapter we saw the defining principles of various controller modes. The selection of the mode to use and appropriate gains depends on many factors involved in the process operation. This decision is made by engineers who are familiar with the process itself and who are aided by process-control technique experts who understand the characteristics of each mode. This chapter will study how modes of controller action are realized using *analog* techniques. The emphasis is on *electronic* techniques, using op amps as the active element because of their widespread use. *Pneumatic* techniques also are discussed because there are many operations where a complete implementation of a process-control loop uses pneumatic methods.

Digital methods of providing controller modes are considered in Chapter 10. Note that the study of analog electronic implementation of controller modes can be considered an application of analog computer techniques because we are, in effect, looking for a solution to controller mode equations.

Specific methods of controller mode realization, using either electronics or pneumatics, are as varied as the manufacturers of this equipment. A detailed analysis of a specific method is limited. For this reason, the material is presented in a general fashion using op amps in electronics and general principles in pneumatics. If a particular manufacturer chooses to use special discrete circuits, for example, then a one-to-one correspondence with op-amp methods will help in understanding the circuit. Because many controllers are now using integrated circuit (IC) op amps, there is a distinct advantage to a specific study of this method.

9.2 GENERAL FEATURES

An analog controller is a device which implements the controller modes described in Chapter 8, using analog signals to represent the loop parameters. The analog signal may be in the form of an electric current or a pneumatic air pressure. The controller accepts a measurement expressed in terms of one of these signals, calculates an output for the mode being used, and outputs an analog signal of the same type. Since the controller does solve equations, we think of it as an analog computer. The controller must be able to add, subtract, multiply, integrate, and find derivatives. It does this working with analog voltages or pressures. In this section we will examine the general physical layout of typical analog controllers.

TYPICAL PHYSICAL LAYOUT

Analog controllers are usually designed to fit into a panel assembly as a slide in/out module, as shown in Figures 9.1. The front displays all necessary information and adjustment capability for the operator. When the unit is pulled out part way, but still connected, other, less frequently required adjustments are available. When the controller is pulled further out, an extension cable can be disconnected, and the entire unit removed from the panel for replacement if necessary.

FRONT PANEL

The front panel of an analog controller displays information for operators and allows adjustment of the setpoint. Figure 9.1 shows a typical front panel. Note that the setpoint knob moves a sliding scale under the fixed setpoint indicator. Thus, a fixed span of measurement above and below the setpoint is visible as indicated by the measurement meter indicator. The error is the difference between the setpoint indicator and the measurement meter. The display is typically expressed in percent-

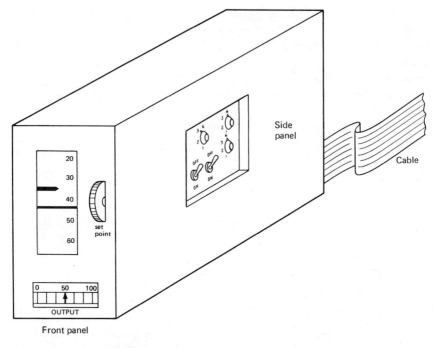

Front panel

FIGURE 9.1 Typical controller layout.

age of span (4–20 mA or 3–15 psi). The lower meter shows the controller output, again expressed in percentage of span. Of course, the output is actually 4–20 mA or 3–15 psi so that 0% would mean 4 mA, for example. There often is a switch on the front panel by which the controller can be placed in a *manual* control, which means that the output can be adjusted independently of the input, using the output adjust knob. In *automatic* mode, this knob has no effect on the output. Connections to the controller are made through electrical or pneumatic cables connected to the rear of the unit.

SIDE PANEL

On the side of the controller, when partially pulled out, knobs are available to adjust operation of the controller modes. On this panel, as shown in Figure 9.1, the proportional, integral (reset), and derivative (rate) gains can be adjusted. In addition, filtering action and reverse/direct operation can often be selected.

9.3 ELECTRONIC CONTROLLERS

In the following treatment of electronic methods of realizing controller modes, emphasis is on the use of op amps as the circuit element. Discrete electronic components are also used to implement this function, but the basic principles are best illustrated using op amp circuits. Note that op-amp circuits other than the ones described can also be developed.

9.3.1 Error Detector

The detection of an error signal is accomplished in electronic controllers by taking the difference between voltages. One voltage is generated by the process signal current passed through a resistor. The second voltage represents the setpoint. This is usually generated by a voltage divider using a constant voltage as a source. An example is shown in Figure 9.2. Here, we assume a two-wire system is in use so that the current drawn from the power supply is the 4 to 20 mA signal current. The signal current is used to generate a voltage, IR, across the resistor R. This is placed in series opposition to a voltage V_{sp} tapped from a variable resistor R_{sp} connected to a constant source V_0. The result is an error voltage $V_E = V_{sp} - IR$. This is then used in the process controller to calculate controller output. Since this voltage is defined opposite to the definition of error used before, we may use an inverting unity gain amplifier to reverse the equation if desired.

9.3.2 Single Mode

The following op-amp circuits illustrate methods of implementing the pure modes of controller action with op-amp circuits.

TWO-POSITION

A two-position controller can be implemented by a great variety of electronic and electromechanical designs. Many household air conditioning and heating systems employ a two-position controller constructed from a bimetal strip and mercury switch as shown in Figure 9.3. Here we see that as the bimetal strip bends due to a temperature decrease, it reaches a point at which the mercury slides down to close

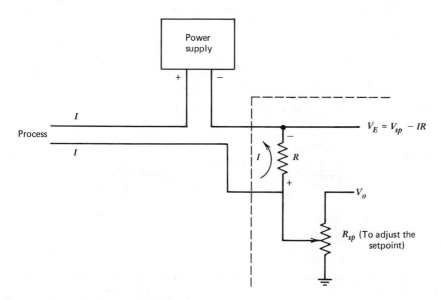

FIGURE 9.2   Typical electronic error detector.

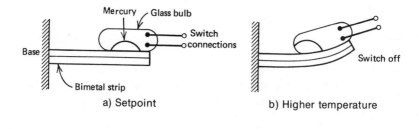

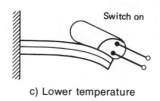

c) Lower temperature

FIGURE 9.3   A mercury switch on a bimetal strip is often used as a two-position temperature controller.

an electrical contact. The inertia of the mercury tends to keep the system in that position until the temperature increases to a value above the setpoint temperature. This provides the required neutral zone to prevent excessive cycling of the system.

A method using op-amp implementation of ON-OFF control with adjustable neutral zone is given in Figure 9.4. Here the controller input signal is assumed to be a voltage level with an "ON" voltage of V_H and an "OFF" voltage V_L, and the

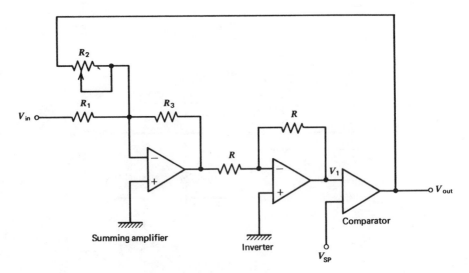

FIGURE 9.4   A two-position controller with neutral zone constructed from op amps and a comparator.

output is the comparator output of zero or V_{out}. The comparator output switches states when the voltage on its input V_1 is equal to the setpoint value V_{sp}. Analysis of this circuit shows that the high (ON) switch voltage is

$$V_H = \left(\frac{R_1}{R_3}\right) V_{sp} \qquad (9\text{-}1)$$

whereas the low (OFF) switching voltage is

$$V_L = \frac{R_1}{R_3}\left[V_{sp} - \left(\frac{R_3}{R_2}\right) V_{out}\right] \qquad (9\text{-}2)$$

Figure 9.5 shows the typical two-position relationship between input and output voltage for this circuit. The width of the neutral zone between V_L and V_H can be adjusted by variation of R_2. The relative location of the neutral zone is made by variation of the setpoint voltage, V_{sp}. The neutral zone is calculated from the difference between Equations (9-1) and (9-2).

EXAMPLE 9.1

Let us design a two-position controller having the circuit of Figure 9.4. The input range is 0–2 volts, the output is 0 or 5 volts, the ON voltage is 1 volt, and the OFF voltage is 0.5 volts.

SOLUTION

We have two requirements as given by Equations (9-1) and (9-2). Because we have four unknowns and only two equations, two of the unknowns may be selected arbitrarily. Let us take $V_{sp} = 5$ volts and $R_3 = 10$ kΩ, then

$$V_H = \left(\frac{R_1}{R_e}\right) V_{sp} \qquad (9\text{-}1)$$

gives

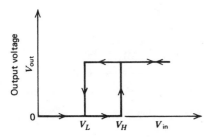

FIGURE 9.5   The circuit of Figure 9.4 shows the characteristic two-position response.

$$R_1 = R_3 \left(\frac{V_H}{V_{sp}}\right) = (10 \text{ k}\Omega) \left(\frac{1}{5}\right)$$

$$R_1 = 2 \text{ k}\Omega$$

And then from

$$V_L = \frac{R_1}{R_3} \left[V_{sp} - \left(\frac{R_3}{R_2}\right) V_{out}\right] \tag{9-2}$$

we get

$$R_2 = R_3 \left[\frac{V_{sp} - \dfrac{R_3}{R_1} V_L}{V_{out}}\right]^{-1}$$

Substituting above values yields

$$R_2 = (10 \text{ k}\Omega) \left[\frac{5 - \dfrac{10 \text{ k}\Omega}{2 \text{ k}\Omega}(0.5)}{5}\right]^{-1}$$

$$R_2 = 20 \text{ k}\Omega$$

The inverter resistors (R in Figure 9.4) can be made any convenient value, say $R = 5 \text{ k}\Omega$. Multiposition controllers can be devised by a similar process of op-amp circuit development. An example for a three-position controller is given in the problems at the end of the chapter.

FLOATING

The floating-type controller can be generated by connecting the output of a three-position controller into an integrator. Such a circuit is shown in Figure 9.6. Here we assume the three-position controller was designed to provide outputs of V_1, zero,

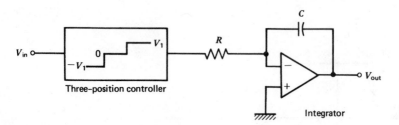

FIGURE 9.6 Construction of a floating controller from a three-position controller and an integrator.

or $-V_1$, depending on input. As an input to the integrator this produces possible outputs of

$$V_{out} = \begin{cases} -\dfrac{V_1}{RC}(t - t_a) + V_a & V_{in} < V_{s1} \\ V_b & V_{s1} < V_{in} < V_{s2} \\ \dfrac{V_1}{RC}(t - t_c) + V_c & V_{in} > V_{s2} \end{cases} \qquad (9\text{-}3)$$

Where

V_{s1} = lower setpoint voltage
V_{s2} = upper setpoint voltage
V_a, V_b, V_c = values of output when the input condition occurs
t_a, t_b, tc = times at which input reaches the setpoints

The *trip* voltage can be set to provide the desired band of inputs producing no output, a positive rate, or negative rate. The actual rate of output change depends on the values of resistor and capacitor in the integrator and the output level of the three-position circuit preceding the integrator. Remember that the output floats at whatever the latest value of output is when the input falls within the neutral zone.

EXAMPLE 9.2

A control signal varies from 0–5 volts. A floating controller, such as that in Figure 9.6, has trip voltages of 2 volts and 4 volts, and a three-position controller has outputs of 0 and ± 2 volts. The integrator consists of a 1 MΩ resistor and a 1 μF capacitor. Plot the controller output in response to the input of Figure 9.7a.

SOLUTION

We can find the output by applying the relations of Equation (9-3) to the input voltage. The result is shown in Figure 9.7b. This is arrived at as follows:

(1) From 0–1 second, the output is zero (an assumed starting point) and remains so because the input is within the neutral zone.

(2) From 1–3 seconds, the lower setpoint has been reached and the output is given by

$$V_{out} = -\frac{V_1}{RC}(t - 1) = -2(t - 1) \text{ V}$$

at $t = 3$ s the output is -4 V.

(3) From 3–4 seconds, the output remains at -4 V because the input is in the neutral zone.

(4) From 4–7 seconds, the input reaches the upper setpoint and the output becomes

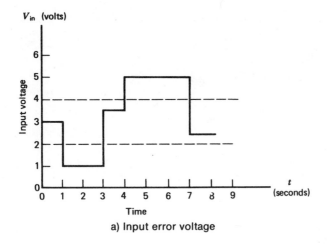

a) Input error voltage

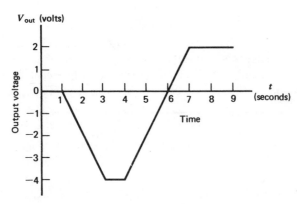

b) Output voltage

FIGURE 9.7 Input and output voltage for the floating controller of Example 9.2.

$$V_{out} = + \frac{V_1}{RC}(t - 4) - 4 \text{ V}$$

$$V_{out} = + 2(t - 4) - 4 \text{ V}$$

At $t = 7$ seconds, the input again falls within the neutral zone and the output becomes

$$V_{out} = +2$$

(5) The output will remain at +2 until the input again hits a setpoint value.

PROPORTIONAL MODE

Implementation of this mode requires a circuit which has a response given by

$$P = K_P E_p + P_o \tag{8-14}$$

Where

P = controller output 0–100%
K_P = proportional gain
E_p = error in percent of variable range
P_o = controller output with no error

If we consider both the controller output and error to be expressed in terms of voltage, we see that Equation (8-14) is simply a *summing amplifier*. The op-amp circuit in Figure 9.8 shows such an electronic proportional controller. In this case, the analog electronic equation for the output voltage is

$$V_{out} = \left(\frac{R_2}{R_1}\right) V_E + V_0 \tag{9-4}$$

Where

V_{out} = output voltage
$K = R_2/R_1$ = gain
V_E = error voltage
V_0 = output with zero error

In practice, Equation (9-4) is an exact analog of Equation (8-14) and implements the proportional controller. The amplifier input and output voltages are conveniently scaled such that a 0–V_{max} amplifier output corresponds to the 0–100% or 4–20 mA controller signal. In a similar fashion, the input error voltage is scaled to match the full range of error signal. The proportional band is adjusted through the gain $= R_2/R_1$ so that the given band of error saturates the amplifier output. When a

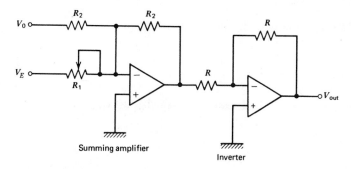

FIGURE 9.8   An electronic proportional controller.

load change occurs, the offset necessary to support a new controller output (without a manual reset through V_0 adjustment) is found from Equation (9-4).

EXAMPLE 9.3

A controller shown in Figure 9.8 with scaling so that 0–10 volts corresponds to a 0–100% output. If $R_2 = 10$ kΩ, and full-scale error is 10 volts, find the values of V_0 and R_1 to support a 20% proportional band about a 50% zero controller output.

SOLUTION

Now, the value of V_0 is simply 50% of 10 volts or 5 *volts* to provide the zero error controller output. To design for a 20% proportional band means that an increase of error by 10% must cause the controller output to go to 100% from 50%. Thus, from

$$V_{out} = KV_E + V_0$$

we note that when the error has changed 10% of 10 volts or 1 volt, we must have full controller output. Thus,

$$K = \frac{V_{out} - V_0}{V_E} = \frac{10 - 5}{1}$$

$$K = 5$$

so that if $R_2 = 10$ kΩ, then

$$R_1 = R_2/K = 2 \text{ k}\Omega$$

EXAMPLE 9.4

If the load in the previous problem changes such that a new controller output of 40% is required, find the corresponding offset error.

SOLUTION

In this case we need a negative error so that the output is 40% of 10 volts = 4 volts.

$$V_{out} = KV_E + V_0$$
$$4 = 5V_E + 5$$

(9-4)

$\therefore V_E = -\frac{1}{5}$ volts and because the full-scale error signal is 10 volts, we have an error of:

$$\frac{-0.2}{10} \times 100 = -2\%$$

Generally, a voltage to current converter is used on the output to convert the output voltages to a 4–20 mA range of current signals to drive the final control element.

INTEGRAL MODE

In the previous chapter we saw that the integral mode was characterized by an equation of the form

$$P(t) = K_I \int E_p(t)\, dt + P(0) \qquad (8\text{-}16)$$

Where

P(t) = controller output in percent of full scale
K_I = integration gain (s^{-1})
$E_p(t)$ = deviations in percent of full scale variable value
P(0) = controller output at $t = 0$

This function is easy to implement when op amps are used as the building blocks. A diagram of an integral controller is shown in Figure 9.9. The corresponding equation relating input to output is

$$V_{out} = K_I \int V_E dt + V_0 \qquad (9\text{-}5)$$

Where

V_{out} = output voltage
K_I = 1/RC = integration gain
V_E = error voltage
V_0 = initial output voltage

The values of R and C can be adjusted to obtain the desired integration time. The initial controller output V_0 is relatively unimportant because the integration output

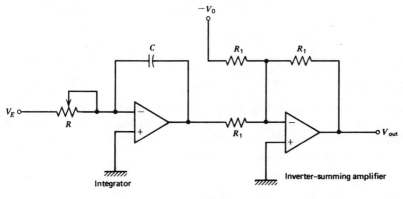

FIGURE 9.9   Electronic integral-mode controller.

floats at values determined by the error history. As we noted earlier the integration time constant determines the *rate* at which controller output increases when the error is constant. If K_I is made too large, the output rises so fast that overshoots of the optimum setting occur and cycling is produced.

EXAMPLE 9.5

Suppose that $V_0 = 0$ in Equation (9-5) for an integral controller. Design specifications require that the controller output should saturate 15 seconds after a constant 5% error is applied. The input range is 0–5 volts and the output is 0–10 volts. Calculate (a) integration gain and (b) select values of R and C to provide this.

SOLUTION

(a) An error of 5% is then 0.25 volts of the full-scale variable range. The controller output saturates at 10 volts, so

$$V_{out} = K_I \int V_E dt + V_0 \qquad (9\text{-}5)$$

becomes, with $V_E = 0.25$ V, $V_0 = 0$

$$10 = K_I 0.25t$$

and for $t = 15$ seconds, we solve for K_I

$$K_I = \frac{10}{(0.25)(15)} = 2.67 \text{ s}^{-1}$$

(b) If we assume $C = 50 \ \mu\text{F}$, and

$$K_I = \frac{1}{RC}, \text{ we solve for } R$$

$$R = \frac{1}{(5 \times 10^{-5})(2.67)}$$

$$R = 7.49 \text{ k}\Omega$$

DERIVATIVE MODE

The derivative mode is never used alone because it cannot provide a controller output when the error is zero (see Section 8.3.6). Nevertheless, we show here how it is implemented with op amps for combination with other modes in the next section. The control mode equation was given earlier as:

$$P = K_D \frac{dE_p}{dt} + P_0 \qquad (8\text{-}17)$$

Where

P = controller output in percent of full output
P_0 = no error derivative controller output
K_D = derivative time constant (s)
E_p = error in percent of full-scale range

This function is implemented by op amps in the configuration shown in Figure 9.10. Here, resistance R_1 is added for stability of the circuit against rapidly changing signals. The response of this circuit for slowly varying inputs is

$$V_{out} = K_D \frac{dV_E}{dt} \tag{9-6}$$

Where

V_{out} = output voltage
$K_D = R_2C$ = derivative time in seconds
V_E = error voltage

The value of R_1 is selected so that the circuit will be stable for high frequencies by setting $2\pi f R_1 C \ll 1$ where f is the frequency in Hz.

The circuits of this section show that the pure modes of controller operation are easily constructed from op amps. As stated in Chapter 8, a pure mode is seldom used in process control because of the advantages of composite modes in providing good control. In the next section, implementation of composite modes using op amps is considered.

9.3.3 Composite Controller Modes

In Chapter 8, the combination of several controller modes was found to combine the advantages of each, and in some cases, eliminate disadvantages. Composite

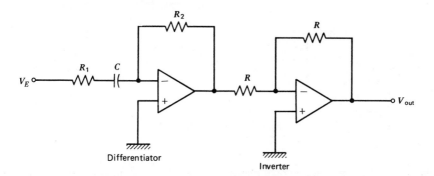

FIGURE 9.10   Electronic derivative-mode controller circuit.

modes are implemented easily using op-amp techniques. Basically, this consists of simply combining the mode circuits introduced in the previous section.

PROPORTIONAL-INTEGRAL

A simple combination of the proportional and integral circuits provides the proportional-integral mode of controller action. The resulting circuit is shown in Figure 9.11. For this case the relation between input and output is most easily found by applying op-amp circuit analysis. We get (including the inverter)

$$V_{out} = + \left(\frac{R_2}{R_1}\right) V_{in} + \frac{1}{R_1 C} \int V_{in} dt$$

Now the definition of the proportional-integral controller includes the proportional gain in the integral term, so we write

$$V_{out} = + \left(\frac{R_2}{R_1}\right) V_{in} + \left(\frac{R_2}{R_1}\right) \frac{1}{R_2 C} \int V_{in} dt \qquad (9\text{-}7)$$

Equation (9-7) has the same form as Equation (8-18) for this mode. The adjustments of this controller are the *proportional band* through $K_P = R_2/R_1$, and the *integration gain* through $K_I = 1/R_2 C$.

EXAMPLE 9.6

Design a proportional-integral controller with a proportional band of 30% and an integration time of 10 seconds. The 4–20 mA input converts to a 0–2 volt error signal and the output is to be 0–10 volts. Calculate values of K_P, K_I, R_2, and R_1, respectively.

SOLUTION

A proportional band of 30% means that when the input changes by 30% of full scale or 0.6 volts, the output must change by 100% or 10 volts. This gives a gain of

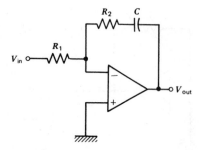

FIGURE 9.11 Electronic proportional-integral controller. Note that an inverter is needed to obtain the response of Equation (9-7).

$$K_P = \frac{R_2}{R_1} = \frac{10 \text{ V}}{0.6 \text{ V}} = 16.67 \text{ V/V}$$

Now, an integration time of 10 seconds implies that

$$K_1 = \frac{1}{R_2 C} = 0.1 \text{ s}^{-1}$$

or

$$R_2 C = 10 \text{ s}$$

As an example of values to accomplish this, we could pick

$$C = 100 \ \mu\text{F which requires}$$

$$R_2 = \frac{10 \text{ s}}{10^{-4}} = 100 \text{ k}\Omega$$

Then, to get the proportional gain, we use

$$R_1 = \frac{100 \text{ k}\Omega}{16.67} = 6 \text{ k}\Omega$$

PROPORTIONAL-DERIVATIVE

A powerful combination of controller modes is the proportional and derivative modes (Section 8.3.2). This combination is implemented using a circuit similar to that shown in Figure 9.12. Analysis shows that this circuit responds according to the equation

$$V_{out} = \left(\frac{R_1}{R_1 + R_3}\right) R_3 C \frac{dV_{out}}{dt} = + \left(\frac{R_2}{R_1 + R_3}\right) V_{in} + \left(\frac{R_2}{R_1 + R_3}\right) R_3 C \frac{dV_{in}}{dt}$$

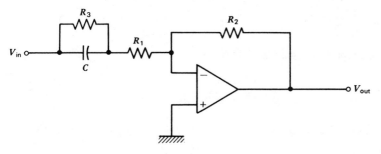

FIGURE 9.12   An electronic proportional-derivative controller. Note that an inverter is required to obtain the response of Equation (9-8).

where the quantities are defined in the figure and the output inverter has been included. We make the derivative coefficient on the left small to eliminate instability. One choice is

$$\frac{R_1}{R_1 + R_3} R_3C = \frac{0.1}{2\pi}T$$

where T is the fastest variable time change to be expected in the process. Then, the equation for the proportional derivative response becomes

$$V_{out} = \left(\frac{R_2}{R_1 + R_3}\right) V_{in} + \left(\frac{R_2}{R_1 + R_3}\right) R_3C \frac{dV_{in}}{dt} \qquad (9\text{-}8)$$

where the proportional gain is $K_P = R_2/(R_1 + R_3)$ and the derivative gain is $K_D = R_3C$. This equation now corresponds to the form given by Equation 8-19 for the proportional-derivative controller. Of course this mode still has the offset error of a proportional controller because the derivative term cannot provide reset action.

EXAMPLE 9.7

A proportional-derivative controller is to have a 20% proportional band and an 18-second derivative time. Input and output are both scaled to 0–10 volts and the fastest expected change time is 1 second. Calculate R_3, K_P, R_1, and R_2, respectively.

SOLUTION

There are, of course, many combinations which can provide this response. Let us first select the capacitor as 50μF because it is easier to assemble odd valued resistors. Then, from $K_D = R_3C = 18$ s, we get

$$R_3 = \frac{18 \text{ s}}{50 \times 10^{-6}\text{F}} = 0.36 \text{ M}\Omega$$

Now, since the input and output share the same scale, we can set

$$K_P = \frac{100\%}{20\%} = 5$$

so that for the proportional gain we have

$$5 = \frac{R_2}{R_1 + 0.36 \text{ M}\Omega}$$

To get R_1 we use

$$\frac{R_1}{R_1 + 0.36 \text{ M}\Omega} (18 \text{ s}) = \frac{(0.1)(1 \text{ s})}{2\pi}$$

which gives $R_1 = 318 \ \Omega$

Then we find R_2 to get the required gain

$$5 = \frac{R_2}{318 \ \Omega + 0.36 \ M\Omega}$$

or $R_2 = 1.8 \ M\Omega$

Putting this all together we get

$$V_{out} = 5 \ V_{in} + 90 \frac{dV_{in}}{dt} \tag{9-8}$$

Recall now that V_{in} is the error voltage.

THREE MODE

The ultimate process controller is that which exhibits proportional, integral, and derivative response to the process error input. In Chapter 8 we saw that this mode was characterized by the equation

$$P = K_P E_p = K_P K_I \int E_p dt + K_P K_D \frac{dE_p}{dt} + P(0) \tag{8-20}$$

Where

P = controller output in percent of full scale
E_p = process error in percent of the maximum
K_P = proportional gain
K_I = integral gain
K_D = derivative gain
$P(0)$ = initial controller output

The zero error term of the proportional mode is not critical because the integral automatically accommodates for offset and nominal setting. This mode can be provided by a straight application of op-amp circuits resulting in the circuit of Figure 9.13. It must be noted, however, that it is possible to reduce the complexity of the circuitry of Figure 9.13 and still realize the three-mode action, but in these cases an interaction results between derivative and integral gains. We will use the circuit of Figure 9.13 because it is easy to follow in illustrating the principles of implementing this mode. Analysis of the circuit shows that the output is

$$V_{out} = \left(\frac{R_2}{R_1}\right) V_{in} + \left(\frac{R_2}{R_1}\right) \frac{1}{R_I C_I} \int V_{in} dt + \left(\frac{R_2}{R_1}\right) R_D C_D \frac{dV_{in}}{dt} \tag{9-9}$$

where R_3 has been chosen from $2\pi R_3 C_D \ll 1$ for stability. Comparison with Equation (8-20) shows that this implements the three-mode controller if

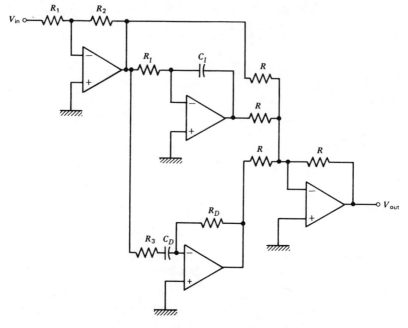

FIGURE 9.13   An electronic three-mode controller. It is possible to implement this mode by other circuits, some with only one op amp.

$$K_P = \frac{R_2}{R_1}, \; K_D = R_D C_D, \; K_I = \frac{1}{R_I C_I}$$

EXAMPLE 9.8

A three-mode controller is to have a 50% proportional band with a 0.2-min integral (reset) time and a 0.5-min derivative time. Find the circuit values for Figure 9.13. Input and output are equally scaled voltages.

SOLUTION

A 50% proportional band implies $K_P = 2$ so that we can pick $R_1 = 1 \text{ k}\Omega$ and $R_2 = 2 \text{ k}\Omega$, for example. The 0.2-minute integration time implies

$$K_I = \frac{1}{12} \, s^{-1}$$

so that if $C_I = 50 \; \mu\text{F}$, then

$$R_I = \frac{12 \, s}{50 \times 10^{-6} \text{F}} = 240 \text{ k}\Omega$$

Finally, for the 0.5-minute derivation time we have 0.5 min = 30 s and thus,

$$R_D C_D = 30 \text{ s or if we use } C_D = 50 \ \mu\text{F},$$
$$R_D = 0.6 \text{ M}\Omega$$

We would pick R_3 for stability.

These circuits have shown that the direct implementation of controller modes can be provided by standard op-amp circuits. It is necessary, of course, to scale the measurement as a voltage within the range of operation selected by the circuit. Furthermore, the outputs of the circuits shown have been voltages which may be converted to currents for use in an actual process-control loop.

Note again that these circuits are only examples of basic circuits which implement the controller modes. Many modifications are employed to provide the controller action with different sets of components.

9.4 PNEUMATIC CONTROLLERS

Historically, the reason for using pneumatics in process control was probably that electronic methods had not been developed to be competitive in cost or reliability. Safety was and still is a factor where the danger of explosion from electrical malfunctions exists. It also is true that the final control element is often pneumatically operated, which suggests that an all pneumatic process-control loop might be advantageous. As of now, it appears that analog or digital electronic methods will eventually replace most pneumatic installations. But we will still have pneumatic equipment for many years until these are depreciated in industry. A good understanding of process-control principles can be applied to either electronic or pneumatic techniques, but it is necessary to consider some special features of pneumatic technology. This section provides a brief description of operations by which controller modes are pneumatically implemented.

9.4.1 General Features

The outward appearance of a pneumatic controller is typically the same as that for the electronic controller shown in Figure 9.1. The same readout of setpoint, error, and controller output appears and adjustments of gain, rate, and reset are available. The working signal is most typically the 3–15 psi standard pneumatic process-control signal, usually derived from a regulated air supply of 20–30 psi. As usual, we use the English system unit of pressure because its use is so widespread in the process-control industry. Eventual conversion to the SI unit of N/m² or pascals will require some alteration in scale (of measurement) to a comparable range.

The pneumatic controller is based on the nozzle/flapper described in Section 7.2.3 as the basic mechanism of operation, much the same as the op amp is used in electronics. The schematic drawings of controller mode implementation are intended to convey the operating principles. The reader is advised that specific designs may vary considerably from the systems shown.

9.4.2 Mode Implementation

In the following discussions, essential features of controller-mode implementation using pneumatic techniques are presented. The equations are stated in general form with units in SI, but the reader should be prepared to work with English units when necessary.

PROPORTIONAL

A proportional mode of operation can be achieved with the system shown in Figure 9.14. Operation is understood by noting that if the input pressure increases, then the input bellows forces the flapper to rotate to close off the nozzle. When this happens, the output pressure increases so that the feedback bellows exerts a force to balance that of the input bellows. A balance condition then occurs when torques exerted by each about the pivot are equal, or

$$(p_{out} - p_0)A_2X_2 = (p_{in} - p_{sp})A_1X_1$$

This equation is solved to find the output pressure

$$p_{out} = \frac{X_1}{X_2}\frac{A_1}{A_2}(p_{in} - p_{sp}) + p_0 \qquad (9\text{-}10)$$

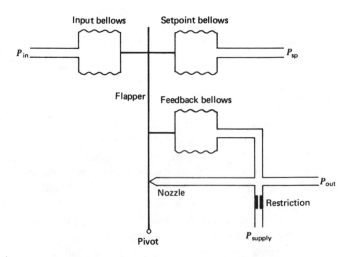

FIGURE 9.14 Pneumatic proportional controller.

Where

p_0 = pressure with no error
p_{in} = input pressure (Pa)
A_1 = input bellows effective area (m^2)
X_1 = lever arm of input (m)
p_{out} = output pressure (Pa)
A_2 = feedback bellow effective area (m^2)
X_2 = feedback lever arm (m)
p_{sp} = setpoint pressure

This relation is based on the notion of torque equaling force time lever arm, and that a pressure in a bellows produces a force which is effectively the pressure times bellows area, much like a diaphragm. Note that Equation (9-10) displays the standard response of a proportional mode in that output is directly proportional to input. The gain in this case is given by

$$K_P = \left(\frac{X_1}{X_2}\right) \left(\frac{A_1}{A_2}\right) \tag{9-11}$$

Because the bellows are usually of fixed geometry, the gain is varied by changing the lever arm length. In this simple representation, the gain is established by the distance between the bellows. If this separation is changed, the forces are no longer balanced and for the same pressure a new controller output will be formed corresponding to the new gain.

EXAMPLE 9.9

Suppose a proportional pneumatic controller has $A_1 = A_2 = 5$ cm^2, $X_1 = 8$ cm and $X_2 = 5$ cm. The input and output pressure ranges are 3–15 psi. Find the input pressures that will drive the output from 3–15 psi. The setpoint pressure is 8 psi and $p_0 = 10$ psi. Find the proportional band.

SOLUTION

First we find the gain from

$$K_P = \left(\frac{X_1}{X_2}\right) \left(\frac{A_1}{A_2}\right) = \left(\frac{8 \text{ cm}}{5 \text{ cm}}\right) \left(\frac{5 \text{ cm}^2}{5 \text{ cm}^2}\right) \tag{9-11}$$

$$K_P = 1.6$$

Now we have

$$p_{out} = K_P (p_{in} - p_{sp}) + p_0$$
$$p_{out} = 1.6 (p_{in} - 8) + 10 \tag{9-10}$$

The low input occurs when $P_{out} = 3$ psi so that

$$3 = 1.6\,(p_L - 8) + 10$$

which gives

$$p_L = 3.625 \text{ psi}$$

The high is found from

$$15 = 1.6\,(p_H - 8) + 10$$

which gives

$$p_H = 11.125 \text{ psi}$$

This means the proportional band (PB) is

$$PB = \left(\frac{11.125 - 3.625}{15 - 3}\right) 100$$

$$PB = 62.5\%$$

Note that this checks with

$$PB = \frac{100}{K_P} = \frac{100}{1.6} = 62.5\%$$

which could be used because the input and output ranges are the same.

PROPORTIONAL-INTEGRAL

This control mode is also implemented using pneumatics by the system shown in Figure 9.15. In this case, an extra bellows with a variable restriction is added to the proportional system as shown. Suppose the input pressure shows a sudden increase. This drives the flapper toward the nozzle increasing output pressure until the proportional bellows balances the input as in the previous case. The integral bellows is still at the original output pressure since the restriction prevents pressure changes from being transmitted immediately. Now as the increased pressure on the output bleeds through the restriction, the integral bellows slowly moves the flapper closer to the nozzle, thereby causing a steady increase in output pressure (as dictated by the integral mode). The variable restriction allows for variation of the *leakage rate* and hence the *integration time*.

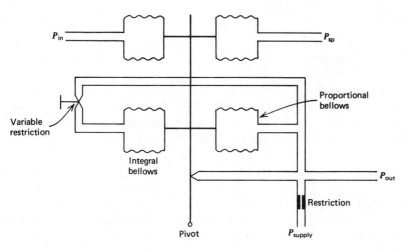

FIGURE 9.15   Pneumatic proportional-integral controller.

PROPORTIONAL-DERIVATIVE

This controller action can be accomplished pneumatically by the method shown in Figure 9.16. Here a variable restriction is placed in the line leading to the balance bellows. Thus, as the input pressure increases, the flapper is moved toward the nozzle with no impedence because the restrictions prevent an immediate response of the balance bellows. Thus, the output pressure rises very fast and then, as the increased pressure leaks into the balance bellows, decreases as the balance bellows moves the flapper back away from the nozzle. Adjustment of the variable restriction allows for changing the derivative time constraint.

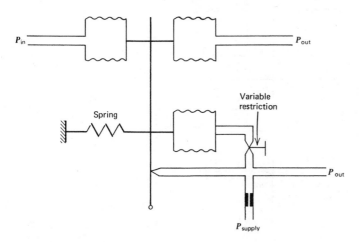

FIGURE 9.16   Pneumatic proportional-derivative controller.

THREE-MODE

The three-mode controller is actually the most common type produced because it can be used to accomplish any of the previous modes by setting of restrictions. This device is shown in Figure 9.17, and, as can be seen, is simply a combination of the three systems presented.

Note that by opening or closing restrictions the three-mode controller can be used to implement the other composite modes. Proportional gain, reset time, and rate are set by adjustment of bellows separation and restriction size.

9.5   DESIGN CONSIDERATIONS

In order to illustrate some of the facets involved in setting up a process-control loop, it would be of value to follow through some hypothetical examples. The following examples assume that a process-control loop is required, and that the controller operation must be provided by electronic analog circuits.

EXAMPLE 9.10

Design a process-control system which regulates light level by outputting a 0–10 volt signal to a lighting system that provides 30–180 lux. The transducer has a transfer function of $-120 \ \Omega$/lux with a 10 kΩ resistance at 100 lux. The setpoint is to be 75 lux, and proportional control with a 75% proportional band has been selected.

SOLUTION

We solve such problems by first establishing the characteristics of each part of the system.

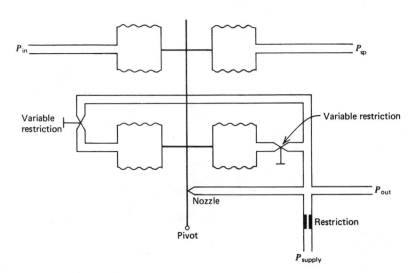

FIGURE 9.17   Pneumatic three-mode controller.

(1) The illumination varies from 30–180 lux. We find the resistance changes according to

$$R = 10 \text{ k}\Omega - 0.12 \text{ k}\Omega \, (I - 100)$$

Where I is the illumination in lux.

(2) This allows us to find the resistance at 30 lux as

$$R = 10 \text{ k}\Omega - 0.12 \, (30 - 100)$$
$$R = 18.4 \text{ k}\Omega$$

and at 180 lux we get **0.4 kΩ**. The setpoint (75 lux) has a resistance of **13 kΩ**.

(3) We can convert this resistance variation to voltage using the photocell in an op-amp circuit. In Figure 9.18 we use an inverting amplifier with a gain of 1 at the setpoint and a constant −1 volt input. The resistance to voltage conversion gives

$$V = - \frac{R}{13 \text{ k}\Omega} (-1 \text{ V})$$

Using this equation, we find the output voltage at 18.4 kΩ to be

$$V = - \frac{18.4 \text{ k}\Omega}{13 \text{ k}\Omega} (-1 \text{ V}) = +1.42 \text{ volts}$$

and at 0.5 kΩ we get

$$V = - \frac{0.5 \text{ k}\Omega}{13 \text{ k}\Omega} (-1 \text{ V}) = 0.038 \text{ volts}$$

(4) Now we use a summing amplifier to find the error as shown in Figure 9.10.

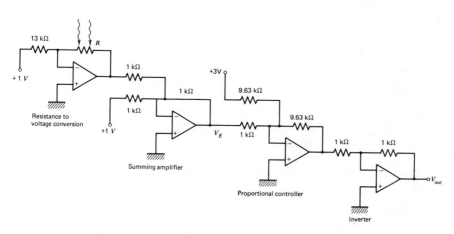

FIGURE 9.18   Circuit for Example 9.10.

$$V_E = \frac{R}{13\ k\Omega} - 1$$

A 75% proportional band controller with a 75 lux setpoint requires a zero error output of

$$V_0 = \frac{75 - 30}{180 - 30}\ 10\ V = 3\ V$$

(5) The 75% band means that when the illumination changes by 75% of $(180 - 30) = 112.5$ lux, the output should swing by 10 volts. Thus, in terms of resistance, this corresponds to 13.5 kΩ and in terms of error voltage it is 13.5 kΩ/13 k or 1.038 volts.

(6) Finally, the gain must be

$$K_P = \frac{10\ V}{1.038} = 9.63$$

This means the overall response is

$$V_{out} = 9.63\ V_E + 3$$

or

$$V_{out} = 9.63 \left(\frac{R}{13\ k\Omega} - 1\right) + 3$$

The rest of the circuit in Figure 9.18 accomplishes this function. Note that when $V_{out} = 0$, $R = 8.9$ kΩ or 90.83 lux and for $V_{out} = 10$ V, $R = 22.4$ kΩ or 203.33 lux so that the output swings 100% as the input swings

$$\frac{203.33 - 90.83}{180 - 30} = 0.75\ \text{or}\ 75\%$$

as required.

The proportional band could not be used to find the gain directly in Example 9.10 because the input and output were not expressed to the same scale, that is, as 0–100% or 0–10 volts, and so on.

EXAMPLE 9.11

A type-J thermocouple (TC) with a 0°C reference is used to control temperature between 100°C and 200°C. Design a proportional-integral controller with a 40%

band and a 0.08-minute reset (integral) time. The final control element requires a 0–10 volt range.

SOLUTION

a. In this problem we must perform the following steps:

(1) Amplify the low TC voltage to a more convenient value than the TC mV output.

(2) Use this amplifier output as input to the proportional-integral controller and pick a proportional gain which swings the output 0–10 V as the input swings 40% of full scale.

(3) Select values to provide a 0.08 min (4.8 s) integral time.

b. The solution is shown in Figure 9.19.

(1) We note that a type-J TC produces a voltage of 5.28 mV at 100°C and 10.77 mV at 200°C. An amplifier with a gain of 100 will convert these to 0.528 and 1.077 volts, respectively.

(2) Now we sum this output to a properly scaled setpoint voltage to get an error signal. The setpoint value is obtained from a voltage divider. To get the proper controller values we note that 40% of the input swing is

$$0.4\,(1.022 - 0.528) = 0.2196\ \text{V}$$

Thus, the proportional gain is

$$K_P = \frac{10\ \text{V}}{0.2196\ \text{V}} = 45.54$$

So values are to be chosen to provide this gain.

(3) The integral time is found from Figure 9.11 and Equation (9-7) as

$$K_1 = \frac{1}{R_2C} = \frac{1}{4.8\ \text{s}}$$

so

$$R_2C = 4.8\ \text{s}$$

Let us nominally pick $C = 10\ \mu\text{F}$ and $R_2 = 480\ \text{k}\Omega$. Then, to get the correct proportional gain, we need

$$R_1 = \frac{480\ \text{k}\Omega}{45.54} = 10.54\ \text{k}\Omega.$$

The overall transfer function for the final circuit shown in Figure 9.19 is found to be

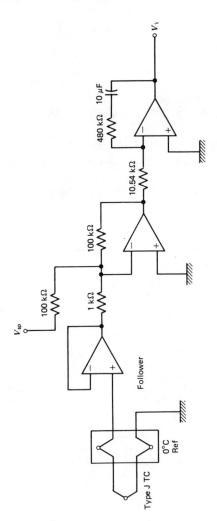

FIGURE 9.19 Circuit for Example 9.11.

$$V_{out} = 45.54 \, V_E - 9.49 \int \dot{V}_E dt$$

where

$$V_E = 100 \, V_{TC} - V_{SP}$$

The output diode and zener limit the swing from 0–10 volts.

EXAMPLE 9.12

A differential pressure gauge is used to measure flow which varies as the square root of the pressure difference (Equation 7-8). The pressure signal is a 0–2 volt range for minimum to maximum flow. A *square root extractor* circuit is available which accepts from 0–10 volts and outputs square root of input. Design a proportional controller with a 15% proportional band having a 0–10 volt output and a nominal (zero error) output of 5 volts.

SOLUTION

The circuit of Figure 9.20 implements this function. The controller input is a 0–3.162 volt signal. A 15% proportional band means that if the input changes by $(0.15)(3.162 \text{ V}) = 0.474 \text{ V}$, the output must change by 10 volts. Thus, the gain is

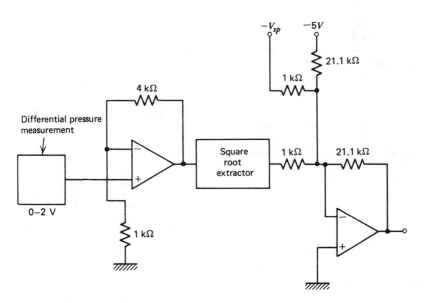

FIGURE 9.20 Circuit for Example 9.12.

$$K_P = \frac{10}{0.474} = 21.1$$

This is provided by the 1 kΩ and 21.1 kΩ resistors.

SUMMARY

This chapter presented numerous methods of implementing the controller function of a process-control loop. It showed typical methods of obtaining controller modes from analog electronic and pneumatic approaches. If the reader understands *these* typical methods, then other specific methods of implementation can be analyzed and understood by analogy.

The topics covered are summarized by the following:

1. Realization of controller modes with op amps is obtained by a straight application of amplifier, integrator, and differentiator circuits using standard op-amp techniques. The gains are found by the external resistors and capacitors used with the op amps.
2. Pneumatic controller mode implementation is made possible by a combination of a flapper/nozzle system, appropriate bellows, and variable flow restrictions. In general, given a three-mode controller, any of the other composite modes is obtained by opening the restrictions.

PROBLEMS

9.1 Derive the proportional-integral response of Figure 9.11 given by Equation (9-7).

9.2 Using the system of Figure 9.4, design a two-position controller which accepts a 10 volt maximum input and provides a 0 or 10 volt output. The setpoint is 4.3 volts and the neutral zone is ±1.1 volts.

9.3 A TC type J with a 0°C reference is used in a proportional mode temperature control system with a 140°C setpoint. The zero deviation output should be 45% and the proportional band 35%. The output is 0–10 volts. Let the full-scale input be 0–1 volt. Design a controller using Figure 9.8.

9.4 Design a proportional-integral controller with an 80% proportional band and a 0.03-minute integration time. Use a 0–5 volt input and a 0–12 volt output. (See Figure 9.10).

9.5 Design a proportional-derivative controller with a 140% proportional band and a 0.2-minute derivative time.

9.6 A three-position controller is shown in Figure 9.21. Show how this circuit implements three-position behavior. Derive equations for the input setpoint voltages at which transition to output states occurs.

9.7 Explain how the setpoint in Example 9-10 can be changed. Implement such a change to provide a setpoint of 90 lux. Show all new values required.

9.8 A liquid-level system converts a 4–10 m level into 4–20 mA. Design a three-mode controller which outputs 0–5 volts with a 50% proportional band, 0.03 minute integral time, and 0.05 minute derivative time. Let the setpoint be determined by an input voltage of 0–5 volts.

9.9 Design a two-position controller which turns lights ON when a silicon photocell cell output reaches 0.22 volts and OFF when this voltage reaches 0.78 volts.

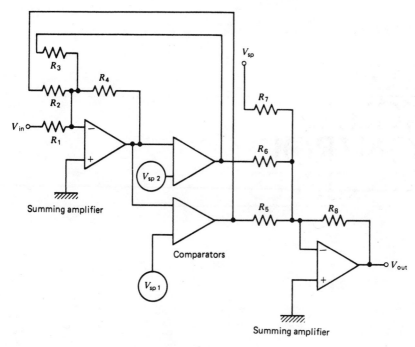

FIGURE 9.21   An electronic three-position controller.

9.10   (a) Design a 45% proportional band controller for motor speed control. The motor speed varies from 100–150 rpm, and the control circuit requires 0–5 volts for this range. A speed transducer linearly changes 2 kΩ to 5 kΩ over the same speed range. A setpoint of 125 rpm is desired. (b) Find the offset for a change in setpoint to 120 rpm.

9.11   A proportional pneumatic controller has equal area bellows. If 3–15 psi signals are used on input and output, find the ratio of distances to pivots which provides a 23% proportional band.

9.12   If the setpoint in Problem 9.11 is 7 psi and the zero error output is 9.2 psi, find the inputs yielding 3 and 15 psi (saturated) outputs.

CHAPTER

10

DIGITAL
CONTROL
PRINCIPLES

INSTRUCTIONAL OBJECTIVES

The specific objectives of this chapter are described by activities which the reader should perform after a thorough study of the chapter. Thus, following completion of a comprehensive study of this chapter, including problems, the reader should be able to:

1. Give examples of how single and multiple variable alarms are implemented in process control.
2. Draw a diagram of a typical data logging system for use in process control.
3. Explain how computer supervisory process-control operations are used in an analog process-control loop.
4. Draw a diagram of a direct digital control system with identification of each element.
5. Explain the effect of ADC time and computer execution time on data sampling rate.
6. Contrast microcomputers, minicomputers, and large-scale computers as applied to process control.
7. Define the effects of noise and aliasing in data sampling systems.
8. Explain how controller modes are implemented in DDC.
9. Determine the computer flow diagram for typical DDC applications in process control.

10.1 INTRODUCTION

Digital electronics received initial impetus from the computer industry to produce smaller, faster, and cheaper digital integrated circuits (ICs) for computers. As ICs were developed, applications extended to many other areas, such as digital watches, TV tuners, and electronic calculators. In process control, there are numerous areas where digital circuits are used *directly*, such as alarms and multivariable interactive control. A few examples of these applications will be presented in this chapter.

The evolution of digital computers, having higher speed, higher reliability, smaller size, and reduced cost has brought about increased use of digital computers in process control. One of the first computer applications was *data logging*, where the computer is used both to store the vast amount of measurement data produced in a complex process and to display the data for review (by process engineers) to determine the condition of the process. Gradually, the computer performed certain kinds of reduction of this data using control equations and even indicated the type of action, if any, which should be taken to tune a process for maximum operating efficiency. All of the loops in the process were still analog and for the most part, independent except for manual adjustment of setpoint under guidance of a process engineer.

A natural extension of this concept led to the development of a technique where the computer itself performs adjustments of loop setpoints and provides a record of process parameters. The loops are still analog, but the setpoints that determine the overall process performance are set by a computer on the basis of equations solved by the computer, using measured values of process parameters as inputs. Such a system is called *supervisory computer control*.

The ultimate result of computer applications in process control has been to use the computer to perform *controller* functions. In such a system, called *direct digital control* (DDC), the only analog elements left in the process-control loop are the measurement function and the final control element. This chapter gives an overall view of these digital electronic and computer applications in process control.

10.2 ELEMENTARY DIGITAL METHODS

There are some instances in process control where simple digital logic circuits can provide the desired regulation. In general, the controller-mode equations studied in Chapter 8 are too complicated to implement by logic circuits, but an exception is simple on/off or two-position control. In this section, we will see how logic circuits can provide this type of control even in cases of interacting variables.

10.2.1 Simple Alarms

One of the simplest digital applications to process control is the implementation of simple alarm circuits. These are very elementary binary processes because we are only concerned about whether a variable is *above* or *below* an alarm *level*. In industrial manufacturing operations, there are many variables, over and above the process-controlled variable, to be monitored. A system may operate without control of pressure, but if the pressure exceeds some preset limit, then an alarm is generated and some corrective action is taken. In this sense, the alarm is similar to two-position controller operation. In digital circuitry, a simple alarm is constructed from a *comparator*, where a voltage level indicates the alarm condition.

EXAMPLE 10.1

Design an alarm that provides a logic high of 5 volts when a liquid level exceeds 42 m. The level has been linearly converted to a 0–10 volt signal for a 0–50 m level.

SOLUTION

The data given show that level L and voltage V are related by 10 V per 50 m or 0.2 V/m, that is,

$$V = 0.2\,L$$

so that the critical level of 42 m is $V_L = \dfrac{0.2\,V}{m}\,(42\ m) = 8.4$ volts.

A simple comparator having a threshold level signal compared to 8.4 volts provides the desired alarm function. Above 8.4 V an alarm signal is generated.

Alarms generally are indicators of trouble in the process and mean that other than normal control procedures are required.

10.2.2 Multivariable Alarms

Some alarms are not predicated on the state of one variable but on the relative states of *several* variables. If each of the variables is expressed through two states, that is, HIGH and LOW, then digital approaches are ideally suited for a multivariable alarm. The general procedure is to express the variable as Boolean parameters and to find the Boolean equation between the variables which gives an alarm (HIGH output). This equation is then implemented using digital logic circuit elements.

EXAMPLE 10.2

In Figure 10.1, a holding tank is shown for which liquid level, inflow A, and inflow B are monitored. These measurements are converted to voltages and then, with comparators, to digital signals that are high when some limit is exceeded. The flow variables FA and FB will be 0 for low flow and 1 for high flow. The level variables are such that $L2$ is 1 if the level *exceeds* the lower limit and $L1$ will be 1 if the level *exceeds* the upper limit. Now the alarm is to be triggered if either of the following conditions occur.

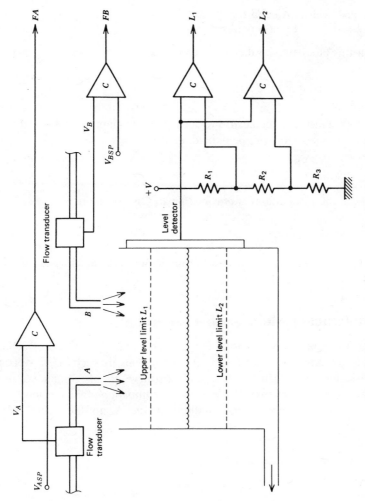

FIGURE 10.1   Holding tank level control system for Example 10.2.

1.   *L2* low **and** neither *FA* **or** *FB* high.
2.   *L1* high **and** *FA* **or** *FB* **or** both high.

Implement this problem with digital logic circuits.

SOLUTION

The variables *FA, FB, L*1, and *L*2 are already Boolean in that they have values of
logic 0 or 1. We first write Boolean equations giving an alarm output $A = 1$ for the
given two conditions. This can be done directly as

1. $A = \overline{L2} \cdot \overline{(FA + FB)}$
2. $A = L1 \cdot (FA + FB)$

Now either of these conditions is provided by an OR operation

$$A = \overline{L2} \cdot \overline{(FA + FB)} + L1 \cdot (FA + FB)$$

Logic gates which can be used to directly implement this equation are shown in
Figure 10.2.

10.2.3 Interactive Multivariable Control

Another area of process control using digital logic circuitry involves interaction of
several variables to produce several control requirements. In such cases, we assume
a set of variables, which can be expressed in two level forms, combine to decide
several types of control actions. This is really a continuation of the previous section
to include multiple outputs as well as multiple inputs. Again, the control require-

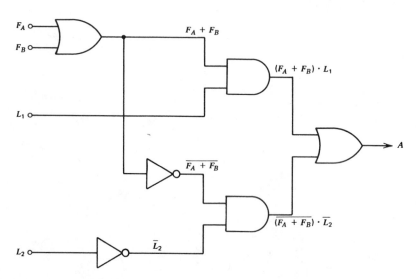

FIGURE 10.2   This logic circuit is one possible solution to Example 10.2.

ments are written in terms of Boolean equations which are implemented with digital logic equations.

EXAMPLE 10.3

The process shown in Figure 10.3 regulates both temperature and level using an outlet valve and heater. The input conditions are for temperature (T) and level (L) to be HIGH or LOW in measurement. The output will be a valve (V) and heater (H) driven HIGH (H) or LOW (L). The requirements are

	Input		Output	
	T	**L**	**H**	**V**
1.	L	L	H	L
2.	H	L	L	L
3.	L	H	H	L
4.	H	H	L	H

Find a logic circuit that implements this control mode.

SOLUTION

The first step is to define the HIGH and LOW states of the input variables using comparators with the input signals. Following this, we write the four requirements as logic equations treating the parameters as Boolean variables. We can implement the requirements directly through the equations

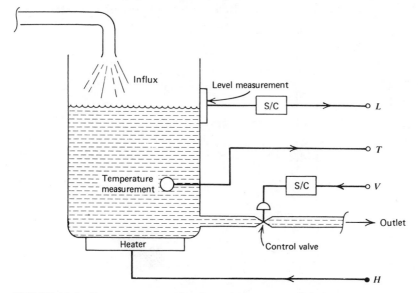

FIGURE 10.3   Temperature and level regulation required for Example 10.3.

$$H = \bar{T} \cdot \bar{L} + \bar{T} \cdot L$$

and

$$V = T \cdot L$$

A direct implementation using AND/OR logic is shown in Figure 10.4. Other solutions can be found and the circuit could be implemented using NAND/NOR logic. The comparators are used to define the LOW or HIGH state of the input parameters. This is an example of two-position, two-variable, interactive control. (The reader should show that there is a much simpler solution for H.)

10.3 COMPUTER DATA LOGGING

The efficient operation of a manufacturing process may involve the interplay of many factors, such as production rates, materials costs, and efficiencies of control. When the process requires implementation of many process-control loops, then the interaction of one stage of the system with another can often be analyzed in terms of the controlled variables of the loops. An example of this would be the rate of production of one loop, expressed as a flow rate, serving as a determining factor in the production rate of a following control system. Historically, an understanding of this type of interaction required analysis, after the fact, of strip chart recordings taken from process parameters during a production run. Such analysis, carried out by trained personnel, may then dictate settings of operational limits of future production runs.

With the development of high-speed digital computers with mass storage, it

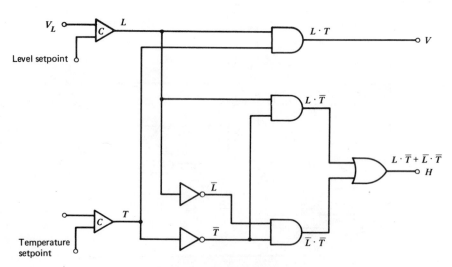

FIGURE 10.4 One possible solution to the problem of Example 10.3.

became possible to record such data continuously and automatically, display the data on command, and perform calculations on the data to reduce it to a form suitable for evaluation by appropriate technical individuals.

The general features of a computer data logging system are shown in Figure 10.5. Let us assume the process is under the control of many analog process-control loops and there is provision for analog process variable measurements to be available as a commonly scaled voltage. Thus, some signal conditioning converts all measurements into a given range, often a specified voltage range as required by a data acquisition system. A brief accounting of the elements of the system is given below.

10.3.1 Data Acquisition System (DAS)

The data acquisition system, discussed in detail in Chapter 3, is the switchyard by which the computer inputs samples of process variable values. The concept of "samples" of these values is an important topic which will be discussed in more detail later in this chapter. The reason for concern over this point is that there are

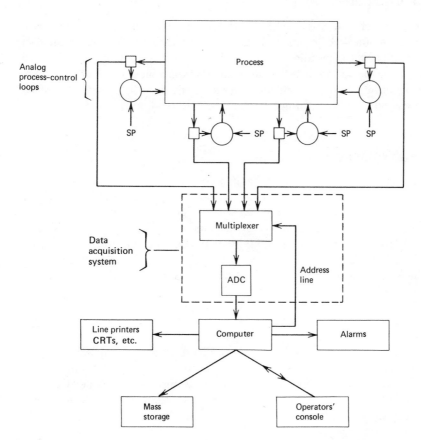

FIGURE 10.5   General features of a data logging system using a computer.

situations where the sample rate can be such that erroneous information about the variable variations results. The rate at which samples of a process variable can be taken depends on how long it takes for the DAS to acquire a value, how long it takes the computer to process the value, and how many other variables are to be sampled. This is illustrated in Example 10.4.

10.3.2 Alarms

An important part of any data logging circuit is an alarm system that monitors inputs from excursions beyond some specific limits. With scan rates of the data as high as 5000 per second, it is possible for a computer to maintain a very tight vigilance over variable values. Every time the computer inputs a particular variable, the value is compared to its preset limits which, if exceeded, triggers an alarm.

10.3.3 Computer

The computer, of course, is the central element in the system. Through programming, the computer accepts inputs and performs prescribed reductions of the data through mathematical operations. The results are evaluated by further programmed tests to oversee the operation of the entire process from which the inputs are taken. Projections of future yields, evaluations of efficiency, deviation trends, and many other operations can be performed and made available to process personnel.

10.3.4 Peripheral Units

The peripheral units are the *support* equipment to communicate computer operations to the outside world. These include the *operator console* where the programs are entered and through which commands can be given to initiate specific actions such as calculations and data outputs by the computer. The console usually has a CRT/keyboard and a typewriter unit for input and outputs. A mass storage system, such as magnetic tape, is used to store data, such as periodically sampled inputs from the process, which can be used in later, more detailed analysis of process performance.

EXAMPLE 10.4

A data logging system such as that shown in Figure 10.5 must monitor 12 analog loops. A small computer requires 4 μs per instruction and 100 instructions to address a multiplexer line and to read in and process the data in that line. The ADC performs the conversion in 30 μs. The multiplexer requires 20 μs to select and capture the value of an input line. Calculate the maximum sampling rate of a particular line.

SOLUTION

The 100 instructions require a time of $(4\,\mu s)\,(100) = 400\,\mu s$, and this must be done for 12 loops. Thus, the total instruction time is $(12)\,(400\,\mu s) = 4800\,\mu s$. The ADC converts in $30\,\mu s$ so that for 12 conversions we have $(12)\,(30\,\mu s) = 360\,\mu s$, and the total time spent in multiplexer switching is $240\,\mu s$. Adding $4800 + 360 + 240 = 5400\,\mu s$ as the minimum time before a particular line can be readdressed. The maximum sampling rate is the reciprocal or **185** samples per second.

10.4 COMPUTER SUPERVISORY CONTROL

A natural extension of a computer data logging system involves computer feedback on the process through automatic adjustment of loop setpoints. As various loads in a process change, it is often advantageous to alter setpoints in certain loops to increase efficiency or to maintain operation within certain precalculated limits. In general, the choice of setpoint is a function of many other parameters in the process. In fact, a decision to alter one setpoint may necessitate the alteration of many other loop setpoints as interactive effects are taken into account. Given the number of loops, interactions, and calculations required in such decisions, it is more natural and expedient to let a computer perform these operations under program control.

Such a system is represented in Figure 10.6 where the effect is shown by the addition of a data output system (DOS). Such a system assumes the controllers of analog loops have been designed to accept setpoint values as some properly scaled voltage. By proper switch addressing, the computer then outputs a signal through the DAC and multiplexer representing a new setpoint to a controller connected to that output line.

It might be helpful in understanding use of this type of control if a hypothetical example is given. Study for a moment the reaction process shown in Figure 10.7. The process specifications are:

1. Reactants A and B combine such that one part A to two parts B produce one part C.
2. Volume production of C varies as the square root of the A and B flow rate product.
3. The operating temperature must be lineary decreased with C volume production rate.
4. For stability, the reaction must occur with the pressure maintained below a critical value.

Now a decision is made to increase production in this operation. The first step to accomplish this is to increase the flow rate of A by a change in setpoint. Let us see the consequences of this in the rest of the process:

1. The setpoint of B flow must be set to twice the A setpoint keeping pressure below p_{max}.
2. The setpoint of C flow must be increased by the square root of the new A and B flow rates.
3. The temperature setpoint must be decreased by a proportion of the new C setpoint.

To accomplish this change in a purely analog system requires monitoring pressure

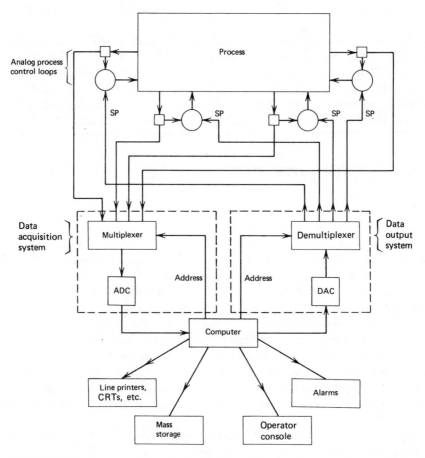

FIGURE 10.6   In computer supervisory control the computer sets the setpoints of analog process-control loops.

constantly while the operations of the three steps outlined above are gradually performed, and the new production rate is finally established. With each new setting of setpoint, we must wait until all parameters have adopted the new setpoints and wait for a safe pressure. To perform this manually requires constant human monitoring as the adjustments are made. In a supervisory control system, the computer performs these operations automatically while still performing other activities in the production.

EVENT FLOW DIAGRAM

In order to describe the steps a computer must go through to operate in some specified manner, we use an event *flow diagram*. This is the same as the flow diagram used by computer programmers, and in fact some required programs can often be written (coded) directly from such a process-operation diagram. For pur-

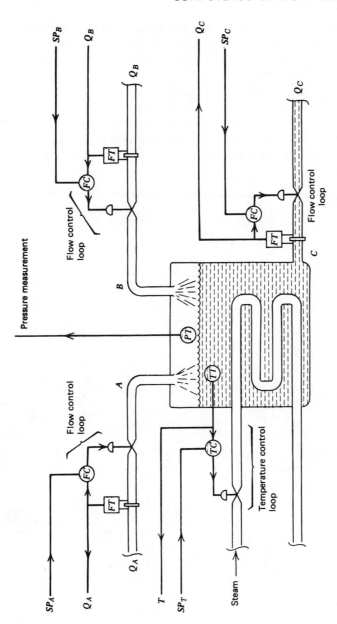

FIGURE 10.7   Computer supervisory control is ideally suited to strongly interacting, nonlinear control problems.

poses of illustration, we will use only three types of symbols to prepare a flow diagram. These are presented in Figure 10.8*a*. In using such a diagram, we do not have to get lost in the details of *how* the input, output, operations, and decisions are made, and we can thus better design the overall solution. The next step would be to consider these details.

The event flow diagram by which a computer might accomplish monitoring is shown in Figure 10.8*b*. Remember now, analog control loops are maintaining the control variables at the setpoint values. The boxes labeled INPUT refer to computer commands to address the input multiplexer to obtain the current values of these

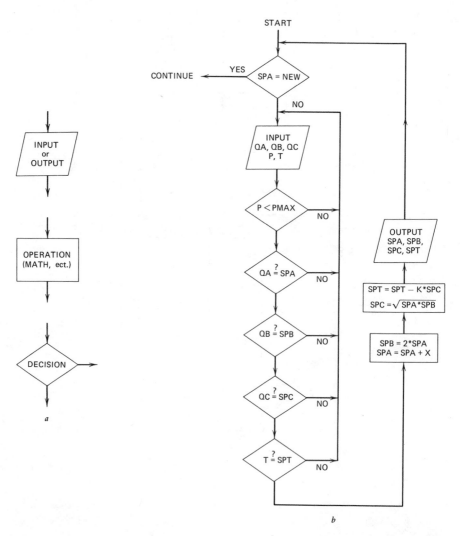

FIGURE 10.8   Flow diagram of setpoint changes in a supervisory process-control system.

parameters. The OUTPUT boxes serve a similar function for the output multiplexer. The notes on the flow diagram indicate the function of each section. An important feature of a computer supervisory control system is that it produces the desired change in operation rate in the minimum possible time. The completion of one run through the instructions in Figure 10.8*b* might typically require less than 100 μs for an average computer. Most of the adjustment time is spent waiting for the loops to stabilize. The instant such stabilization occurs, the next increment of setpoints is made by the computer via the controller.

10.5 DIRECT DIGITAL CONTROL (DDC)

In our discussions of analog control implementation, it was stressed that the circuits actually represented an analog computer. This is evident in the operations of summing, integrating, differentiating, and so on, which are required to implement the various control modes. The digital computer can also perform these math operations on analog data which is first encoded into the binary format required for digital computations. In the previous section, we saw that the data associated with a process can be provided as input and output of a digital computer through a multiplexer and AD or DA converters. A natural consequence is that, given the variable inputs from a process, we let the computer compare these internally to the setpoint values, solve the controller mode equations, and output any necessary signals to the final control elements. When this is done it is called *direct digital control* (DDC).

DDC is a natural extension of the data logging and supervisory computer control of Section 10.4. In Figure 10.9, a diagram of DDC shows that the analog loop is gone and the setpoint is now established as a programmed value, for comparison with the measured dynamic variable value.

10.5.1 General Description

Although there are many configurations used in DDC, certain common features remain. In the following paragraphs the important features of a typical DDC system are presented. The next section describes some of the configurations employed.

VARIABLE INPUT

Variables that are either under control or simply monitored are input through the same type of data acquisition system, described in Section 10.3. Although a random selection of input can be made through a computer address command to the multiplexer, it is important to note that a *regular pattern of reading* is necessary in DDC. This is necessary because the computer is also performing the controller function and must therefore have a steady input of these variables under control for use in solving mode equations. Typically, however, a single computer instruction is executed in 3–5 μs. A greater time is required for ADC operation so that a single sample can be taken in about 5 μs. In a process with 100 control loops, it might require more than 5 ms to sample values from every loop, that is, the time between

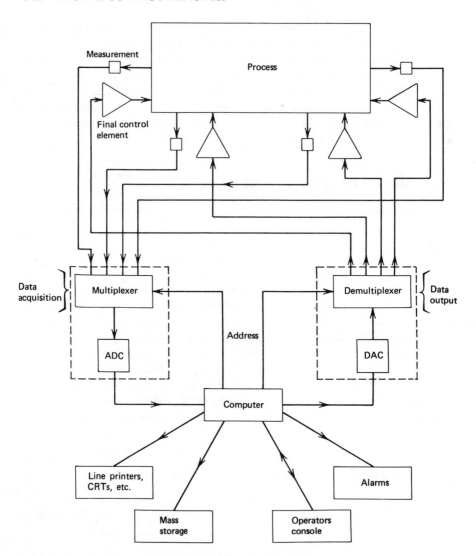

FIGURE 10.9 In direct digital control (DDC) the analog process-control loop is replaced by digital computer operations.

successive samples could be 5 ms. Of course, many machine instructions must be executed for each loop to solve the control mode equation and perform other functions so that the computer cannot return to new samples every ms. Even if we assume that 100 instructions per input are necessary, however, the delay is only on the order of (100 instructions) × (100 loops) × (5 μs per instruction) = 500 ms so that a new set of samples could be taken every 500–1000 ms. Because most industrial process reaction times are measured in minutes, it is clear that sample times are not typically a problem. Generally, a given number (typically 10) of samples of each variable are taken and averaged before use in computations associated with the control function.

CONTROL OUTPUT

The result of a solution to the mode equation for a particular variable is the updated setting of the final control element for that loop. This is provided by addressing the output multiplexer for that particular loop and outputting the calculated value. If the final control element is a digital device such as a stepping motor, then the output is direct. If it is an analog device, the DAC will be used to convert the digital information into a properly scaled analog signal. Note that if the final control element requires constant excitation to hold a particular state, as a control value, then a latch must be provided to hold the excitation when the computer is not outputting to that loop. In this instance, the output of the computer serves to update or refresh the state of the final control element.

PERIPHERAL DEVICES

The DDC computer often is used for data logging and data reduction in addition to the control operations. In any event, numerous peripheral devices are typically employed with the DDC system. These include

1. The operator console with CRT and teletype or keyboards.
2. A line printer for rapid output of bulk process data.
3. A bulk storage system, such as magnetic tape for process data.
4. Monitoring CRTs with teletypes.

ALARMS AND INTERRUPTS

An important feature of any DDC system is its alarm and interrupt facilities. The *alarm* has the same meaning as in supervisory control where a signal is generated to notify process personnel that a parameter has exceeded some preset limits. Such an alarm is often both an audio and a visual signal to attract attention to the condition. Often, the computer will have a programmed set of operations to perform under an alarm condition such as an orderly shutdown of some facet of the process. In any case, however, such an alarm in DDC is usually a computer's cry for help in a situation beyond its direct control.

The *interrupt* feature enables the operator or an external process condition to halt the normal computer operation and initiate some other procedure. Thus, boiler pressure may be constantly monitored by the computer on an interrupt line which goes HIGH when pressure exceeds a preset limit. The computer never directly addresses the line to see if the pressure is HIGH, but if it does happen, internal mechanisms cause the computer to immediately stop its present execution and perform some other operation such as turning the heater off. In this case, it is similar to an alarm except the computer never addresses and examines the condition specifically. The interrupt also is used to terminate normal operations so that new data, setpoints, and so on, can be input by an operator. Interrupts are often provided on a *priority* basis. The levels of priority may be such that high priority requires immediate cessation of present excecution and low priority allows completion of a particular program instruction set before the interrupt condition is given attention.

___

EXAMPLE 10.5

Diagram a DDC system to implement the control function of Figure 10.7 showing required inputs and outputs.

SOLUTION

We must provide multiplexing of five inputs for the three flow rates, temperature, and the pressure. Figure 10.10 shows a diagram of such a system.

___

10.5.2 DDC Configurations

The implementation of DDC into a process is not typically a straightforward assembly of the system of Figure 10.9. There are numerous other factors that must be considered, including methods of failure recovery, number of loops which can be handled by a computer, effects of peripheral access on loop performance, com-

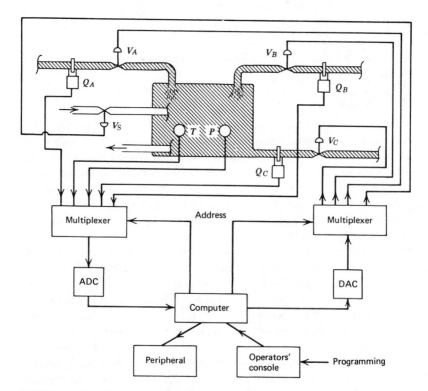

FIGURE 10.10   Figure for Example 10.5. Signal conditioning must be provided to scale the input and output for the multiplexer and converter requirements.

munication with other computers, and many other factors. In this section we will consider several of these and the typical methods of organizing DDC systems.

FAILURE PROTECTION

In analog process control, a failure of one of the elements in the loop is usually corrected by removing the failed device and replacing it with a spare. In some cases, where delays in control are critical, a backup (redundant) unit may be installed in parallel. If a failure occurs, the backup is either manually or automatically switched into operation. The defective unit can now be removed and repaired and another unit installed in its place to provide backup again.

In modern DDC, this substitution procedure can still be used since microcomputer controllers are as small and easy to replace as analog controllers. When larger computers are used for control, a second (backup) computer is often used.

In most cases, the *spare* (or dual) computer is used to perform many routine calculational chores and program development work as well as monitoring the control computer operation. In this sense, the dual computer is not simply idling and is not a complete redundancy to the main control computer. In any event, some sort of failure-recovery system is an important consideration, particularly when a process shutdown is serious because of expense, damage, and restart.

FOREGROUND-BACKGROUND

One method of DDC implementation with dual computers lets one computer perform the control operations while the other is used for data logging, program development, and other operations. In this mode of operation, the *control* computer is referred to as the *foreground* computer and the other as *background*.

MICROCOMPUTER DDC

A microcomputer is a complete computer reduced to a single printed circuit board measuring perhaps 20 cm by 20 cm. The heart of the microcomputer is a single large-scale integrated circuit (LSI) which contains the instruction set, registers, arithmetic logic unit (ALU), and computer control unit. To this chip is added solid state memory in the form of both Random Access Memory (RAM), which is volatile since it erases when power is off but can be both read from and written into, and Read Only Memory (ROM) which is permanent in that it is retained when power is off but it must be preprogrammed and cannot be altered by the computer. When input/output ICs are added, we have a complete computer.

Figure 10.11 illustrates the block diagram of a microcomputer. Such devices are available with data words of 4 bits, 8 bits, and 16 bits at the present time. The computer operates from a program which has been implanted into the ROM. The RAM is used for temporary storage during program execution. When connected to one or more process-control loops, the unit performs all required controller operations according to the program stored in ROM and independent of outside influence unless commands are given to display some variable, change a setpoint, and so on. Often, a microcomputer operates on a *real-time* basis where an internal real-

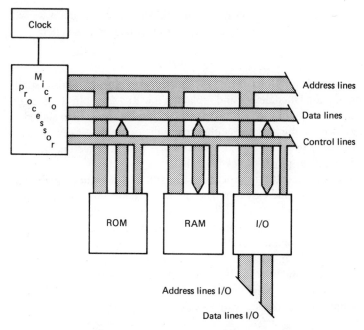

FIGURE 10.11   Block diagram of microcomputer elements.

time clock signal is available, and certain operations on the process must be done at specified real-time intervals. Thus, a process system may require purging at 5:00 PM daily.

In general, a microcomputer will be used to control from one to ten process-control loops. Numerous factors determine how many loops are to be controlled by one microcomputer. Some of these factors are:

1.   The more loops under control, the longer a time between samples of any one loop. This is a function of the computer speed, the data acquisition and output speed, and the extent of calculations to be done for each loop.
2.   The degree of interaction between loops can determine how many and which loops must be under control of one microcomputer. Therefore, strongly interacting loops should be controlled by one microcomputer which has the programming to handle the interactive effects.

Microcomputer controllers can be housed directly in the plant area or in central control rooms.

MINICOMPUTERS

The minicomputer is a unit which will typically have data word sizes of 16 to 32 bits and a larger addressable memory than the microcomputer. These units also can handle more complicated calculations than the microprocessor-based systems. The dual computer configurations, such as foreground/background, most commonly employ this size of unit. These computers not only control more loops per machine but also are used for data logging and for calculations of production efficiency and other operational quantities. Because of the fear of plant shutdown from computer failure, the use of minicomputers for DDC is giving way to microcomputers. Minicomputers are then used to oversee the operation of numerous microsystems in a network as discussed later.

LARGE-SCALE COMPUTERS

A large-scale computer usually has a word size of 64 to 128 bits, large and fast memory access, and vast mass tape or disk storage. DDC usually is not performed using such computers because the expense of the machine far overshadows its use in such a fashion. A more typical application uses these computers for the financial and administrative operations of a process facility. They may also be used by process experts to analyze process operations using historical data with complex equations and to determine conditions for improved operation.

NETWORKS

Perhaps the most successful deployment of computers in process control involves a network consisting of all three computer sizes. In a network, the microcomputer is used to control one to perhaps ten loops in a process. Several such units would provide computer control of all loops requiring computer control in the process. Each of the microcomputers is supervised by a minicomputer which also provides the required mass storage and evaluation of overall process performance. In a large plant with a number of major processes, the network would include a large-scale computer in communication with the minicomputer supervisors. Such a system is shown in block diagram in Figure 10.12.

10.6 PRACTICAL CONSIDERATIONS OF DDC

In this section we will consider some of the practical details involved in direct control of a process by computer. Since the front-line control is often performed by microcomputers, much of the discussion will be directed toward the use of that type of computer for control. Obviously, the purpose of the discussion below is to make the reader familiar with the features of control by computer. If expertise in micro-computer applications is desired, a course in microcomputer hardware and software would be required.

10.6.1 General Description

A computer control system can be most easily discussed in two parts: *hardware* and *software*.

HARDWARE

This term is used to describe the computer, data acquisition and output systems, and all the real equipment necessary to operate the computer control system.

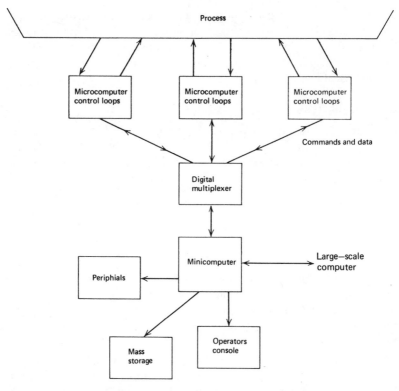

FIGURE 10.12   An entire process can be controlled by microcomputer control loops connected to a minicomputer which oversees the operation.

SOFTWARE

This term refers to the program which the computer executes in order to provide control. The controller modes, linearization, and many other functions are all provided by the software. We use the word *algorithm* to describe the mathematical and logical procedures which must be coded into the program to provide control. The program consists of instructions to be executed by the computer, and, as such, it is stored in a part of the computer's memory which is not altered by the running of the program (ROM).

10.6.2 Input/Output

The input/output part of the computer is the means by which the controlled variable value is brought into the computer and by which the command to the controlling variable (final control element) is sent from the computer. Generally, a data acquisition system (DAS) and a data output system (DOS) are used to translate signals from the real world to the digital domain and vice versa. There are several methods by which microcomputers read inputs from the DAS and send outputs to the DOS.

I/O PORTS

Some computers have special lines, called *I/O ports*, connected to the data lines within the computer. The address lines of the computer are used to activate or *select* the specific port, using a device select code. When this happens, any digital data which resides in the port is placed on the data lines of the computer. Thus, this information can now be processed by the computer. This system is illustrated in Figure 10.13*a*.

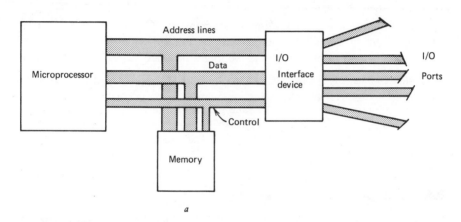

a

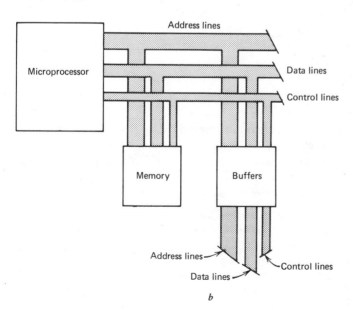

b

FIGURE 10.13   Two methods of I/0 to microcomputers are typical.

MEMORY MAPPED

Another popular I/O system simply assigns some memory locations to the DAS or DOS data lines. When the computer executes a command to fetch the data in a memory address which happens to be a DAS data line, then the computer actually *reads* from that memory location in an external data input. In the same fashion if the computer *writes* to a memory location which is actually the DOS input, then some output to a final control element will be transmitted. Figure 10.13*b* illustrates this system.

SOFTWARE CONTROL

It should be noted that data input and output is accomplished by software; that is, a program instruction is executed to input data from an I/O port or a data memory location. In general, several instructions may have to be executed to actually perform the input. For example, the steps we might need are:

1. Output a command to DAS via I/O port to select a specific analog data channel (actually addressing the analog multiplexer).
2. Go into a wait mode until the DAS sends a signal that the data is ready. (Waiting to acquire the analog signal and the AD conversion of the signal requires a simple program loop while inputing from the I/O port and looking for a "data ready" signal.)
3. Input from the I/O port the required data.

10.6.3 Sampled Data Processing

The use of computers to perform process-control functions has some very important results with respect to the measurement of process variables. It is important to note that the variable value is not known continuously but rather on a sampled basis. Usually, the value of a controlled variable is sampled on a regular basis, that is, with approximately the same time interval between samples. In the paragraphs below, several consequences of the use of a computer for sampling of data are discussed.

SIGNAL NOISE

Typically, the data acquisition system can acquire a data channel and perform a sample *hold* in only a few microseconds. This means that the measurement may be overresponsive to noise on the variable signal. Figure 10.14 shows how the periodically sampled values of a process variable can erroneously indicate variations that do not exist. This type of error can be eliminated in two ways:

1. Filters can be used in the data acquisition system to eliminate the higher frequency spikes. This will work, if the frequency of the noise is much greater than real variations of the process variable.
2. If the samples are taken at a much higher rate than is needed by the computer for determination of an output update, the computer can be programmed to average the variable values. In this way, the noise signal, which is random, will disappear from the averaged signal.

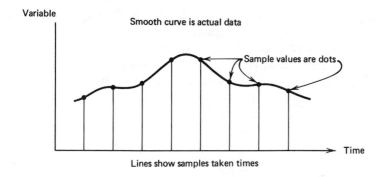

Variable

Smooth curve is actual data

Sample values are dots.

Time

Lines show samples taken times

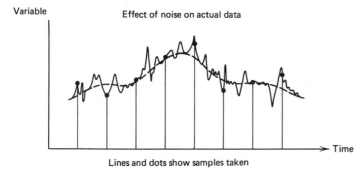

Variable

Effect of noise on actual data

Time

Lines and dots show samples taken

FIGURE 10.14   The presence of noise can cause the sampled data to feed erroneous data to the computer.

AVERAGING

In most cases, the computer can take samples of the controlled variables at a much higher rate than would be practical for calculating an output to the final control element. For example, if a microcomputer controlled 12 flow loops and sampled data (as in Example 10.4), then a flow update would be available every 5.4 ms. It would be of no value in most cases to correct a valve opening every 5 to 10 ms. For this reason, as well as the noise reduction referred to above, the computer often works with sample averages. A *simple average* of n samples is often employed, by which the computer simply accumulates (sums) a number of samples, n (where n is often 10 or more), and then divides by the number of samples taken when the time comes to process that loop.

$$<x> = \frac{x_1 + x_2 + x_3 + x_4 \cdots + x_n}{n} \tag{10-1}$$

When processing is finished using the controller mode equations and a new final control element signal has been outputted, a new accumulation of n samples is started. A *running average* of *n*-samples is where the computer always has the *n* latest values of the variable. With every new sample input, the oldest one is dropped off and the latest is added. In this case, *n*-sample average is available at all times, whereas in the previous case the computer must wait for the accumulations of *n* samples to get an average. Care must be taken in averaging not to take so many samples that important signal information is lost in the averaging.

EXAMPLE 10.6

The graph of Figure 10.15 represents typical flow over a 160 s interval in some process. Assuming a sample every five seconds, contrast by plotting: (a) straight data samples, (b) average determined every three samples, (c) running three sample averages.

SOLUTION

The results of the three approaches are shown in Figure 10.16 *a*, *b*, and *c*. Notice that the three sample straight average only updates every 15 seconds and that all the small-scale variation is eliminated. In the straight data samples, there is substantial scatter of the data due to small-scale variations. The running average presentation retains some of the small-scale variation and still clearly shows the large-scale variation. Furthermore, the data points are updated and available every sample period, that is, every 5 seconds. In most cases, this latter presentation would be preferred unless the small-scale variation is purely noise and of no interest.

LINEARIZATION

Use of a computer as part of the process-control loop makes the problem of non-linear measurement transducers much less important because in nearly all cases of

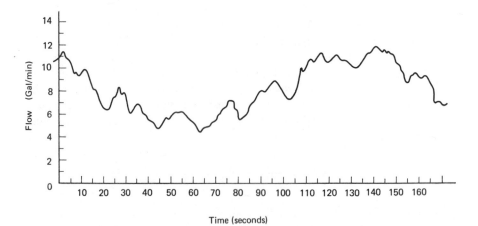

FIGURE 10.15 Figure for Example 10.6.

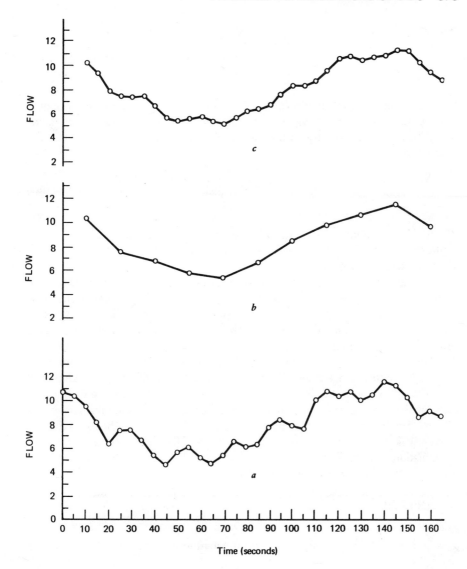

FIGURE 10.16   Solution for Example 10.6.

nonlinear response of a transducer to the controlling variable, the computer can "linearize" the response via software, that is, programs. There are two common ways to perform such linearization: equations or tables.

In the first case, consider as an example a pressure-to-voltage transducer for which the output voltage varies with pressure according to:

$$V = 0.5\sqrt{p\text{-}12} \text{ volts} \tag{10-2}$$

Where:

$$V = \text{output volts}$$
$$p = \text{pressure in pascals}$$

Thus, the voltage varies (nonlinearly) as the square root of the pressure. Suppose that a DAS converts the voltage to a digital signal which is inputted by the computer. To perform analysis with the pressure, the computer simply calculates the pressure by

$$p = 12 + \left(\frac{V}{0.5}\right)^2 \qquad (10\text{-}3)$$

We think of Equations (10-2) and (10-3) as inverses of one another. So when the nonlinear response is *known*, the inverse equation is programmed into the computer to linearize the data.

The table method is used when an analytical expression for the nonlinearity is not known. Thus, if a type-J TC is used, the computer uses a table (stored in memory) of voltage and temperature for this thermocouple. A voltage reading is inputted to the computer which then uses a table search and interpolation algorithm to find the corresponding temperature.

A table search routine has the known values of voltage and temperature pairs from the table stored in memory. Thus, pair 1 gives $V(1)$ and $T(1)$, pair 2 gives $V(2)$ and $T(2)$, and so on to the last entry in the table. The search routine is simply to compare the input voltage to the voltage of each pair in sequence until a pair voltage is larger. Suppose this happens at pair I, that is, $V(I)$ is larger than the input voltage. Then you know that the temperature is between $T(I\text{-}1)$ and $T(I)$. Any standard interpolation routine, such as Equation (4-14) can be coded to find the temperature. This same process can be used with any tabulated transfer function.

EXAMPLE 10.7

An RTD has a quadratic approximation of $R = 360\Omega$ at 30°C, $\alpha_1 = 0.003/°C$, and $\alpha_2 = -1 \times 10^{-6}/(°C)^2$. Find the inversion equation for temperature.

SOLUTION

From Equation (4-12), we can write the equation for resistance as

$$R = 360[1 + 0.003(T - 30) - 4 \times 10^{-6}(T - 30)^2] \ \Omega$$

We need to solve this for T. Let us first get a standard quadratic equation for the quantity $T - 30$, this is

$$4 \times 10^{-6}(T - 30)^2 - 0.003(T - 30) + R/360 - 1 = 0$$

Now, the standard equation for the roots of a quadratic can be used to find the values of $(T - 30)$ which satisfy this equation

$$(T - 30) = \frac{0.003 \pm \sqrt{(0.003)^2 - 4(4 \times 10^{-6})(R/360 - 1)}}{2 (4 \times 10^{-6})}$$

Simplifying this equation gives the desired result

$$T = 405 - 125 \sqrt{25 - 16R/360} \text{ °C}$$

We used the negative sign from the two solutions because only that would give $T = 30°C$ at $R = 360 \, \Omega$ as was required.

ALIASING

The fact that computer control systems operate by using periodic samples of the process variables can itself lead to a special type of error called *aliasing*. This refers to the fact that it is possible for *variations* in the process variable *and* the *sample rate* to *interact* so that *false* information about the variable results. This can easily be seen from Figure 10.17.

In Figure 10.17*a*, we see that the process variable is varying periodically, that is, *oscillating*. Such a condition is very serious in many processes, and the controller action must stop the oscillation. Note that here, due to the rate at which samples are taken, the computer is *not* aware that the oscillation is occurring! An extension of this is shown in Figure 10.17*b*. Here, the frequency of the actual variation is not the same as that which the samples imply (an alias) to the computer. Clearly such aliasing can have serious effects on the ability of the computer to regulate the value of the controlled variable in the process. The aliasing can be eliminated by using a sample rate which is faster than any reasonably expected variation of the actual data. Thus, suppose one determines that the actual process variable can never oscillate at a rate faster than 10 cycles per minute, which is a period of 0.1 minutes or 6 seconds. If we sampled at 10 times that rate or 100 samples/minute (0.6 seconds between samples), then the effects of aliasing would be prevented.

10.6.4 Controller Modes

In DDC, the operation of the controller is entirely taken over by software within the computer, that is, by programs. The process-controlled variable has been measured in samples and provided to the computer, the value of the setpoint has been inputted by the computer operator via keyboard or other input device, and a program now calculates the error and then the required controller output to the final control element. It is interesting that the control algorithm, that is, the calculations required as controller operation, are the same three *modes* used in analog controllers: *proportional, integral (reset)*, and *derivative (rate)*. In this section we discuss

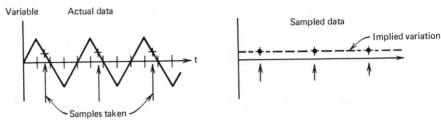

a   samples fail to show oscillation

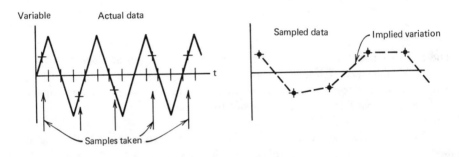

b   Samples display false frequency

FIGURE 10.17   Effects of sampling, called aliasing, create false impressions of the data variation.

briefly what kinds of algorithms are used to solve the controller mode equations by computer.

ERROR

In the computer, error can be represented by the same equation used to define error as percentage of range in Chapter 8.

$$E_p = \frac{C_m - C_{sp}}{C_{max} - C_{min}} \times 100 \tag{8-3}$$

PROPORTIONAL MODE

In Chapter 8 we saw that the proportional mode controller action is defined by a term that is directly proportional to the error. The equation was:

$$P = K_P E_p + P_0 \tag{8-14}$$

Where

K_P = proportional gain
E_p = error
P_0 = controller output with no error

The gain is expressed as percent controller output per percent error. The concept of proportional band (PB) was defined as $1/K_P$ and represents the percentage error which will cause a 100% change in controller output. This mode is easily implemented by the computer in the form of an algorithm which simply calculates Equation (8-14) directly.

INTEGRAL MODE

The integral or reset mode calculated a controller output which depended upon the history of the controlled variable error. In a mathematical sense, history is measured by an integral of the error,

$$P = K_I \int_0^t E_p dt + P(0) \qquad (8\text{-}16)$$

Where

K_I = integral gain in percent controller output per second per percent error (or, more commonly, per minute)

In order to use this mode in computer control, we need a way of evaluating the integral of error. There are many algorithms which have been developed to do this, all of which are only approximate, since only samples of the error in time are available. The simplest is called *rectangular* and is often accurate enough for use in process control. To see how this works, you should note that the integral in Equation (8-16) is merely the net area of the E_p curve from 0 to t. This is shown in Figure 10-18.

$$\int_0^t E_p dt = \text{net area} = (\text{area of } E_p > 0) - (\text{area of } E_p < 0)$$

In rectangular integration, we simply use the periodic samples of E_p to construct a series of rectangles of height equal to the sample error and width equal to the time between samples. The integral (or area) is then approximately equal to the sum of the rectangle areas. This is shown in Figure 10.18b. In an equation, rectangular integration specifies that

$$\int_0^t E_p dt \approx [S + E_{pi}] \Delta t \qquad (10\text{-}4)$$

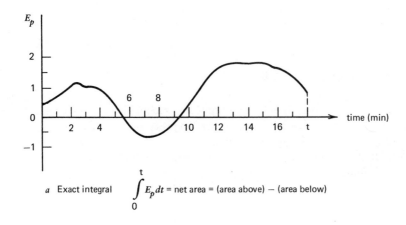

a Exact integral $\displaystyle\int_0^t E_p\,dt$ = net area = (area above) − (area below)

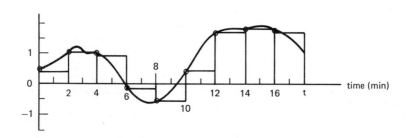

b Approximate integral = sum of rectangle areas

FIGURE 10.18 Rectangular integration approximation.

Where

Δt = time between samples

$S = E_{p_1} + E_{p_2} + \cdots$ (sum of errors calculated from previous variable samples)

E_{pi} = last sample taken at time t specified in the integral.

It should be clear that the smaller the time between samples, the more closely the approximate answer will approach the actual integral.

EXAMPLE 10.8

Find the approximate integral of E_p in Figure 10.18a from 0 to 14 minutes. Do this for the sample time shown in the figure (2 minutes) and again for a sample time of

1 min. What percentage change in the value of the integral results from the difference in sample time?

SOLUTION

For the sample time of 2 minutes, we find the integral from the rectangular integral procedure as

$$\text{Integral}_{2 \text{ min}} = 2(0.4 + 1.1 + 1 - 0.2 - 0.6 + 0.4 + 1.6)$$
$$\text{Integral}_{2 \text{ min}} = 7.4\%\text{-min}$$

For a sample time of 1 minute, we find the integral as

$$\text{Integral}_{1 \text{ min}} = 1(0.4 + 0.6 + 1.1 + 1.1 + 1 + 0.4 - 0.2 - 0.6 - 0.6$$
$$- 0.4 + 0.4 + 1.1 + 1.6 + 1.7)$$
$$\text{Integral}_{1 \text{ min}} = 7.6\%\text{-min}$$

This gives a percentage change in integral value between the two approaches as

$$\text{change} = \frac{7.6 - 7.4}{7.6} \times 100 = +3.7\%$$

DERIVATIVE MODE

The derivative controller mode, also called rate, derives a controller output which depends on the instantaneous rate of change of the error,

$$P = K_D \frac{dE_p}{dt} + P_0 \tag{8-17}$$

Where

K_D = derivative gain

$\dfrac{dE_p}{dt}$ = rate of error change in percent per second (or minute)

P_0 = controller output with no error change

The gain expresses the percent controller output for each percent/second change in error. This mode is implemented in computer control by calculating an approximate derivative of the error from the data samples. A derivative is defined as the rate at which a quantity is changing at an instant in time. We can calculate only the rate at which it is changing over the sample period, Δt, which is therefore only an approximation. In terms of an equation, this is

$$\frac{dE_{pi}}{dt} \approx \frac{E_{pi} - E_{pi-1}}{\Delta t} \tag{10-5}$$

Where

E_{pi} = present error sample
E_{pi-1} = previous error sample
Δt = time between samples

Figure (10.19) shows that this process results in a derivative which is not the actual derivative. Notice that as the time between samples is made smaller, the error will become less.

EXAMPLE 10.9

Determine an approximate value of the derivative of E_p at a time of 12 minutes from Figure 10.18a, using samples every 2 minutes and every 1 minute. Compare the results.

SOLUTION

For 2-minute samples, we get the derivative by using the sample at 10 minutes and at 12 minutes.

$$\text{Derivative}_{2 \text{ min}} = \frac{1.6 - 0.4}{2} = 0.6\%/\text{min}$$

For samples every minute, we use a sample at 11 minutes and at 12 minutes.

$$\text{Derivative}_{1 \text{ min}} = \frac{1.6 - 1.1}{1} = 0.5\%/\text{min}$$

This means there is a difference of about 17%.

THREE-MODE CONTROLLER

A three-mode controller is accomplished in computer control by combining the three terms given in the previous paragraphs and using the three-mode equation given in Chapter 8 to determine the controller output.

$$P = K_P E_p + K_P K_I \int_0^t E_p dt + K_P K_D \frac{dE_p}{dt} + P(0) \tag{8-20}$$

The integral and derivative terms are calculated by some algorithm such as those given and the Equation (8-20) for P is then calculated, all by a program.

10.6.5 Program Design Considerations

Once a controller mode design has been developed, it is necessary to code the algorithm into a set of computer instructions, the *program,* by which the control

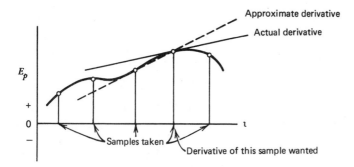

FIGURE 10.19   Approximate calculation of the derivative from sampled data.

can be accomplished. There are a variety of ways this can be done depending on the type and size of computer being used and other factors. In this section we will consider some of the necessary steps.

GENERAL FLOW DIAGRAM

Once the control algorithm has been established, a diagram is designed by which the sequence of steps necessary to provide control is displayed. In general, the standard symbols of computer programming flow charts are used. To illustrate this concept, we will employ three symbols defined in Figure 10.8a. At this stage, we are not concerned with the details of how an input operation, for example, is to be performed. Thus, we do not specify I/0 ports, data ready status, and so on, but merely indicate that an input is to occur. To illustrate this concept, Figure 10.20 shows the flow diagram of proportional-integral (PI) controllers. This is a continuously running loop with the time between samples, determined by the timer started immediately after a sample is taken. Of course, we must be sure the total time spent in the indicated operation does exceed the timer. Assuming that this is true, the *wait loop* at the top delays taking another sample until the timer indicates that a new sample should be taken.

DETAIL FLOW DIAGRAM

Each of the blocks in Figure 10.20 may itself be specified by a more detailed flow diagram, which may depend on the type of computer and data input/output system being used. To illustrate this, suppose the variable V of Figure 10.20 is on channel 5 of a multichannel data acquisition system. Then, the input block may be written, as shown in Figure 10.21, as several operations depicted by diagram symbols. As the total flow diagram becomes more detailed, we are closer to the actual coding of blocks into computer instructions.

MACHINE LANGUAGE

The final objective is the coding of the control algorithm into a set of computer-executed instructions called a *machine language program*. All computers from

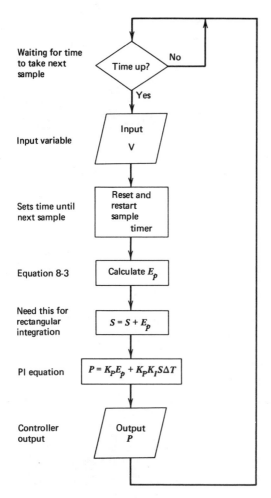

Waiting for time
to take next
sample

Time up?   No

Yes

Input variable

Input
V

Sets time until
next sample

Reset and
restart
sample
timer

Equation 8-3

Calculate E_p

Need this for
rectangular
integration

$S = S + E_p$

PI equation

$P = K_P E_p + K_P K_I S \Delta T$

Controller
output

Output
P

FIGURE 10.20   General flow diagram of a PI controller algorithm using rectangular integration.

large-scale to microcomputers have a built-in (hardware wired) set of instructions by which they perform specified operations. These instructions are in the form of one or more binary words which, when executed by the computer, cause some specified operation to occur. The number and type of instructions vary from computer to computer. As an example, in a Z-80 microprocessor-based microcomputer, the binary, 8-bit word 10000101 causes the computer to take the *contents* of an 8-bit register, L, add it to the *contents* of an 8-bit register, A, and deposit the result in register A, that is, $A \leftarrow A + L$. It is quite common to express 8-bit binary numbers or instructions by two hexadecimal (base 16), abbreviated as hex, numbers so that the previous instruction would be 85 h. The construction of all the coding necessary for even a simple process-control algorithm in machine language would be much

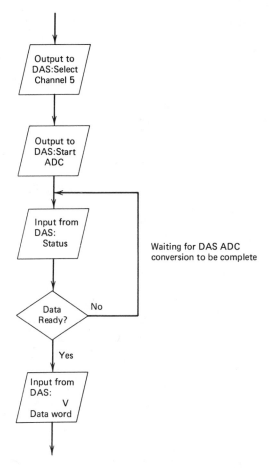

FIGURE 10.21   Flow diagram of details of a section of the general diagram of Figure 10.20.

too difficult by hand so other means have been developed as described in the next section.

INTERRUPTS

The controller program also must allow for servicing of interrupts which may be initiated by a condition in the process or by a human operator. An interrupt is an input to the computer not normally provided for by the program, which forces the computer to cease the normal flow of the program to service the interrupt. In most cases, the interrupt causes the computer to leave (in an orderly fashion) the normal program and switch to another program stored in memory. An operator may interrupt the program to insert different gains into the controller equation, or some external condition, such as a high temperature in some part of the process, may require the computer to take some action. Of course, the program is designed to ensure that the process will not suffer severe errors in the time taken to service the interrupt.

MULTIPLE-LOOP CONTROL

In most cases, a computer in process control is expected to control several loops. This introduces a number of complications into the software design required for the computer. Specifically, the problems are:

1.   The measurements from different loops may require special programming action such as linearization.
2.   The different loops may need different control modes.
3.   The gains in each loop for optimum control will most certainly be different even if the control algorithm is the same.
4.   The sampling rate and/or averaging interval of different loops may not be the same.

For these reasons, the design of multiple-loop process control by one computer requires very carefully written software which takes advantage of every possible loop similarity.

NUMBER REPRESENTATIONS

It should be noted that in machine language all numbers are considered to be integers. Thus, the content of a data word is treated as if it were just a whole number in all math operations. This is called a fixed-point system, where the decimal point is assumed to be fixed just to the right of the LSB of the data word. This is of no consequence when whole numbers are all that is used. For example, suppose we are adding 25 and 5 to get 30. We would simply input 25 into the computer as **00011001** (assuming an 8-bit data word) and enter the 5 as **00000101**. Now the add instruction from the instruction set of the machine would get as a result **00011110** which is just 30. But suppose we wanted to add 25 to 0.5? There is no way to directly enter a fractional number into the data word. In this case, we might enter the 25 as a 250 (**11111100**), and then the 5 as a **00000101**, and add to get **11111111** which is 255 as required. But we would have to use some programming technique to remember that the decimal point had been shifted so that the answer is really 25.5. It should be clear, since even this simple example becomes complicated, that when we consider trying to solve controller-mode equations, provide linearization, and other difficult math, the use of fixed-point representation of numbers is very impractical.

The way out of this is to use a representation of numbers called the *floating-point representation*. This requires that a number be represented by more than one data word, but it makes the math operations necessary much simpler to perform. In floating-point representations, numbers are written as a fixed-point number called the *mantissa* times a base raised to some power. This is just like scientific notation except that the base does not have to be 10. The general form of a floating-point number is,

$$N = n \cdot nn - - n \times B^{yyyy} \qquad\qquad (10\text{-}6)$$

Where

$$\begin{aligned}
N &= \text{number to be represented} \\
n \cdot nn - - n &= \text{fixed point part (mantissa)} \\
yyyy &= \text{exponent part} \\
B &= \text{base of exponent}
\end{aligned}$$

In this representation, the fixed-point number occupies at least one data word and the exponent part another data word. Some computers use floating point with the base of 10, just like scientific notation, while others use 2, 8, or even 16 as the base.

In some computers, floating-point data representation is provided by hardware, that is, as a set of basic instructions, hardwired into the computer. Thus, the computer might have an instruction ADD P,Q which adds the contents of memory locations P and Q as fixed-point data words. Another instruction might be FADD P,Q which adds the contents of symbolic memory locations P and Q as floating-point numbers, assuming that the contents of P and Q have been properly prepared as floating-point numbers. Other computers make such floating-point operations available as special software routines which are called by numonics similar to the FADD above. Generally, the software routines are much slower than hardware systems. Process-control systems implemented by computer generally use floating-point representations.

MEASUREMENT DATA REPRESENTATIONS

It is important for you to understand the exact relationship between the actual process variable and the data word or words, which are inputted to the computer to represent that variable. Often, the control algorithms are easier to code if the process variable has been transformed back to its correct numeric magnitude after inputting. Unless the relationship between process variable and data word contents are known, such reconstruction of the variable value is impossible. In most cases, the computer will function with a floating-point package so at least the problem of fixed-point number manipulations are not faced. There are other factors to consider in the reconstruction, however. For example, suppose an 8-bit ADC with a 10-volt reference inputs 6.75 volts from the process. Then, the resulting data word inputted by the computer will be **10101100**. But this is *not* 6.75! If you recall the way the ADC works, the data word inputted is a fixed-point number with the decimal point assumed to the left of the MSB. In other words the number is actually 0.675. Remember? The ADC converts *fractional* binary; that is, the fraction of the input to the ADC compared to 10 volts. This means the input section of the program must properly pack the inputted number into the floating-point package, so that the result is 6.75, for example. We now have an added part of the I/0 as the necessity of preparing the inputted data words for proper introduction into the floating-point package. Assuming the data word has been expressed in floating point, we must now take this number mathematically back through all the signal conditioning to find the value of the process variable. The following example illustrates this process.

EXAMPLE 10.10

Given a data channel like that shown in Figure 10.22, give the program evaluation necessary to find the pressure from the data word input. Assume the data word has been expressed in floating point.

SOLUTION

Let's just work back through the signal conditioning shown in Figure 10.20. Let D = the data expressed in the computer in floating point. Then D is a fractional number, as outputted from the ADC.

1. Multiply D by 2.5 to get the actual analog input to the ADC.
2. Divide this result by 4 to get the amplifier input.
3. Divide this by 0.005 mV/psi to get the pressure.

Combining all these operations we find,

$$p = \frac{(2.5)(D)}{(4)(0.005)}$$

$$p = 125\,D$$

Thus, the program should have the operation of multiplying the value of D by 125 to get the pressure in psi.

10.6.6 Programming Languages

The last section described the process by which a desired control system is developed from the control objective description, to a control algorithm, to a flow diagram, and finally to a machine language program. To facilitate efficient coding of the flow diagram into machine code, special languages have been developed which use the computer itself to *translate* statements similar to human language into machine

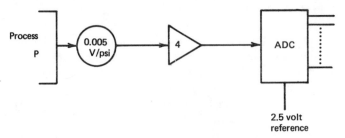

FIGURE 10.22   Data acquisition for Example 10.10.

language instructions. In this section, we describe the basic features of these languages.

ASSEMBLY LANGUAGE

In its simplest form, this type of translator allows the programmer to express the computer instructions by alphabetic and numeric numonics. For example, the Z-80 instruction given earlier in machine language to add register L to register A and place the contents in register A, becomes simply ADD A,L in assembly language. In general, an assembly language instruction must be written for each machine language instruction so that there is no savings in the number of instructions used in the code. The number of such instructions can be very large for even a modest control algorithm and involves keeping track of many branches, subroutines, labels, and variables. It would not be unreasonable to expect a PI controller like that in Figure 10.20 to require 100 or more assembly language instructions. If the algorithm requires linearization, averaging, other controller modes, and multiple loop routines, then the number of instructions can become thousands. Another problem with using assembly language directly is the complexity of the mathematics required to manipulate the data in control equations. Thus, simple calculations, such as finding the sine of an angle or multiplying an input by a scale factor, can become quite complicated in assembly language because special routines must be written to calculate each math operation. The problems of assembly language have been eased somewhat by the development of even more sophisticated translation languages.

HIGHER LEVEL LANGUAGES

In higher level languages, the coding of required computer operations is made, using instructions very near to normal human language. A single statement in one of these languages often translates into many machine language instructions, and then it is much easier for the programmer to write the flow diagram into a set of higher level language statements and let the computer translate this into the machine language. There are many higher level languages in use throughout the world. Most have been designed for particular purposes in science, business, and other disciplines.

In process control, we need languages which can handle complicated mathematical operations but which also allow for easy input and output of data with the real world. Although some special process-control languages have been developed which have controller modes built in as single statements, most applications rely on one of the standard languages, perhaps with some modification to ease input and output. The most common languages are FORTRAN and BASIC. Another, PASCAL, has shown promise of being of value in efficient process-control coding. Generally, a program written in one of these languages looks very similar to the general flow diagram shown in Figure 10.20. In fact, the coding of that PI controller into BASIC might look like the following:

```
500 T=UDF(0)
510 IF T=0 THEN 500
520 V=UDF(5)
530 T=UDF(1)
540 E=(V-S)/(V2-V1)
550 S=S+E
560 P=K1*(E+K2*S*D)
570 P=UDF(10)
580 GOTO 500
```

Notice in this hypothetical example that the coding can be done almost directly from the flow diagram. The function UDF stands for a User Defined Function, which the language lets the user employ for the construction of special functions. In this case, we use the function to write input and output routines for the process data. Thus, in the example above $V = UDF(5)$ might direct the computer to a machine language routine similar to that given in Figure 10.21 to input data from DAS channel 5. User defined routines such as this are not part of the standard language definition and must be obtained as modifications to make the language more suitable for process control.

Higher level languages, such as these, have common math functions available as built-in routines. Thus, $Z = SQRT(X)$ will calculate the square root of a number assigned to variable name X and assign that value to variable Z. This same kind of operation is provided for the trig functions and other math functions. If the signal processing necessary to linearize or otherwise process the data requires such functions, the higher level languages are often used instead of assembly language. Where this cannot be done, it is necessary to construct assembly language routines which calculate the math functions and use them in conjunction with the DDC routines.

10.7 DDC EXAMPLES

The following examples illustrate some of the difficulties which arise from seemingly simple control problems implemented digitally. It is not the purpose of this chapter or text to give you all the expertise to fully design a DDC system, but merely to give you an understanding of the steps of such a design and how some of the features of the design are actually built into the system.

EXAMPLE 10.11

A proportional control system with a 50% PB controls conveyor speed by using weight measurement. Specifications are

1. Weight range: 50 to 100 lb.
2. Weight setpoint: 75 lb.
3. Signal conditioning already available converts 0 to 100 lb to a voltage scaled at 0.05 volts/lb.

4.   An 8-bit ADC converts an input of 0 to 5 volts to 00h to FFh, where 5 volts input produces FFh.

5.   Conveyor speed is regulated by a motor control circuit which operates directly from the output of an 8-bit DAC. There are 256 speeds selected according to the following: 00h is off, FFh is full speed, and 80h is the speed corresponding to the setpoint of 75 lb.

Design the DDC control system to implement the proportional control in accordance with the above specifications. Take the design to the detailed computer flow diagram. Assume data input and output routines are already available.

SOLUTION

Let us start the design by writing the equations which must be implemented to perform the control function and work out a general block diagram.

1.   Assume we input the weight as a variable W. Then, the error term can be calculated from Equation (8-3) as

$$E_p = \left(\frac{W - 75}{100 - 50}\right) \times 100$$

$$E_p = 2(W - 75) \text{ in percent} \tag{10-7}$$

2.   The controller-mode equation is given by Equation (8-14) where a 50% PB means that the proportional gain is $K_P = 100/50 = 2$ so that the equation is

$$P = 2E_p + 50 \text{ in percent} \tag{10-8}$$

where we have set $P_0 = 50$ because 80h is 50% of the span from 00h to FFh.

These are the only two equations necessary to form the controller action. Thus, the block diagram of Figure 10.23 shows a direct application of Equations (10-7) and (10-8).

Let us now see how the details of a coding would be worked out to actually perform the simple block diagram of Figure 10.23. We will do this block by block.

INPUT W

This would involve addressing an input channel waiting for the data ready signal and inputting the result of ADC operation. The encoding is such that 0 to 100 lb will correspond to 00h to FFh.

$E_p = 2(W - 75)$

To perform this math, we must express the 75 in the same encoding as W. The 100 lb is distributed in 256_{10} states of the 8-bit word. Thus, we find the 75 is given by $(75/100) \times 256 = 192_{10}$ which is C0h. The operation of multiplication by 2 in a binary system is accomplished by a left shift on the digital word. Thus, the equation is implemented by two steps:

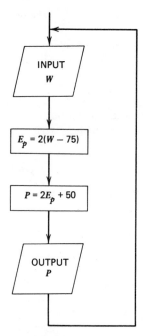

FIGURE 10.23   Figure for Example 10.11.

1.   Form $W - C0h$
2.   Left shift the result once to get E_p

$$P = 2E_p + 50$$

This equation calculates P as 0 to 100%. Since E_p is expressed from the operation above in hex, we must form this equation in hex also, that is, by simply writing the 50 in hex, or 32h. Again, multiplication by 2 is a left shift. Thus, the mode equation is accomplished by

1.   Left shift E_p
2.   Add 32h to the result to get P

OUTPUT P

Of course, the output is provided by addressing an output channel, loading the data, P, and outputting. There is a problem here, however, in that P varies from 0 to 100_{10} or 0 to 64h as calculated above. Yet the motor controller wants 256 steps which would be provided by an output of 0 to 256_{10} or 0 to FFh, so we must multiply P as calculated by a scale factor of $(256/100) = 2.56$ before outputting. It is not always possible to form a perfect binary representation of a decimal fraction, as was noted in Chapter 3, so we simply come as close as possible. In this case, we proceed as follows

1. Save the original P.
2. Left shift P once and save the result. This multiplies P by 2 from the required 2.56.
3. Get the original P and right shift, save the result. This multiplies the value of P by 0.5 out of the required 2.56.
4. Get the original P and right shift four times and save the result. This actually multiplies P by 0.0625 instead of by 0.06 required from the 2.56. We could get a little closer by a more complicated shifting, but let us accept this degree of error.
5. We add the results of (2), (3), and (4) to get the required output (actually 2.5625P).

The result of Step 5 is now outputted to the motor control DAC. If the error term is zero, you can show that the output value of P from Steps 1 through 5 will be 80h as required. The next step in this problem would be to begin coding the detailed operations above into the assembly language of the computer employed in the DDC. In Figure 10.24, the detailed flow diagram is shown from which the decoding would begin.

You can see from this example that the development of actual coding from the original DDC specification involves consideration of many factors including the encoding of numerical data and binary operations of math. For more complicated operations, such as finding averages, linearization, integration, and finding derivatives, the math operations in binary become even more complicated. In general, routines are written in a top-down fashion, starting with a general statement of the problems and then working into the details as shown in this example. The following example illustrates a more complicated DDC problem. The problem takes the control algorithm only to the stage of a rather general flow diagram, from which the detailed flow diagram would be developed.

EXAMPLE 10.12

A DDC system, shown in Figure 10.25, will use PI control with a 40% PB, 0.5%/min integration time, and a 0.75-minute sample time. The input is to be the average of five temperatures, and the output will be a control signal to a valve. Specifications of the system are:

1. A temperature measurement system is available which provides a voltage from the temperature transducer with the following relationship

$$V = 0.075\,T + 0.0004\,T^2 \tag{10-9}$$

2. Temperature range is 0°C to 90°C with a 57°C setpoint.
3. For initial startup, the output signal to the valve should be 5 volts. The valve is driven directly from a DAC, adjusted so that 0h is 0 volts and FFh is 10 volts.
4. A data acquisition system is available with five channels and a ADC which converts 0 volts to 0h and 10 volts to FFh.
5. The computer has floating-point hardware so that inputs are converted automatically to floating-point numbers for use in the programs. Thus, 0h input produces 0.0, and FFh produces 225.0_{10}.

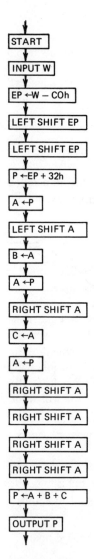

FIGURE 10.24 Figure for Example 10.11.

Construct the general flow diagram for the control algorithm using rectangular integration.

SOLUTION

We find the solution by working through the system from input to output in steps. For convenience, we will express the temperature error in terms of percent.

1. *Find the temperature.* The input section will provide a value from 0 to 255 which is the voltage representing the temperature. We can find the voltage by dividing by 25.5. Knowing

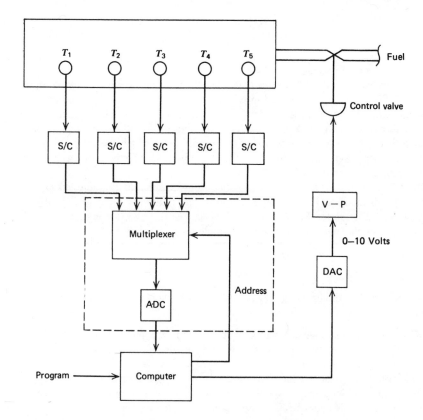

FIGURE 10.25   Process system for Example 10.12.

the voltage, we find the temperature by inversion of the relationship between voltage and temperature, Equation (10-9). First we write it in a quadratic form,

$$0.0004T^2 + 0.075\,T - V = 0$$

Now, from the quadratic root equation we find

$$T = \frac{-0.075 \pm \sqrt{(0.075)^2 - 4(-V)(0.0004)}}{2(0.0004)}$$

This relation reduces to

$$T = -93.75 + \sqrt{8789 + 2500V} \qquad (10\text{-}10)$$

where we have used the positive sign because we know T must be 0 when the voltage is 0.

2. *Find the average temperature.* We can find the average by simply inputting all five temperatures, by addressing the proper input channels, and then summing and dividing by five.

3. *Find the error.* The error will be found in percent, as usual, using Equation (8-3)

$$E_p = \left(\frac{AT - 57}{90 - 0}\right) \times 100 \tag{10-11}$$

where AT = temperature average.

4. *Proportional gain*. A 40% PB means that when the error changes by 40% the output changes by 100%. Since E_p is expressed in percent, we have $K_P = 100/40 = 2.5\%/\%$.

5. *Integral term*. The integral gain is the 0.5%/min given in the specification of the problem. This means that for a 1% error the term contributes 0.5% change in proportional-mode controller output every minute. Since the samples from which the integral is constructed are taken every 0.75 minutes, the contribution actually will be (0.5%/min)(0.75 min/sample) or 0.375%/sample.

6. *Output*. The output signal can be constructed from Equation (8-18), adjusted so that 100% corresponds to 255 so that an output of FFh is sent to the DAC when $P = 100\%$.

The final flow diagram for the controller is given in Figure 10.26. It has been assumed here that the input channels are identified by a single integer, 1 through 5.

SUMMARY

This chapter presented a discussion of the important features of digital and computer applications in process control. General features are:

1. Application of digital methods started with the simple application of digital logic methods to simple control problems and the use of computer installations to provide data logging services in process control.

2. In data logging, advantages of the computer to oversee the operation of a process through high-speed sampling of loop data proved the computer's value in process control. ADC and multiplexer systems allow rates of 5000 samples per second to be taken.

3. Supervisory control lets the computer adjust process-loop setpoints for optimum process performance even though standard analog control loops are still used for control.

4. Direct digital control (DDC) eliminates the analog controller and replaces the mode implementation by programs within the computer.

5. DDC is most efficiently implemented using minicomputers to control several loops or many microcomputers to control a fewer number of loops each.

6. The microcomputers can be used to provide local closed loop control.

PROBLEMS

10.1 A force transducer has a transfer function of 2.2 mV/N. Design an alarm that triggers at 1050 N.

10.2 A type-J TC with a 0°C reference monitors a process temperature. Design a two-alarm system at 100°C and 150°C.

10.3 Assume that a conveyor system speed S, load L, and loading rate R have been converted to binary signals by comparison to preset levels. What Boolean equation provides an alarm A when (1) S is low and L is high and R is high, or (2) S is high and L is low?

10.4 Implement the equation of Problem 10.3 with digital circuit elements.

10.5 A multiple-variable digital control uses a reaction vessel for which level, temperature, and pressure determine valve input open/close and valve output open/close. Valve open is a digital high (**H**) and valve closed is a digital low (**L**). The following conditions must be satisfied.

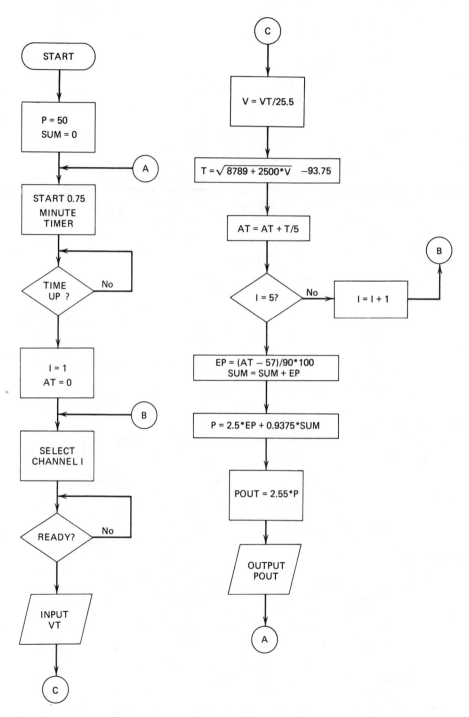

FIGURE 10.26   Figure for Example 10.12.

Pressure	Temp	Level	Valve In	Valve out
H	H	H	L	H
L	H	H	L	H
L	H	L	L	L
L	L	L	H	L
H	L	L	H	L
H	L	H	L	L

Find a Boolean equation to provide this operation.

10.6   A data-logging system must take samples of 40 variables at 100 samples per second What is the maximum signal acquisition and procesing time in microseconds?

10.7   A computer must sequentially sample 100 process parameters. It requires 14 instructions at 5.3 μs/instruction for the computer to address and process one line of data. The multiplexer switching time is 2.3 μs and the ADC conversion time is 34 μs. Find the maximum sampling rate of one line.

10.8   Develop a block diagram similar to Figure 10.6 of a supervisory computer control system for the process illustrated in Figure 10.27. This is a firing operation for ceramic articles. The conveyor motor speed is determined by the feed rate.

10.9   For the process of Problem 10.8, to obtain an increase of conveyor speed by 1%, the preheat setpoint is increased 0.9%, the oven setpoint is increased by 2.2%, and the cooler setpoint is decreased by 1.4%. Draw a flow diagram to show the computer steps necessary to increase the motor speed by 5% in steps of 0.5%.

10.10   Diagram a flow loop which inputs a voltage that varies with light intensity (W/m^2) as

$$V = 0.8 \, \ell n \, I$$

and provides an intensity error signal when the setpoint is 400 W/m^2.

10.11   A type-J TC with a 20°C reference will be used to measure temperature from 200° to 280°C in a DDC application. The input system makes the voltage measured directly available in the program. Devise the program flow diagram of a software system to use reference correction and an interpolation routine to find the temperature in °C. Assume the table values in 5°C increments for a type-J TC with a 0°C reference are available in memory.

10.12   A strain gage is used in bridge circuit as in Figure 5.11a with $R_1 = R_2 = 120\Omega$. The strain gages have $R = 120\Omega$, $GF = 2.05$, $V = 10$ volts. Assume the bridge offset voltage is inputted to a computer. Find the equation the program must use to get the strain from the offset voltage.

10.13   Suppose the setpoint for the flow in Figure 10.15 is 8 gal/min. Find the approximate integral from 10 to 100 seconds using the three sample results of Figure 10.16.

10.14   A DDC process is to operate under PID with a 60% PB, 1.2 minute integration time, and 0.05 minute derivative time. If the error is available as percent full range, show a program flow diagram of controller action. Sample time is 2 minutes.

10.15   A DDC proportional-integral controller is required for the following specifications:

1.   Temperature is to be controlled from 50° to 150°C, with a setpoint of 100°C.
2.   A transducer provides temperature as a voltage given by

$$V = (T - 100)/10 + 5$$

DIGITAL CONTROL PRINCIPLES

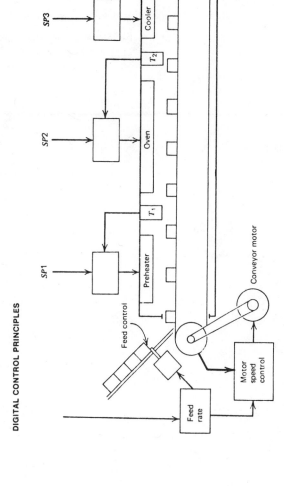

FIGURE 10.27   System for problems.

3. An 8-bit ADC provides this voltage adjusted as 0h is 0 volts, and FFh is 10 volts.
4. We use a 70% PB and an integral constant of 1.5%/min.
5. Samples are taken every 0.5 minutes
6. The output is through an 8-bit DAC with a 10 volt reference.

Develop a general flow diagram, assuming a floating point package is available.

10.16   Repeat Problem 10.15 if the detailed flow diagram of coding is to be developed as in Example 10.11.

11

CONTROL
LOOP
CHARACTERISTICS

INSTRUCTIONAL OBJECTIVES

This chapter provides the process-control technologists with a general background in the practical considerations of process-control loop implementations. After reading the chapter the reader should be able to:

1. Explain the characteristics of single variable, compound, cascade, and multivariable control.
2. Define three standard measures of quality in a control system.
3. Describe the control loop stability criteria with respect to a Bode plot.
4. Describe the open loop transient disturbance method of loop tuning.
5. Describe the Ziegler-Nichols method of process-control loop tuning.
6. Define phase and gain margin.
7. Explain how the frequency response method can be used to tune a process-control loop.

11.1 INTRODUCTION

In the previous chapters of this text, the detailed elements of process control have been described. The instrumentation used to perform the process-control function has been presented in some detail. At this point, the reader is similar to an individual who has acquired detailed knowledge of all the elements of an airplane. Such a person, however competent in knowledge of the airplane parts, is certainly *not* considered competent to *pilot* the aircraft. In process control, there also is a remarkable difference in knowing the elements of a process-control loop and being able to tune and operate a process-control loop installation.

In this chapter, consideration is given to the various ways of setting up a process-control loop system and tuning this system for optimum performance. A word of caution: A complete understanding of such *optimization* requires detailed mathematical study of stability theory, a substantial depth of knowledge of the process, and probably many years of experience. The general intent of this chapter is merely to make the reader familiar with the general concepts.

The objectives of this chapter have been carefully selected to provide the reader with as much background as can be expected without more extensive mathematical background. The emphasis is not on the theoretical development of process-control loop characteristics but on the interpretation and employment of the results of such theory. Many volumes are devoted to studies of loop stability, optimization techniques, multiloop DDC operations theory, and a host of other topics that could never be included in this brief chapter. In short, this chapter provides the process-control technologists with a general background in the practical considerations of process-control loop implementations.

11.2 CONTROL SYSTEM CONFIGURATIONS

We consider first the types of control-loop configurations that are encountered in typical industrial applications. The control-loop concept as outlined in the previous chapters is a correct one, but necessarily oversimplifies many of the actual configurations used in industry. The decision of the arrangement to select is made by the process designers based on the goal of the process, relative to production requirements, and the physical characteristics of operations under control.

11.2.1 Single Variable

The elementary process-control loop (which has formed the basis for most of our discussion in previous chapters) is a *single variable* loop. The loop is designed to maintain control of a given dynamic variable by manipulation of a controlling variable, regardless of the other process parameters.

INDEPENDENT SINGLE VARIABLE

In many process-control applications, certain regulations are required regardless of other parameters in the process. In these cases a setpoint is established, controller action is started, and the system is left alone. Thus, in Figure 11.1, a flow-control system is used to regulate flow into a tank at a fixed rate determined by the setpoint. This system then makes adjustments in valve positions, as necessary following a load change, to maintain *flow rate* at the setpoint value.

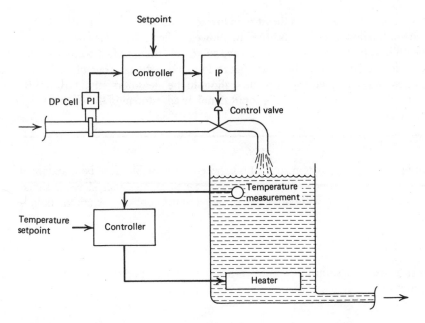

FIGURE 11.1 Two variable process-control loops with interaction.

INTERACTIVE SINGLE VARIABLE

A second single variable control loop, shown in Figure 11.1, regulates the *tempera-ture* of liquid in the tank by adjustment of heat input. This also is a single variable loop that maintains the liquid temperature at the setpoint value. Under nominal conditions, the flow into the tank is held constant and the temperature is also held constant, both at their respective setpoint values. Note, however, that a change in the setpoint of the flow-control system appears as a *load change* to the temperature control system because the fluid level in the tank or rate of passage through the tank must change. The temperature system now responds by resetting the heat flux to accommodate the new load and bring the temperature back to the setpoint.

We say then that these two loops *interact*. Almost *any* process where several variables are under control shows such *interactive behavior*. Note that any cycling or other instability of the flow control loop *causes* cycling in the temperature system because of this interaction.

COMPOUND VARIABLE

In some cases, a *single* process-control loop is used to provide control of the rela-tionship between *two or more* variables. This can be accomplished by using mea-surements from, say, two transducers as input to the process controller. Here, a signal-conditioning system must *scale* the two measurements, and add them prior to input to the controller for evaluation and action. The analysis of such systems can become quite complicated.

A common example is when the *ratio* of two reactants must be controlled. In this case, one of the flow rates is measured but allowed to float, that is, is not regulated, and the other is both measured and adjusted to provide the specified constant ratio. An example of this system is shown in Figure 11.2. Here the flow rate of reactant A is measured and added, with appropriate scaling, to the measurement of flow rate B. The controller reacts to the resulting input signal by adjustment of the control valve in the reactant B input line.

EXAMPLE 11.1

In a compound control system, the ratio between two variables is to be maintained at 3.5 to 1. If each has been converted to a $0-5$-volt range signal, devise a signal conditioning system that will output a zero signal to the controller when the ratio is correct.

SOLUTION

In this case, we use a summing amplifier (Section 2.5.2). Then the output is related to the input by

$$V_{out} = -\frac{R_f}{R_1}V_1 - \frac{R_f}{R_2}V_2$$

If we make $V_{out} = 0$, then the voltage ratio is

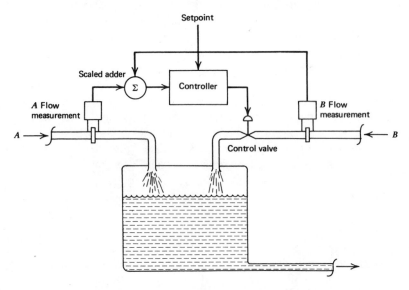

FIGURE 11.2   A compound system for which the ratio of two flow rates is controlled.

$$\frac{V_1}{V_2} = -\frac{R_1}{R_2}$$

This means that one input voltage must be negative (because we cannot use negative resistance!) and since the ratio of the resistance should also be 3.5 to 1,

$$R_1 = 3.5\,R_2$$

Because the gain is unspecified, we can use $R_f = R_1$, for example. Then the circuit of Figure 11.3 accomplishes the desired function, where

$$V_{out} = -\left(\frac{3.5\ k\Omega}{3.5\ k\Omega}\right)(-V_1) - \left(\frac{3.5\ k\Omega}{1\ k\Omega}\right)V_2$$

or

$$V_{out} = V_1 - 3.5\,V_2$$

Thus, whenever $V_1 = 3.5\,V_2$, the output is zero.

In some cases, the compound control system is known as cascade, although it differs from true cascade systems which will be described below.

11.2.2 Cascade Control

The inherent interaction that occurs between two control systems in many applications is sometimes used to provide better overall control. One method of accom-

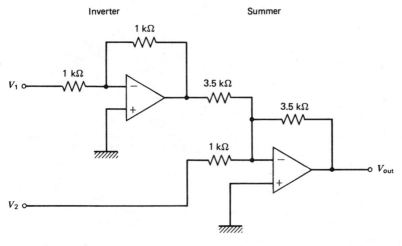

FIGURE 11.3   Circuit for Example 11.1.

plishing this is for the *setpoint* in one control loop to be determined by the *measurement* of a different variable for which the interaction exists. A block diagram of such a system is shown in Figure 11.4. Note that two measurements are taken from the system and each used in its own control loop. In the outer loop, however, the controller output is the *setpoint* of the inner loop. Thus, if the outer loop controlled variable changes, the error signal that is input to the controller effects a change in setpoint of the inner loop. Even though the measured value of the inner loop has not changed, the inner loop experiences an error signal and thus new output by virtue of the setpoint change. Cascade control generally provides better control of the outer loop variable than is accomplished through a single-variable system.

An example of a cascade control system not only shows how it works, but suggests how control is improved. Consider the problem of controlling the level of liquid in a tank through regulation of the input flow rate. A single-variable system to accomplish this is shown in Figure 11.5a. Here, a level measurement is used to adjust a flow control valve as a final control element. The setpoint to the controller establishes the desired level. In this system, upstream load changes cause changes in flow rate that result in level changes. The level change is, however, a second stage effect here. Consequently, the system cannot respond until the level has actually *been changed* by the flow rate change.

Figure 11.5b shows the same control problem solved by a *cascade* system. Here, the flow loop is a single-variable system as described earlier, but the setpoint is determined by a measurement of *level*. Note that upstream load changes are *never*

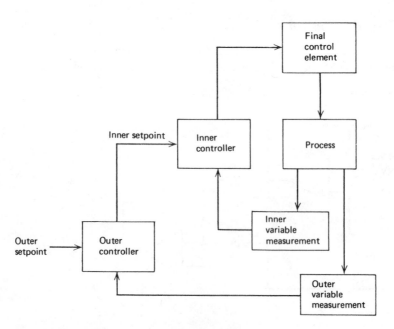

FIGURE 11.4   General features of a cascade process-control system.

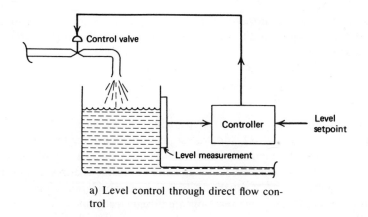

Control valve

Controller

Level setpoint

Level measurement

a) Level control through direct flow control

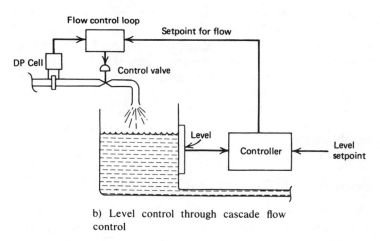

Flow control loop

Setpoint for flow

DP Cell

Control valve

Level

Controller

Level setpoint

b) Level control through cascade flow control

FIGURE 11.5   Cascade control often provides better control than direct methods.

seen in the level of liquid in the tank because the flow control system regulates such changes before they appear as substantial changes in level.

11.3 MULTIVARIABLE CONTROL SYSTEMS

One could correctly say that any reasonably complex industrial process is multivariable because many variables exist in the process and must be regulated. In general, however, many of these are either noninteracting or the interaction is not a serious problem in maintaining the desired control functions. In such cases, either single variable or cascade loops suffice to effect satisfactory control of the overall process. The use of the word *multivariable* in this section refers to those processes wherein many, *strongly* interacting variables are involved. Such a multivariable system can have such a complex interaction pattern that the adjustment of a single setpoint causes a profound influence on many other control loops in the process. In some cases, instabilities, cycling, or even runaway result from the indiscriminant adjustment of a few setpoints.

11.3.1 Analog Control

When analog control loops are used in multivariable systems, a carefully prepared instructional set must be provided to the process personnel regarding the procedure for adjustment of setpoints. Generally, such adjustments are carried out in very small increments to avoid instabilities that may result from large changes. Let us try to give you an idea of the kind of situation which we are considering. Suppose we have a reaction vessel in which two reactants are mixed, react, and the product is drawn from the bottom. We are now concerned with controlling the reaction rate. It is also important, however, to keep the reaction temperature and vessel pressure below certain limits and, finally, the level is to be controlled at some nominal value. If the reactions are exothermic, that is, self-sustaining and heat-producing, then the relation between all of these parameters can be critical. If the temperature is low, then an indiscriminant increase in steam-flow setpoint could cause an unstable runaway of the reaction. Perhaps, in this case, the level and reaction flow rates must be altered as the steam-flow rate is increased to maintain control. The necessary steps are often empirically determined or from numerical solutions of complicated control equations.

11.3.2 Supervisory and Direct Digital Control

The computer is ideally suited to the type of control problem presented by the multivariable control system. The computer can make any necessary adjustments of system operating points in an *incremental* fashion, according to a predetermined sequence, while monitoring process parameters for interactive effects. The problem in such a system is determining the *algorithm* which the computer must follow in providing the control function of setpoint change sequence. In some cases, control equations are used to accomplish this. Usually, in complex interactions, these relations are not analytically known. In some cases *self-adapting* algorithms are used, causing the computer to sequence through a set of operations and letting the result of one operation determine the next operation. As an example, if the temperature is slightly raised and the pressure rises, then drop the temperature, and so on. The computer can sequence through a myriad of such microadjustments of setpoints looking for the optimum adjustment path.

EXAMPLE 11.2

A process requires adjustment of setpoints to increase production. A particular sequence must be followed to provide the increase. SP1, SP2, and SP3 are the setpoints, P and PCR are the pressure and a critical pressure, respectively, and T and TCR are the temperature and critical temperature, respectively. Develop a flow diagram that increases the setpoints as follows:

1. Increase SP1 by 1%.
2. Wait 10 s, test for pressure compared to critical.
3. If the pressure is less than critical then
   a. decrease SP2 by ½%
   b. increase SP3 by ¾%
   c. wait for T < TCR
   d. increase SP2 by 1%
   e. go to step 1
4. If the pressure is above critical,
   a. decrease SP1 by ½%
   b. decrease SP2 by ¼%
   c. go to step 2

SOLUTION

To implement this with either DDC or supervisory systems requires a flow diagram from which the program instructions are developed. Such a flow diagram is shown in Figure 11.6. The computer sequences through these steps in an optimum fashion according to the stipulations of the problem. If the sequence is performed manually, several operators would be required to monitor equipment and make necessary adjustments.

11.4 CONTROL SYSTEM QUALITY

When a manufacturing concept is to be implemented, the ultimate goal is to develop a product that satisfies present criteria on what that product should be. If the product is crackers, they should have certain color, flavor, salinity, size, and so forth. If it is gasoline, it should possess certain octane, lead, viscosity, and so forth. The manufacturing process depends on the operation of a set of process-control loops to impart the desired characteristics to the product. The ultimate gauge of control system quality is, therefore, whether such control provides a product that is within specifications. The operation can not even get started unless that *level* of quality is provided. We assume that in any practical situation it *is* satisfied. There are, however, other levels of quality that represent a deeper study of the process-control system, and these are the subject of the present discussion. In brief, given that a control system *can* provide a product which meets specifications we ask, how well does it perform this job, what variation in parameters exists, what percentage of rejected product occurs, and so on. To answer these questions, we must first describe measures of quality in a control system and then analyze how the loop characteristics affect these measures. It is most important to note that no *absolute* answers exist. What is considered good control in one manufacturing process may not be satisfactory in another. Some of the general criteria that are applied are discussed below.

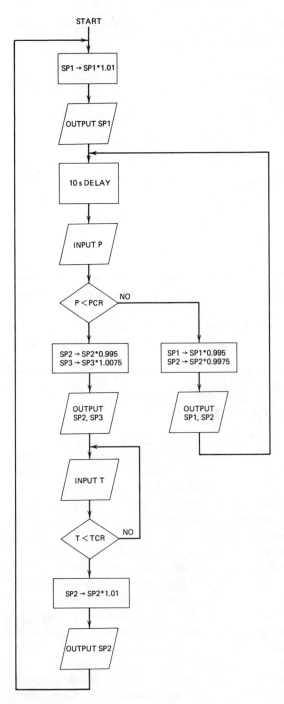

FIGURE 11.6   Flow diagram for an interactive control example.

11.4.1 Definition of Quality

Let us consider for a moment one control loop in a manufacturing process. We need a control loop because the variable under control is dynamic, changes from many influences, and therefore needs regulatory operations. It is not possible for a control loop to regulate this variable to *exactly* the setpoint. Let's face it, the variable must *change* before the loop can generate a corrective action to oppose the change. Then, in considering quality, we must accept from the outset that *perfect* control is not possible and that some inevitable deviations of variables from the optimum values will occur. In fact, then, a definition of quality is concerned with these deviations and their interpretation in terms of the ultimate product. A process-control technologist is not necessarily in a position to evaluate how deviations of a loop variable from the setpoint actually (1) affect the specific properties of the final product, (2) cause it to be rejected, or (3) how such deviations may affect the cost efficiency of the manufacturing process. In order to provide a link of communication between the process experts and the product experts, a set of measures or criteria have been devised. These provide a *common language* so that a product can be evaluated in terms of the dynamic characteristics of a specific loop and serve, therefore, as our measures of quality. To understand the measures, we must first define quality in terms of the process-control loop.

LOOP DISTURBANCE

Because we are interested in a measure of control, we first assume that a condition arises requiring control action. To be specific, two disturbance conditions are used, both of which introduce a step-function change into the loop. A step-function change in *setpoint*, as shown in Figure 11.7a, is an *instantaneous* change of the loop setpoint from an old value to a new value. The second possible disturbance is a step-function change in *process load*, as suggested in Figure 11.7b, that also occurs instantaneously in time. The load change can come from the sudden change of any of the process parameters that constitute the process load.

In order to provide measures of quality, we evaluate how the system responds to either of these sudden changes.

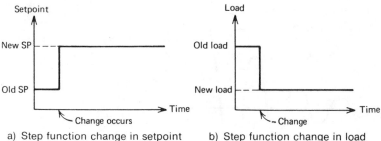

a) Step function change in setpoint   b) Step function change in load

FIGURE 11.7   Loop disturbances can occur from intentional setpoint changes or changes in the process load.

STABILITY

The most basic characteristic in defining process-loop quality is that it provides for stable regulation of the dynamic variable. Stable regulation means that the dynamic variable does not grow without limit. In Figure 11.8, two types of unstable responses are shown. In one case, a disturbance causes the dynamic variable to simply increase without limit. In the other, the variable begins to execute growing oscillations, where the amplitude is increasing without limit. In both cases, some nonlinear breakdown (such as an explosion or other malfunction) eventually terminates the increase. We will consider stability in the next section, but for a measure of quality we assume stable operation has been achieved. Note that a dynamic variable in some process may be stable and still cyclic. This is the case, for example, in two-position control where the controlled variable oscillates between two limits under nominal load conditions. A change in load may change the period of oscillations, but the amplitude swing remains essentially the same; hence, the variable is under stable control.

MINIMUM DEVIATION

If a process-control loop has been adjusted to regulate a dynamic variable at some setpoint value, then an obvious definition of quality is the extent to which a disturbance causes a deviation from that setpoint. Where a disturbance is a change

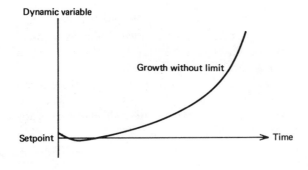

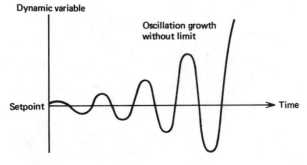

FIGURE 11.8   Instability in a process-control loop refers to the uncontrolled growth of the dynamic variable.

in setpoint, this can be considered as any *overshoot* or *undershoot* of the dynamic variable in achieving the new setpoint. In general, we want to *minimize* any deviation of the dynamic variable from the setpoint value.

MINIMUM DURATION

If a disturbance occurs, we can conclude that some deviation will occur. Another definition of quality is the *length of time* before the dynamic variable regains or adopts the setpoint value or at least falls within the acceptable limits of that value.

Thus, the quality of a process-control loop is defined through an evaluation of *stability, minimum deviation,* and *minimum duration* following a disturbance of the dynamic variable.

11.4.2 Measure of Quality

In general, it is not enough to simply state that we will design or adjust the process-control loop to provide for stable, minimum deviation, minimum duration operation. For example, the achievement of minimum deviation may result in a less than minimum duration (it usually occurs). The final product also may favor a less than absolute minimum adjustment to provide for a faster production rate at an acceptable degradation in product specification. To accommodate such circumstances, we distinguish several *measures of quality* by which we can convey the degree to which we have approached the ideals.

Assume stable operation has been achieved. There are three possible responses to a disturbance which a dynamic variable in a process-control loop can execute. The specific response depends on the controller gains and lags in the process. Referring to Figure 11.9a for a load change and Figure 11.9b for a setpoint change, we have the following definitions.

OVERDAMPED

The loop is *overdamped* in case A of Figure 11.9; the deviation approaches the setpoint value smoothly (following a disturbance) with no oscillations. The duration is *not* a minimum in such a case. For that matter, the deviation itself usually is not a minimum either. Such a response is safe, however, in assuring that no instabilities occur and that certain maximum deviations never occur.

CRITICALLY DAMPED

Careful adjustment of the process-control loop brings about curve *B* of Figure 11.9. In this case, the duration is a minimum for a noncycling response although the deviation may be larger. This is the optimum response for a condition where no overshoot is desired in a setpoint change or no cycling, in general, is desired.

UNDERDAMPED

The natural result of further adjustments of the process-control loops is cyclic response, where the deviation executes a number of oscillations about the setpoint.

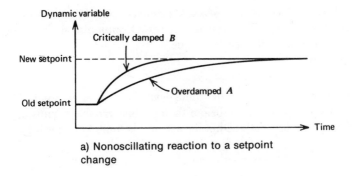

a) Nonoscillating reaction to a setpoint change

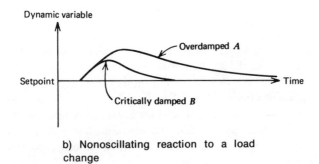

b) Nonoscillating reaction to a load change

FIGURE 11.9   In some cases the reaction of a process-control loop to setpoint or load changes is adjusted for no oscillations.

It is possible that this response gives minimum deviation and minimum duration in some cases. If the cycling can be tolerated, then such a response is preferred.

Two specialized measures of control are used when *none* of the above conditions serve to define the measure of control desired in a process.

QUARTER AMPLITUDE

When a process-control loop has a damped cyclic response to a disturbance, a criteria is sometimes used which is *neither* minimum deviation nor minimum duration. This measure of quality is found by adjusting the loop until the deviation from a disturbance is such that each deviation peak is down by one quarter from the preceding peak as shown in Figure 11.10. In this case, the *actual* magnitude of the deviation is *not* included in the measure nor is the time between each peak. In this sense, neither duration nor magnitude of the deviation are directly involved in a quarter amplitude criterion.

EXAMPLE 11.3

Suppose the deviation following a step function disturbance is a 4.7% error in the dynamic variable. If a quarter amplitude criterion is used to evaluate the response, find the error of the second and third peak.

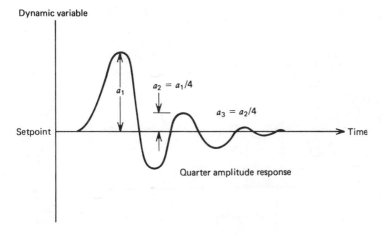

FIGURE 11.10   In one type of cyclic response, the system is adjusted to make each peak down by one quarter of the previous peak.

SOLUTION

The amplitude of each peak must be one-quarter of the previous peak. In this case the second peak is 4.7%/4 or 1.18%. The third peak is an error of 1.18%/4 or 0.30%.

MINIMUM AREA

In cases of *cyclic* or underdamped response, the most critical element is sometimes a combination of *duration* and *deviation*, which must be minimized. Thus, if minimum deviation occurs at one loop setting and minimum duration at another, then *neither* is optimum. One type of optimum measure of quality in these cases is to minimize the net area of the deviation as a function of time. In Figure 11.11 this is shown as the sum of the shaded areas. Analytically, this can be expressed as

$$A = \int |C_M - C_{SP}| \, dt \qquad (11\text{-}1)$$

Where

A = area of deviation
C_M = measured value
C_{SP} = setpoint value

Another way of representing this is to use the full-scale percent error (Section 8.3.1) to write

$$A_P = \int |E_P| \, dt \qquad (11\text{-}2)$$

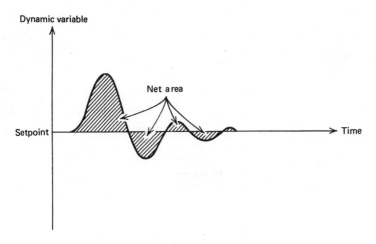

FIGURE 11.11   The minimum area criterion for cyclic response adjusts the process-control loop until the net shaded area is a minimum.

Where

E_P = full-scale error in %
A_P = area as % − s

By adopting this criteria of loop response, we are keeping the extent and duration of deviation, and thus product rejections, to a minimum.

EXAMPLE 11.4

Given the two response curves of Figure 11.12 for deviation versus time, following an initial disturbance, find the response preferred using the minimum area criterion.

SOLUTION

We find the area by application of

$$A_P = \int |E_P|\, dt \qquad (11\text{-}2)$$

In these cases we find the areas geometrically by finding the net area of each curve. For curve 1 we have

$$A_1 = 13\% - \text{minute}$$

and for curve 2 we get

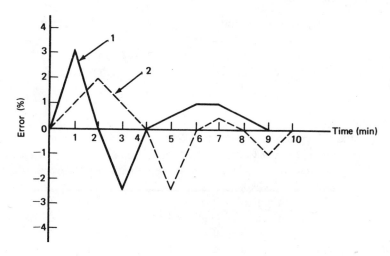

FIGURE 11.12   Figure for Example 11.4.

$$A_2 = 10.5\% - \text{minute}$$

Thus, it is clear that curve 2 is preferred under the minimum area test.

11.5 STABILITY

Earlier in this chapter, we assumed that the loop response is stable as a prerequisite for use in a practical control application. In fact, a great deal of effort is expended in the design and development of process-control loops to achieve this stability. Although a detailed treatment of stability theory in process control is beyond the intent of this book, it is of value to discuss some of the important considerations of such studies.

11.5.1 Why Instability?

Before we discuss how to make a process-control loop stable, let us address the question of why a process-control loop would ever be expected to be unstable. The fact of the matter is that whenever you use feedback, as we do in the loop, the threat that your control system may *cause* an instability is always present. Usually, the instability will be in the form of a growing oscillation of the controlled variable at some specific oscillation frequency. In some instances, the oscillation is so severe that the final control element is driven rapidly from one extreme to the other even though the process is too slow to respond. In any case, one is not regulating the controlled variable.

The fact that the instability usually occurs at some specific frequency suggests that the cause may be frequency dependent, and it is. To see how this comes about, we will develop a simple model of a control system and process. In Figure 11.13 we have lumped the entire process and controller into one box, leaving out only the error detector. An error may come from either a change of setpoint or from the system itself.

TRANSFER FUNCTION

The term $T(\omega)$ in the system box represents the effects of all the components of a process-control loop from controller to process and back. We are interested in the value of this transfer function with respect to an oscillation frequency, f, expressed in Hertz. It is common practice to express this as angular frequency, ω, which is equal to $2\pi f$ and is measured in radians per second (rad/s). The function $T(\omega)$ is normally expressed as a *gain* of magnitude $|T(\omega)|$, and a *phase* expressed as $\angle T(\omega)$. All this means is that if you input an error of frequency ω and magnitude E, then the output will have a *magnitude* of $|T(\omega)| \cdot E$ and will be shifted in phase by $\angle T(\omega)$ degrees. It is common for transfer function gain and phase to vary with frequency. In fact, that variation is at the very heart of instability.

EXAMPLE 11.5

A process-control loop model has a transfer function of 5 $\angle -90°$ at 4 Hz and 0.5 $\angle -180°$ at 10 Hz. Express the frequency in rad/s and find the output for oscillation errors of magnitude 2 at 4 Hz and then again at 10 Hz frequency.

SOLUTION

First we find the angular frequencies as $\omega_1 = (2\pi)(4 \text{ Hz}) = 25.1$ rad/s and $\omega_2 = (2\pi)(10 \text{ Hz}) = 62.8$ rad/s. The effects of the transfer function on the error are:
At 4 Hz: Output is $(2)(5) = 10$ shifted in phase by $-90°$
At 10 Hz: Output is $(2)(0.5) = 1$ shifted in phase by $-180°$

INSTABILITY

Instability is caused by a condition where for some frequency the transfer function is such that feedback to the error detector actually *increases* the error because of the gain and phase shift. Now, if there is *any* frequency for which this condition exists,

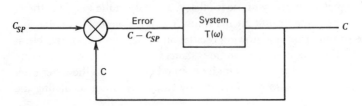

FIGURE 11.13   Process control loop reduced to a single block and the error detector.

then oscillations will always start and grow at that frequency. To illustrate this, we have assumed that a small oscillation is introduced from external sources as an error term. Assume that at this frequency the gain is greater than one, say two, and the phase shift is −180°, that is, a lag of 180°. Now let us study the result, instant by instant. In Figure 11.14a, the original input is shown and the output. Note that we are using the model of Figure 11.13 for the process-control loop. The error detector

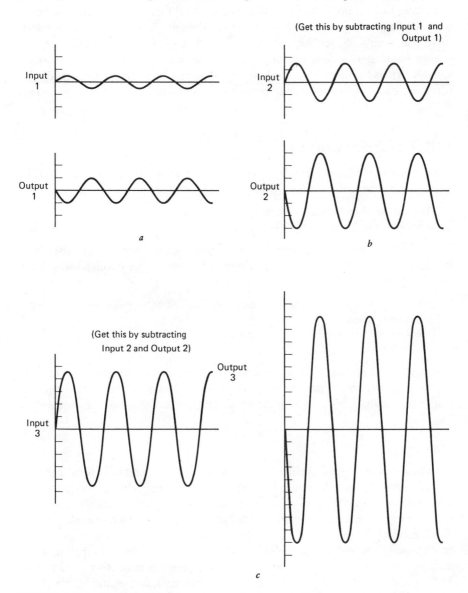

FIGURE 11.14   The instant-by-instant growth of an oscillation in an unstable process-control loop.

subtracts the output from the input, giving the input of the next instant in Figure 11.14*b*, and the output with the gain of two and phase shifted by −180°. Again, we subtract output from input, getting the input of the next instant, as in Figure 11.14*c*, and the corresponding output. Thus, you can see that the error is actually growing! The control system is forcing the oscillating error to increase, instant by instant. Of course, in actuality, it happens smoothly and the output would look something like Figure 11.8, for oscillating growth. If there is *any* frequency where such conditions for growth exist, the system is unstable and something like random noise will eventually set the system into growing oscillation. When a process-control installation is designed, one has the objective of regulating the controlled variable *without* instability in the loop. Stability can be assured by designing the controller gains so that oscillation growth is never favorable, according to certain criteria.

11.5.2 Stability Criteria

We can derive stability criteria by determining just what conditions of system gain and phase lag can lead to an enhancement of error. We are led to two conclusions:

1.   The gain must be greater than one.
2.   The phase shift must be −180°(lag).

Thus, if there is any frequency for which the gain is greater than one and the phase is −180°, then the system is unstable. From this argument, we develop two ways of specifying when a system is *stable*. These rules are:

□   **Rule 1:** A system is stable if the phase lag is less than 180° at the frequency for which the gain is unity (one).
□   **Rule 2:** A system is stable if the gain is less than one (unity) at the frequency for which phase lag is 180°.

The application of these rules to an actual process requires evaluation of the gain and phase of the system transfer function $T(\omega)$ for all frequencies to see if Rules 1 and 2 are satisfied. This is easier to do if a plot of gain and phase versus frequency is used.

BODE PLOT

A particular type of graph, called a Bode plot or diagram, normally is used to plot the gain and phase of the control system versus frequency. In this case, the frequency is plotted along the abscissa (horizontal) on a log scale and expressed commonly as rad/s. Remember, if you want to know the actual frequency in Hz you just divide the rad/s by 2π. Use of a log scale permits a very large range of frequency to be displayed. The ordinate actually consists of two parts: gain is plotted on a log scale in one part, and phase is plotted linearly as degrees in another part. Figure 11.15 shows the Bode plot with a gain/phase plotted for an example. We have labeled the gain as *open loop gain*, since we display only the $T(\omega)$ and not the effect of actually feeding that signal back into the system.

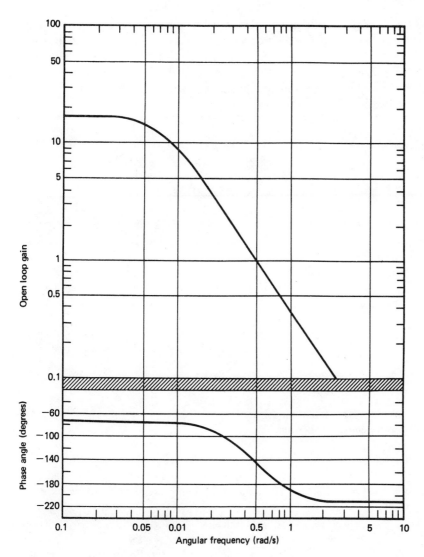

FIGURE 11.15 Bode plot with an example of process-loop frequency response.

EXAMPLE 11.6

Determine if the system of Figure 11.15 is stable according to the rules given earlier.

SOLUTION

By *Rule 1*, we find the frequency for which the gain is unity (one). The system is stable if the phase lag is less than 180°. Unity gain occurs at 0.5 rad/s; this is $0.5/2\pi$

= 0.08 Hz, or about 5 cycles per minute. Such frequencies are not uncommon in many industrial processes. For this frequency, the phase is about $-140°$ so the lag is less than 180°, and the system is **stable** by *Rule 1*. For *Rule 2*, we find the frequency for which the phase lag is 180° and check the gain, which should be less than one for stability. The phase lag is 180° at 0.8 rad/s, for which the gain is 0.5. Thus, the system is **stable** by *Rule 2*.

In some cases, a modified form of the Bode plot given in Figure 11.15 is used, in which the gain is plotted in dB. This gain is then plotted on a linear scale where

$$\text{Gain (dB)} = 20 \log_{10}(|T(\omega)|) \tag{11-3}$$

Thus, a gain of 100 gives 40 dB, unity gain is 0 dB, and a gain less than one gives a negative dB, as 0.1 gives -20 dB. The stability rules would be revised to read 0 dB instead of unit gain.

If the transfer functions of several elements are known, the transfer function of a series of such cascaded elements is found by a *product* of the *magnitudes* and a *sum* of the *phases* or

$$|T(\omega)| = |T_1(\omega)| \cdot |T_2(\omega)| \cdot |T_3(\omega)| \tag{11-4}$$
$$\angle T(\omega) = \angle T_1(\omega) + \angle T_2(\omega) + \angle T_3(\omega) \tag{11-5}$$

Where

$T(\omega)$ = system transfer function
T_1, T_2, T_3 = element transfer function

Note that on log-log scales the product of Equation (11-5) becomes addition. The transfer function for an entire process-control loop therefore involves summation of the transfer functions of all the elements of the loop, *including the process itself.*

$$log|T(\omega)| = log|T_1(\omega)| + log|T_2(\omega)| + log|T_3(\omega)| \tag{11-6}$$

In most cases, an exact solution cannot be found using Bode plots because the *process transfer function* is too complicated or unknown. Stability can still be assured through adjustment of the proportional gain, derivative time, and integral time. The final adjustment of these quantities *tunes* the system for stable operation.

11.6 PROCESS LOOP TUNING

The last aspect of process-control technology we consider refers to the actual startup and adjustment of a process-control loop. We have seen how the various settings of the controller can have a profound effect on loop performance. Now, the most natural question is how to select these settings. There are in fact *many* methods for determination of the optimum mode gains, depending on the nature and complexity of the process. We consider here three common tuning methods to give a basic idea of how optimum adjustments are found. Two of the methods given are semiempirical in that they depend on measurements made on the system to determine factors used in the adjustment formulas. The last is more analytical in that it is based on both a known transfer function of the process and loop.

11.6.1 Open Loop Transient Response Method

This method of finding controller settings was developed by Ziegler and Nichols and is sometimes referred to as a *process reaction* method. The basic approach is to open the process-control loop so that no control action (feedback) occurs. This usually is done by disconnecting the controller output from the final control element. All of the process parameters are held at their nominal values.

At some time, a transient disturbance is introduced by a small, manual change of the controlling variable using the final control elements. This change should be as small as practical for making necessary measurements. The controlled variable is measured (recorded) versus time at the instant of and following the disturbance.

A typical open loop controller response is shown in Figure 11.16 where the disturbance is applied at t_1. We have expressed the deviation as a percent full scale as usual, and we assume the final control element disturbance also is expressed as a percentage change. A tangent line, shown as a dashed line, is drawn at the *inflection point* of the curve. The inflection point is defined as that point on the curve where the slope stops increasing and begins to decrease. Where the tangent line crosses the origin, we get

$$L = \text{lag time in minutes} \tag{11-7}$$

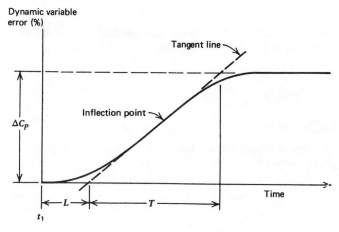

FIGURE 11.16   Process reaction graph for loop tuning.

as the time from disturbance application to the tangent line intersection as shown in Figure 11.16. Also from the graph we get

$$N = \frac{\Delta C_p}{T} \qquad (11\text{-}8)$$

Where

N = reaction rate in %/min
ΔC_p = variable change in %
T = change time in minutes

The quantities defined by Equations (11-7) and (11-8) are used with the controlling variable change ΔP to find the controller settings. The following sections give the stable control definitions for the various modes as developed by Ziegler and Nichols and corrections developed by Cohen and Coon (when the quarter amplitude response criterion is indicated). In the latter case, a log ratio is used, defined by

$$R = \frac{NL}{\Delta C_p} \qquad (11\text{-}9)$$

Where

R = log ratio (unitless)

PROPORTIONAL MODE

For the proportional mode, the proportional gain setting K_p is found from

$$K_P = \frac{\Delta P}{NL} \qquad (11\text{-}10)$$

Corrections to the value of K_p are sometimes used to obtain the quarter amplitude criterion of response. One given by Cohen and Coon is shown bracketed as

$$K_P = \frac{\Delta P}{NL}\left[1 + \frac{1}{3}\frac{NL}{\Delta C_p} \right] \qquad (11\text{-}11)$$

PROPORTIONAL-INTEGRAL MODE

When the controller mode is proportional-integral, the appropriate settings for proportional gain and integration time are

$$K_P = 0.9\,\frac{\Delta P}{NL} \qquad (11\text{-}12)$$

$$1/K_I = T_I = 3.33L$$

If the quarter-amplitude criterion is used, the gain is

$$K_P = \frac{\Delta P}{NL} \left[0.9 + \frac{1}{12} R \right] \tag{11-13}$$

$$T_I = \left[\frac{30 + 3R}{9 + 20R} \right] L$$

THREE MODE

For the three-mode controller, we find the appropriate proportional gain, integration time, and derivative time from

$$K_P = 1.2 \frac{\Delta P}{NL}$$

$$T_I = 2L \tag{11-14}$$
$$T_D = 0.5L$$

If the quarter amplitude criterion is used, these equations are corrected by

$$K_P = \frac{P}{NL} \left[1.33 + R/4 \right]$$

$$T_I = \left[\frac{32 + 6R}{13 + 8R} \right] L \tag{11-15}$$

$$T_D = L \left(\frac{4}{11 + 2R} \right)$$

EXAMPLE 11.7

A transient disturbance test is run on a process loop. The results of a 9% controlling variable change give a process reaction graph as shown in Figure 11.17. Find settings for three-mode action.

SOLUTION

By drawing the inflection point tangent on the graph, we find a lag $L = 2.4$ minutes, and a deviation change time of 4.8 minutes. The reaction rate is

$$N = \frac{\Delta C_p}{T} = \frac{3.9\%}{4.8 \text{ min}} = 0.8125 \text{ \%/min} \tag{11-8}$$

The controller settings are found from Equation (11-14)

$$K_P = 1.2 \frac{\Delta P}{NL} = 1.2 \frac{9\%}{(0.8125)(2.4)}, \text{ yielding } K_P = 5.54$$

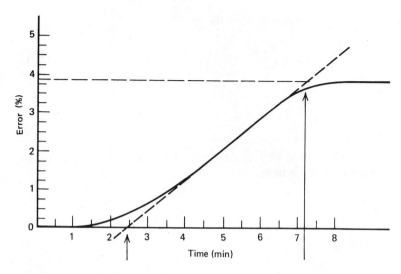

FIGURE 11.17   Process reaction graph for Example 11.7.

or a proportional band of

$$\frac{100}{K_P} = \frac{100}{5.54} = 18\%$$

$$T_I = 2L = (2)(2.4 \text{ min}) = \textbf{4.8} \text{ min}$$
$$T_D = 0.5L = (0.5)(2.4 \text{ min}) = \textbf{1.2} \text{ min}$$

EXAMPLE 11.8

For Example 11.7 find the three-mode settings for a quarter amplitude response.

SOLUTION

Here we use Equation (11-15) together with the lag ratio

$$R = \frac{NL}{\Delta C_p} = \frac{(0.8125)(2.4)}{3.9} \tag{11-9}$$

$$R = \textbf{0.50}$$

Then we get

$$K_P = \frac{\Delta P}{NL} (1.33 + R/4)$$

$$K_P = \frac{9}{(0.8125)(2.4)} (1.33 + 0.50/4) \tag{11-15}$$

$$K_P = \textbf{6.72}$$

or a **14.9%** proportional band.

$$T_I = L\left(\frac{32 + 6R}{13 + 8R}\right)$$

$$T_I = 2.4\left(\frac{32 + 6(0.5)}{13 + 8(0.5)}\right) \qquad (11\text{-}15)$$

$$T_I = 4.94 \text{ min}$$

$$T_D = L\left(\frac{4}{11 + 2R}\right)$$

$$T_D = (2.4)\frac{4}{11 + 2(0.5)} \qquad (11\text{-}15)$$

$$T_D = 0.80 \text{ min}$$

11.6.2 Ziegler-Nichols Method

Ziegler and Nichols also developed another method of controller setting assignment which has come to be associated with their name. This technique, also called the *ultimate cycle method*, is based on adjusting a loop until steady oscillations occur. Controller settings are then based upon the conditions that generate the cycling.

The particular method is accomplished through the following steps:

1. Reduce any integral and derivative actions to their minimum effect.
2. Gradually begin to increase the proportional gain while providing periodic small disturbances to the process. (These are to "jar" the system into oscillations.)
3. Note the critical gain K_c at which the dynamic variable just begins to exhibit steady cycling, that is, oscillations about the setpoint.
4. Note the critical period T_c of these oscillations measured in minutes.

Now, from the critical gain and period, the settings of the controller are assigned as follows:

PROPORTIONAL

For the proportional mode alone, the proportional gain is

$$K_P = 0.5 K_c \qquad (11\text{-}16)$$

A modification of this relation is often used when the quarter-amplitude criterion is applied. In this case, the gain is simply adjusted until the dynamic response pattern to a step change in setpoint obeys the quarter-amplitude criterion. This also results in some gain less than K_c.

PROPORTIONAL-INTEGRAL

If proportional-integral action is used in the process-control loop, then the settings are determined from

$$K_P = 0.45 \, K_c \qquad (11\text{-}17)$$
$$T_I = T_c/1.2$$

In case the quarter-amplitude criterion is desired, we make

$$T_I = T_c \qquad (11\text{-}18)$$

and adjust the gain for that necessary to obtain the quarter-amplitude response.

THREE MODE

The three-mode controller requires proportional gain, integral time, and derivative time. These are determined for nominal response as

$$K_P = 0.6 \, K_c$$
$$T_I = T_c/2.0 \qquad (11\text{-}19)$$
$$T_D = T_c/8$$

For adjustment to give quarter-amplitude response, we set

$$T_I = T_c/1.5$$
$$T_D = T_c/6 \qquad (11\text{-}20)$$

and adjust the proportional gain for satisfaction of the quarter-amplitude response.

EXAMPLE 11-9

In an application of the Ziegler-Nichols method, a process begins oscillation with a 30% proportional band and an 11.5-min period. Find the nominal three-mode controller settings.

SOLUTION

First, a 30% proportional band means the gain is (Section 8.4.5)

$$K_c = \frac{100}{\text{PB}} = \frac{100}{30} = 3.33$$

Then, from Equation (11-19) we find settings of

$$K_P = 0.6 \, K_c = (0.6)(3.33) \qquad (11\text{-}19)$$
$$K_P = 2$$

$$T_I = T/2 = 11.5/2 \qquad (11\text{-}19)$$
$$T_I = 5.75 \text{ min}$$

$$T_D = T/8 = 11.5/8$$
$$T_D = 1.44 \text{ min}$$

(11-19)

11.6.3 Frequency Response Methods

The frequency response method of process controller tuning involves use of Bode plots for the process and control loops. The method is based on an application of the Bode plot stability criteria given in Section 11.5.4 and the effects which proportional gain, integral time, and derivative time have on the Bode plot.

GAIN AND PHASE MARGIN

The stability criteria given in Section 11.5.4 represent *limits* of stability. For example, if the gain is slightly less than one when the phase lag is 180°, the system is *stable*. But if the gain is slightly greater than one at 180°, the system is *unstable*. It would be well to design a system with a margin of safety from such limits to allow for variation in components and other unknown factors. This consideration leads to the *revised stability criteria* or, more properly, *a margin of safety* provided to each condition. The exact terminology is in terms of a *gain margin* and *phase margin* from the limiting values quoted. Although no standards exist, a common condition is

1.   If the phase lag is less than 140° at the unity gain frequency, the system is stable. This, then, is a 40° *phase margin* from the limiting value of 180°.
2.   If the gain is 5 dB below unity (or a gain of about 0.56) when the phase lag is 180°, the system is stable. This is a 5 dB *gain margin*.

EXAMPLE 11-10

Determine if the system with a Bode plot of Figure 11.18 satisfies both the gain and phase margin conditions.

SOLUTION

We first examine the phase at unity gain. Unity gain occurs at an angular frequency of 0.15 rad/s, and the phase is $-120°$. Thus, the first stability condition is satisfied with a 60° phase margin.

Now a 180° phase occurs at 0.3 rad/s angular frequency where the gain is 0.7, which is too high. Thus, the 0.56 or less gain margin is *not* satisfied, and the controller gain will have to be reduced slightly.

TUNING

The operations of tuning using the frequency response method involve adjustments of the controller parameters until the stability is proved by the appropriate phase and gain margins. If the process and control elements transfer functions are known, the correct settings can be determined analytically. If not, the Bode plot can be deter-

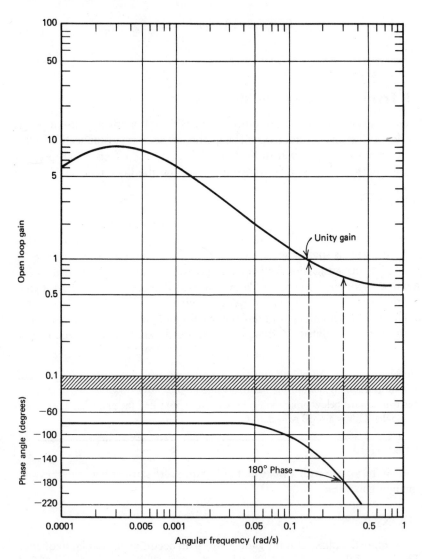

FIGURE 11.18   Bode plot for Example 11.10.

mined experimentally by opening the loop and providing a variable frequency disturbance of the controlling variable. If measurements of phase and gain are made, then the Bode plot can be constructed. From this the proper settings can be determined. Note that the significance of the unity gain crossover in frequency is that the system can correct any disturbances of frequency less than that of the unity gain frequency. Any disturbance of *higher* frequency has no effect on the control system.

The tuning operation is based on the fact that the gains of each mode have a particular effect on the system Bode plot. By adjusting these gains, we can alter the Bode plot until it satisfies the gain and phase margins of safety for a stable system.

Remember that on the Bode plot, the gains appear as products and phases of the modes simply add algebraically.

PROPORTIONAL ACTION

This mode of the controller simply multiplies the gain curve by a constant independent of frequency and has no phase effect at all. Thus, if a system gain curve is found for a porportional gain of 2 and we increase this to 4, then the entire gain curve is multiplied by 2. This term can be used to move the intact gain curve up (increased gain) or down (decreased gain). Generally, moving the curve up extends the range of frequency which the system can control, provided the stability margins are maintained.

INTEGRAL ACTION

In its pure form, the integral mode contributes

$$\text{integral gain} = \frac{K_I}{\omega}$$

$$\text{integral phase} = -90° \text{ (lag)}$$

Thus, we see that the integral mode contributes more gain at lower frequency and less gain at higher frequency because of the inverse dependence on the frequency ω. The phase shift is made more lagging by 90°. The integral *mode* gain K_I can raise or lower the entire curve. In practical form, the integral mode has some lower frequency (called the *breakpoint*) at which its effects cease, and the gain curve becomes flat while the phase lag goes to zero. This is shown in Figure 11.19 where $K_I = 1$ has been assumed.

DERIVATIVE ACTION

Because of problems of instability at higher frequencies, the derivative mode is almost never used in its pure form. If it were, the gain and phase would be

$$\text{derivative gain} = K_D \omega$$
$$\text{derivative phase} = +90° \text{ lead}$$

Since the gain increases with frequency, it must be limited at some upper frequency. In Figure 11.19, this mode action is shown (for a mode gain $K_D = 10$) with the characteristic upper limit frequency at which the gain becomes constant, and the phase shift is zero. In general then, this mode can be used to increase the gain at higher frequency, but more importantly it can be used to drive the phase away from the $-180°$ (lag) shift where instability can begin.

EXAMPLE 11.11

Find the fractional decrease in proportional gain which would make the system of Example 11.10 stable.

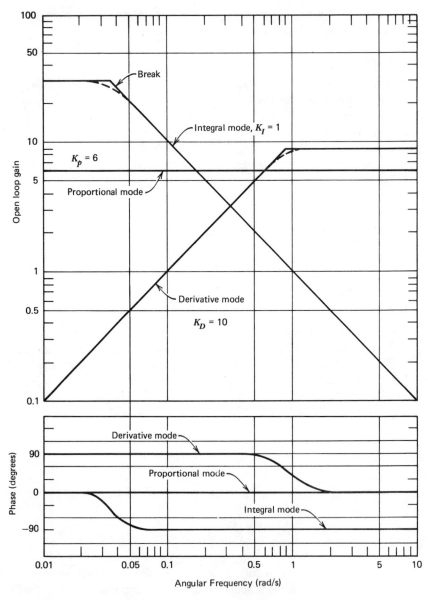

FIGURE 11.19   Bode plots of proportional, integral, and derivative controller modes.

SOLUTION

The system failed to satisfy the 5 db gain margin at $-180°$, since the gain was 0.7 instead of 0.56 or less. Thus, if the proportional gain is reduced by a factor of $0.7/0.56$ or about 0.8, then the gain margin will be satisfied. The phase margin will not be affected.

EXAMPLE 11.12

Consider the Bode plot of Figure 11.20 for some hypothetical process and controller with a proportional gain of 10. Show that the system is unstable and find a proportional gain for a 2.5 db gain margin.

SOLUTION

Evaluation of the curve shows that the phase lag at unity gain is 170°, which, while stable, does not satisfy the required 40° phase margin. The gain at 180° phase lag is

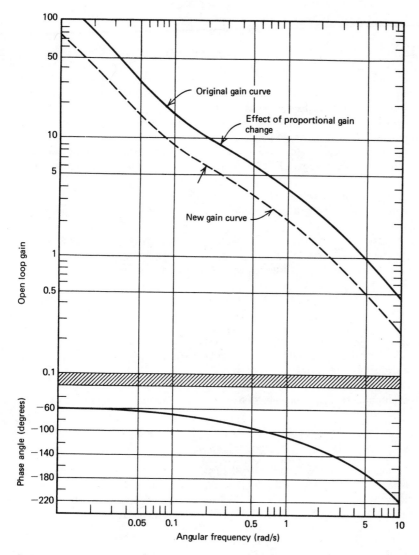

FIGURE 11.20   Bode plot for Example 11.11. Note that changes in proportional gain do not change the phase.

approximately 0.8, which, although less than one, does not satisfy the stipulated gain margin. Notice that if the entire gain curve is reduced or moved down by a constant amount, as indicated, with no effect on the phase lag, then the stability criteria can be satisfied. This is done by changing the *proportional gain* because this affects gain only. In particular, the gain must be derated by a factor of 1/2 (note difference at 5 rad/s) to achieve the required response. Thus, the new proportional gain should be 10/2 or ·5.

In general, the derivative, integral, and proportional gains are adjusted until the system satisfies the stability criteria and the specified unity gain frequency.

SUMMARY

This chapter has presented an overview of operational characteristics of the entire process-control loop. The following items were specifically considered:

1.   Although the single variable control loop is widely used in process control, we often must account for interactive effects between variables.

2.   In many cases, a cascade control loop is used where the setpoint of one loop is determined by the controller output of another loop.

3.   Multivariable control systems are used where strong interactions between variables exist. In many ways, supervisory or DDC techniques are best suited for these control problems.

4.   The quality of control-system response is defined by characterizing the response of the system to a disturbance. The stability, deviation magnitude, and duration are important to this quality.

5.   The *minimum area* and *quarter amplitude* criteria are often used to measure quality together with damping characteristics.

6.   The stability of a process-control loop can be studied with a Bode plot, consisting of gain and phase lag versus frequency.

7.   The tuning of a process-control loop consists of finding the optimum settings of controller gains for good control.

8.   The open loop transient method of tuning relies on a semiempirical technique using the reaction of the open loop to a transient disturbance.

9.   The Ziegler-Nichols method finds the conditions that generate steady oscillation and derates the settings for optimization.

10.   The frequency-response method relies on the stability conditions determined from a Bode plot. Adjustments of gains provide for the phase and gain margin and frequency band width.

PROBLEMS

11.1   A compound control system dictates the ratio of pressure (in N/m^2) to temperature (in K) be held at 0.39, a constant ratio. (a) Diagram a control loop to provide this control. (b) Identify the function of each element in the loop necessary to accomplish the control.

11.2   Suppose that for Problem 11.1 the temperature is available at 2.75 mV/K and the pressure at $11.5 \frac{mV}{N/m^2}$. Design signal conditioning that outputs zero when the ratio is correct. (First convert each to a scale of 1 volt per K and 1 volt per N/m^2.)

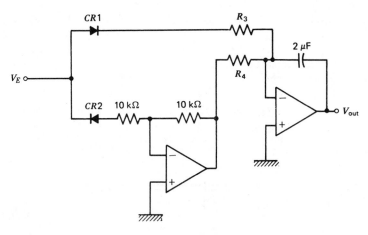

FIGURE 11.21   Circuit for Problem 11.6.

11.3   Draw a block diagram and an operation flow diagram to show how Problem 11.1 is implemented by DDC.

11.4   Draw a diagram of a cascade control system which has an inner loop of temperature measurement as input to a flow control system, and an outer loop of viscosity to determine the flow controller setpoint.

11.5   Draw a block diagram and flow diagram to show how the cascade system of Problem 11.4 is implemented by DDC.

11.6   Assume we want an electronic means of finding the area of the error-time graph following a disturbance. (a) Show that the circuit of Figure 11.21 does this. (b) Find values of R_3 and R_4 so that the output is the area of Equation (11-2) in V-s.

11.7   An open loop transient test provides the process reaction graph given in Figure 11.22

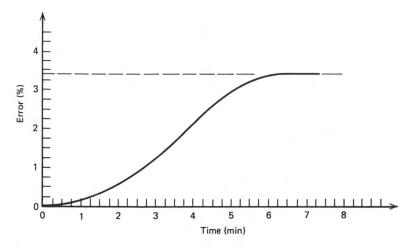

FIGURE 11.22   Process reaction graph for Problem 11.7.

for a 7.5% disturbance. (a) Find the standard proportional-integral settings. (b) Find the three-mode quarter amplitude settings.

11.8   In the Ziegler-Nichols method, the critical gain was found to be 4.2 and the critical period was 2.21 min. Find the standard settings for (a) proportional control (b) proportional-integral control (c) three-mode control.

11.9   Specify the gain and phase margins for Figure 11.23. Is the system stable?

11.10   If the nominal proportional gain for the process in Figure 11.23 was 11.5, determine the gain which will just satisfy a gain margin of at least 0.56 and a phase margin of at least 40°.

11.11   Draw the resultant Bode plot of the series combinations of the proportional, integral, and derivative modes shown in Figure 11.19.

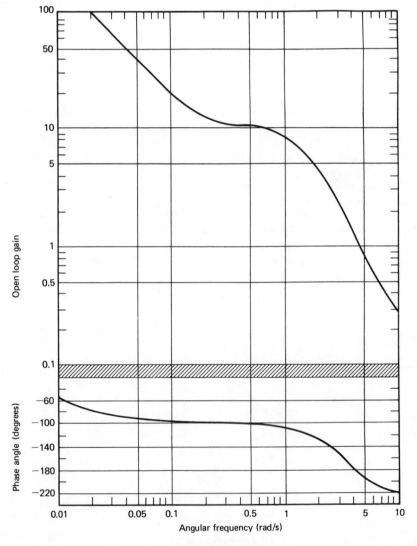

FIGURE 11.23   Bode plot for problems.

11.12   A process and control system has components with transfer functions at 10 rad/s of 2<0°, 1.4<−90°, 0.05<90°, and 3.2<85°. Find the transfer function at 10 rad/s.

11.13   Figure 11.24 shows the Bode plots of a process, the proportional mode and the integral mode of an associated controller. Find the composite transfer function Bode plot and evaluate the stability.

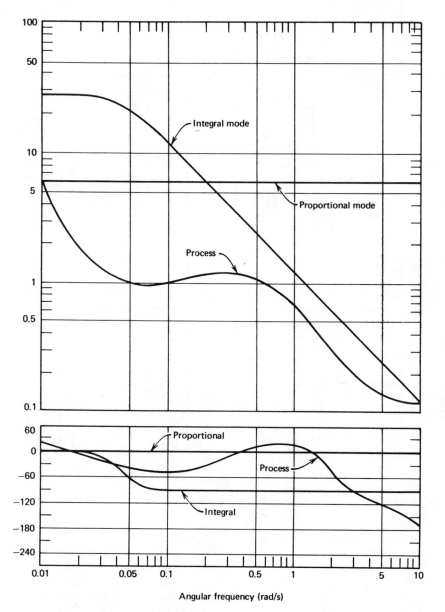

FIGURE 11.24   Figure for Problem 11.13.

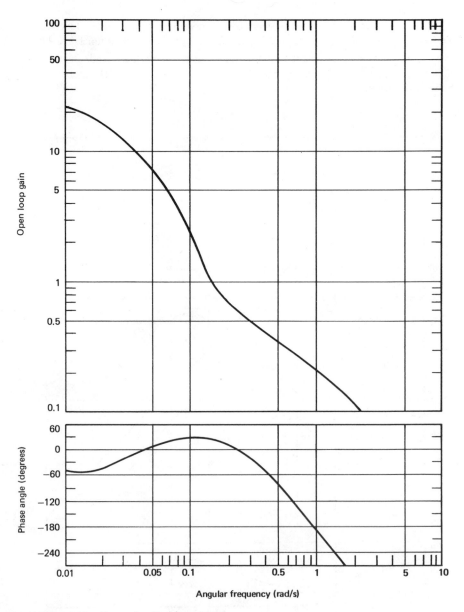

FIGURE 11.25 Figure for Problem 11.15.

11.14 Show how adding derivative action to the system of Problem 11.13 would change the composite Bode plot. Discuss the stability of the system if derivative action is added.

11.15 The composite Bode plot of a system is shown in Figure 11.25. This is for proportional action only with $K_P = 1$. How much proportional gain will drive the system into instability? The critical gain of the ultimate cycle or Ziegler-Nichols method is that which will drive the system to instability. What would be the proper gain setting for ultimate cycle tuning? What phase and gain margin does this provide?

1

UNITS

A.1.1 SI UNITS

The system of units which is gaining acceptance throughout the world is called the Systéme International d'Unités (SI). This system was adopted by the 1960 Conférence Genérale des Poids et Measures. The basic set of these units is given in Section 1.6.1. The description of other physical quantities can be described in terms of *derived* units which are expressible in terms of the basic set. The following list shows the more common derived units and their descriptors.

Quantity Symbol	Quantity Definition	Unit Name	Unit Symbol	Unit Definition
f	Frequency	hertz	Hz	s^{-1}
W	Energy	joule	J	$kg \cdot m^2/s^2$
F	Force	newton	N	$kg \cdot m/s^2$
R	Resistance	ohm	Ω	$kg \cdot m^2/(s^3 \cdot A^2)$
V	Voltage	volt	V	$A \cdot \Omega$
p	Pressure	Pascal	Pa	N/m^2
ω	Angular frequency	radians per second	rad/s	rad/s
E	Illuminance	lux	lx	lm/m^2
Q	Charge	coulomb	C	$A \cdot s$
L	Inductance	henry	H	$kg \cdot m^2/(s^2 \cdot A^2)$
C	Capacity	farad	F	$s^4 \cdot A^2/kg\text{-}m^2$
G	Conductance	siemen	S	Ω^{-1}
Φ	Luminous flux	lumen	lm	cd/sr
	Luminous efficacy	lumen per watt	lm/w	lm/w

A.1.2 OTHER UNITS

Many systems of units are employed in the world, although a unified effort is being made to adopt the SI units universally. Two widely employed unit systems are the English and centimeter-gram-seconds.

English System

The system of units that has been historically employed in the United Kingdom and United States is the English system. There are actually several types of English systems, but the most common one for technological practices is the engineering system. It is inevitable that some quantity in a design specification which has been employed in several countries for many years may be expressed in English units. Until this awkward system is eliminated, it will be necessary to perform conversion between the SI and English systems of units. The following table of conversion factors are provided by showing the equivalent number of English units for particular SI units.

The measure of force and mass in the English system of units is somewhat confusing because of differences in definitions among users of the system. We will define and use what is considered the engineering definitions although it is well to note that these are not the legal definition.

The English unit of *force* is the *pound* (lb_f). The pound is best defined by its relation to the SI unit of force through the conversion 0.2248 lb/N. Thus, a force of 16 lb becomes

$$(16 \text{ lb}) \left(\frac{\text{N}}{0.2248 \text{ lb}} \right) = 71.2 \text{ N}$$

Measurement of *mass* in the English system is in the form of a unit called the *slug*. A slug is defined to be that mass which accelerates at 1 ft/s^2 when acted on by a force of 1 lb. The conversion between the slug and SI unit of mass is 14.59 kg/slug.

To complete the picture of English units we note now several other definitions that are sometimes employed which differ from those already presented.

(1) POUND MASS

The pound was originally designed as a mass measurement. When used in this sense it is best defined in relation to the SI kg from the conversion 1 lb (mass) = 0.454 kg.

(2) POUNDAL (PDL)

The poundal is a unit of force defined as that force which gives the pound mass, defined above, an acceleration of 1 ft/s^2. In terms of the SI force unit, the poundal conversion is 1 pdl = 0.1383 N

Centimeter-Gram-Second (CGS) System

The CGS system of units is often employed in scientific work and appears in journal articles and on some scientific measurement equipment. This system is based on the centimeter (cm) of length, the gram (g) of mass, and second (s) of time. Transformations may be made using the scale factors listed in the following table.

Quantity	SI	CGS	English
length	1 meter	100 centimeters	3.28 feet
mass	1 kilogram	10^3 grams	0.0685 slugs
time	1 second	1 second	1 second
force	1 newton	10^5 dynes	0.2248 pounds
energy	1 joule	10^7 ergs	0.7376 ft-lb
pressure	1 pascal	10 dyne/cm^2	1.45×10^{-4} lb/in^2

A.1.3 STANDARD PREFIXES

The SI decimal multiple and submultiple designations are shown in the table below:

Multiple	SI Prefix	Symbol
10^{12}	tera	T
10^9	giga	G
10^6	mega	M
10^3	kilo	k
10^2	hecto	h
10	deka	da
10^{-1}	deci	d
10^{-2}	centi	c
10^{-3}	milli	m
10^{-6}	micro	μ
10^{-9}	nano	n
10^{-12}	pico	p
10^{-15}	femto	f
10^{-18}	atto	a

2

DIGITAL REVIEW

A.2.1 NUMBER SYSTEMS

The base 10 or decimal number system is normally employed in the analog description of information. In digital applications, a base 2 or binary counting system is used, often with the octal or base-8 system for ease in writing numbers. The hexadecimal system or base 16 has also become popular. It is important that an individual involved in digital applications be prepared to translate numbers between these systems. This can be accomplished by the following operations and equations.

Binary

A mathematical relationship can be constructed between a base 10 number and its binary equivalent using the equation

$$N_{10} = a_n 2^n + a_{n-1} 2^{n-1} + \cdots a_3 2^3 + a_2 2^2 + a_1 2^1 + a_0 2^0 \qquad \text{(A.2-1)}$$

Where

$$N_{10} = \text{base 10 number}$$
$$a_n a_{n-1} \cdots a_3 a_2 a_1 a_0 = \text{base 2 number}$$
$$n + 1 = \text{number of digits in the binary number}$$

The reverse procedure for finding the binary equivalent of a decimal number can be represented in several ways. The easiest way is to use a set of conversion tables.

Lacking that, a process of successive division-by-two is employed. In any division, the remainder will always be zero or one half, and this determines the value of that binary digit. The first division yields the value of the least significant bit (LSB), which is the a_0 term, and the last remainder gives the most significant bit (MSB), which is a_n.

EXAMPLE A.2.1

Find the binary equivalent of 259_{10}.

SOLUTION

We use successive division by 2 as follows

$$
\begin{array}{lll}
259/2 = 129 + 1/2 & \quad a_0 = 1 \\
129/2 = 64 + 1/2 & \quad a_1 = 1 \\
64/2 = 32 + 0/2 & \quad a_2 = 0 \\
32/2 = 16 + 0/2 & \quad a_3 = 0 \\
16/2 = 8 + 0/2 & \quad a_4 = 0 \\
8/2 = 4 + 0/2 & \quad a_5 = 0 \\
4/2 = 2 + 0/2 & \quad a_6 = 0 \\
2/2 = 1 + 0/2 & \quad a_7 = 0 \\
1/2 = 0 + 1/2 & \quad a_8 = 1 \\
\end{array}
$$

Thus, the answer is that 259_{10} is 100000011_2.

Octal

Since it is very frequently employed in digital work, we need to consider a number base formed of 8 counting states. A principle reason for using this numbering system is that it becomes both easy and convenient to represent binary numbers in the octal system. This can be done by grouping the binary numbers in groups of three digits from the right and noting that these three digits represent the full range of octal numbers. Thus, 000_2 corresponds to 0_8, and 111_2 corresponds to 7_8. The small subscript 2 and 8 denote base 2 and base 8, respectively. Thus, 101111_2 can be grouped as 101_2 followed by 111_2 and written 57_8 in the octal system. The correspondence between octal and decimal could be accomplished through conversion first to binary. The direct method, however, is to use the equation

$$N_{10} = d_n 8^n + d_{n-1} 8^{n-1} + \cdots + d_2 8^2 + d_1 8^1 + d_0 8^0 \qquad (A.2\text{-}2)$$

Where

$$N_{10} = \text{decimal number}$$
$$d_n d_{n-1} \cdots d_1 d_0 = \text{octal number digits}$$
$$n = \text{one less than the number of digits in the octal number}$$

EXAMPLE A.2.2

Find the binary and decimal equivalent of the octal number 33_8.

SOLUTION

We can find the binary most easily by noting that

$$3_8 = 011_2 \text{ and thus}$$
$$33_8 = 011011_2 = 11011_2$$

The decimal number could be found by either of Equations (A.2-1) or (A.2-2). Using

$$N_{10} = a_n 8^n + a_{n-1} 8^{n-1} + \cdots + a_1 8^1 + a_0 8^0 \qquad \text{(A.2-2)}$$

we have $n = 1$ so

$$N_{10} = (3) \, 8^1 + (3) \, (8^0)$$
$$N_{10} = 27_{10}$$

To convert from decimal to octal, we can either convert first to binary and then to octal or use tables or use the procedure of successive division where now the divisor will be eight instead of two. As before, the least significant value is found by the first division.

EXAMPLE A.2.3

Find the decimal number 356_{10} as both octal and binary numbers.

SOLUTION

Let us use the successive division procedure to find the octal number first and then find the binary from that

$$\frac{356}{8} = 44 + \frac{4}{8} \qquad \text{so } d_0 = 4$$

$$\frac{44}{8} = 5 + \frac{4}{8} \qquad \text{so } d_1 = 4$$

and

$$d_2 = 5$$

We find then that 356_{10} is 544_8 and from the octal to decimal conversion of 5_8, 4_8, and 4_8 we get a binary number of 101100100_2.

Hexadecimal

The rapid development of 4-bit, 8-bit, and 16-bit microcomputers has brought about the common use of base 16 number representations. This is called the hexadecimal or hex system. Any binary system with 4, or a multiple of 4, bits to the word can be conveniently expressed in hex because a hex number is formed from the counting states of 4 binary numbers. Since we must go beyond 9 before the hex set is complete, it is necessary to have new symbols to complete the set. The accepted standard is to use the letters A through F for the remaining counts. Instead of a subscript 16 to denote hex, it also has become common practice to use a small letter "h" to indicate that a number is hex. Thus, from 0_{10} through 9_{10} we have the hex numbers Oh through 9h, and from 10_{10} through 15_{10} we have the hex equivalents of Ah, Bh, Ch, Dh, Eh, Fh. Conversion of a hex number into the decimal equivalent can be made through the equation

$$N_{10} = c_n 16^n + c_{n-1} 16^{n-1} + \cdots + c_2 16^2 + c_1 16^1 + c_0 16^0 \qquad \text{(A.2-3)}$$

Where

$$
\begin{aligned}
N_{10} &= \text{ decimal number (base ten)} \\
c_n c_{n-1} \cdots c_1 c_0 &= \text{ hex number} \\
n &= \text{ one less than the number of hex digits}
\end{aligned}
$$

Conversion of a decimal number to hex can be made by successive division by 16 as was done for base 2 and base 8 systems. The following examples illustrate these conversions.

EXAMPLE A.2.4

Find the decimal equivalents of 47h, 30Dh, and A2Fh.

SOLUTION

We use Equation (A.2-3) directly, where for the actual evaluation the individual hex digits are expressed in their respective base ten equivalents.
For 47h, we have $c_0 = 7$ and $c_1 = 4$ so that

$$
\begin{aligned}
N_{10} &= 4 \times 16^1 + 7 \times 16^0 = 64 + 7 \\
N_{10} &= 71_{10}
\end{aligned}
$$

For 30Dh, we have $c_0 = Dh = 13_{10}$, $c_1 = 0$ and $c_2 = 3$

$$
\begin{aligned}
N_{10} &= 3 \times 16^2 + 0 \times 16^1 + 13 \times 16^0 = 3 \times 256 + 13 \\
N_{10} &= 781_{10}
\end{aligned}
$$

For A2Fh, we have $c_0 = Fh = 15$, $c_1 = 2$, and $c_2 = Ah = 10$

$$N_{10} = 10 \times 16^2 + 2 \times 16^1 + 15 \times 16^0$$
$$N_{10} = 2607_{10}$$

EXAMPLE A.2.5

Find the hex equivalent of binary 10110101_2 and octal 422_8.

SOLUTION

The simplest way to convert binary to hex is to group the binary number into 4-bit sets and find the hex of each set. Thus, 10110101_2 is 1011_2 and 0101_2. But 1011_2 is 11_{10} which is Bh and 0101_2 is 5_{10} which is 5h. Thus, the hex is B5h. The octal number can be converted easily to binary and then to hex. Thus, 422_8 is 100010010_2 which can be grouped into three groups of 4 bits (using leading zeros) as 0001_2, 0001_2, and 0010_2. The hex is then 112h.

EXAMPLE A.2.6

Find the hex equivalents of 29_{10}, 175_{10}, and 3412_{10}.

SOLUTION

Conversion of decimal to hex is accomplished by successive division by 16. This finds the least significant digit first.

$$29/16 = 1 + 13/16 \qquad c_0 = 13 = \text{Dh}$$
$$1/16 = 0 + 1/16 \qquad c_1 = 1 = \text{1h}$$

Thus, the hex is 1Dh.
For the next decimal number, we have

$$175/16 = 10 + 15/16 \qquad c_0 = 15 = \text{Fh}$$
$$10/16 = 0 + 10/16 \qquad c_1 = 10 = \text{Ah}$$

Thus, the answer is AFh.
The last decimal number is converted as follows:

$$3412/16 = 213 + 4/16 \qquad c_0 = 4$$
$$213/16 = 13 + 5/16 \qquad c_1 = 5$$
$$13/16 = 0 + 13/16 \qquad c_2 = 13 = \text{Dh}$$

Thus, the hex number is D54h.

Negative Numbers

The representation of negative numbers can be accomplished in both binary and octal by preceding the number with a negative sign as in decimal. In many cases, however, it is desirable to let a binary number itself carry information about the sign. This is accomplished in binary by one of several artificial techniques, for example:

SIGN AND MAGNITUDE

In this method, the binary number has another term added to it. There is no convention here and in some cases, a 1 represents a negative, and a 0 represents a negative. Thus, we would write the binary number -1011 as 11011 if our convention choice was to make the highest bit 1 for a negative number.

2s COMPLEMENT

In this method, an extra digit is (again) provided which is always 1 for a negative number but the code of the number is altered. This method is very popular because of the ease with which math operations can be performed. In this method, a negative number is formed by the relation

$$(-N_2) = (2^{n+1})_2 - N_2 \qquad (3\text{-}3)$$

Where

$(-N_2) =$ the 2s complement binary number
$\quad n =$ highest digit in the number
$(2^{n+1})_2 =$ a binary number 2 digits larger than N_2

Thus, the number -1011_2 becomes in 2s complement with $n = 4$.

$$100000_2 - 1011_2 = 10101_2 \quad \text{(2s complement)}$$

A simple rule for finding the 2s complement of a binary number can be stated as follows:

To find the 2s complement of a binary number, copy all zeros from LSB up to and including the first 1 then replace all remaining 0s by 1s and 1s by 0s. Finally, place a 1 before the MSB.

EXAMPLE A.2.7

Find the 2s complement of 1101010_2.

SOLUTION

Using the rule above, we leave the 0 and 1 on the right alone, change all the rest, and include a final 1.

$$10010110 \text{ (2s complement)}$$

A.2.2 BOOLEAN ALGEBRA

The application of digital techniques to the solution of control problems often involves the use of complex logical operations. Boolean algebra is a mathematical method which is very useful for setting up and solving logical equations.

The (only) values of variables in the logical math of Boolean algebra are those of true or false. Thus, the variables are binary in that they can take on only two values which are assigned as 1 for a true state and 0 for a false state. As an example, if the level of liquid in a tank is the variable, we may let the variable A denote this level. Then, if the level is *higher* than that desired, we set $A = 1$ (a logical true) and if *lower*, $A = 0$ (a logical false).

There are four math operations in Boolean algebra which relate how combinations of variables may operate on one another and be related to one another.

Equality "="

Equality between two variables is defined by the same symbol employed in traditional algebra and has the same meaning. If we write $A = B$ and know $A = 1$, then $B = 1$ also.

Complement "$\bar{A}$"

The complement operation is denoted by a bar over a variable and indicates a variable of the opposite value as the original. If we know $B = 1$, then $\bar{B} = 0$. The complement operation is sometimes referred to as a *NOT*.

AND "·"

The AND operation is denoted by a dot "·" between two variables. We note that if $C = A \cdot B$, then C will be 1 if and only if $A = 1$ and $B = 1$; otherwise $C = 0$. Note that in some publications the dot will be omitted such that $A \cdot B = AB$.

OR, "+"

The plus symbol is employed to define the OR operation between two variables. Thus, we note that if $C = A + B$, the C will be 1 if A or B or both are 1 and will be 0 otherwise.

These math operations may be combined into many combinations for the description of some required logical operation. Thus, we may have $C = \overline{A \cdot B}$ which would mean C was equal to the complement of A AND B or NOT A AND B. In many cases, a particular logical combination can be represented by a different logical combination but giving the same result. Thus, we can show that the previous $C = \overline{A \cdot B}$ is exactly the same as the operation $\bar{A} + \bar{B}$, that is, NOT A OR NOT B. Note that in both cases, C takes on the same logic state for the same logic states of

TABLE A.2.1

$A \cdot 0 = 0$	$A + A = A$
$A + 0 = A$	$A \cdot A = A$
$A \cdot 1 = A$	$A + \bar{A} = 1$
$A + 1 = 1$	$A \cdot \bar{A} = 0$

A and B. There are several theorems that aid in changing the form of Boolean math operations without resorting to a truth table in each case.

The basic theorems of Boolean algebra are separated into single variable and multivariable categories. Table A.2.1 shows the single variable theorems and Table A.2.2 the multivariable theorems.

The theorems can be of great value in reducing a logical equation into another equivalent form that may not be intuitively obvious and may be much simpler.

EXAMPLE A.2.8

Find a simpler logic expression for the equation

$$D = A \cdot B + C \cdot (A \cdot B + \bar{C})$$

SOLUTIONS

Let us apply the theorems in steps to simplify this equation. First, by distribution, we can write

$$D = A \cdot B + C \cdot A \cdot B + C \cdot \bar{C}$$

TABLE A.2.2

$A + B = B + A$ $A \cdot B = B \cdot A$	Commutation
$A + (B + C) = (A + B) + C$ $A \cdot (B \cdot C) = (A \cdot B) \cdot C$	Association
$A \cdot (B + C) = A \cdot B + A \cdot C$ $A + B \cdot C = (A + B) \cdot (A + C)$	Distribution
$A + A \cdot B = A$ $A \cdot (A + B) = A$	Absorption
$\overline{A \cdot B} = \bar{A} + \bar{B}$ $\overline{A + B} = \bar{A} \cdot \bar{B}$	De Morgan

But
$C \cdot \overline{C} = 0$ by Table A2.1 and by applying association we can write

$$D = A \cdot B + A \cdot B \cdot C$$

Then, from distribution again

$$D = A \cdot B \cdot (1 + C)$$

But $1 + C = 1$ so that

$$D = A \cdot B + C \cdot (A \cdot B + \overline{C}) = A \cdot B$$

So the rather complex expression becomes a simple AND between A and B and is independent of C.

A.2.3 DIGITAL ELECTRONIC BUILDING BLOCKS

The building blocks for digital electronic circuits are a set of black boxes which implement the basic Boolean math operations for electrical signals. These elements are referred to as *gates* and each has a characteristic symbol which has been adopted for universal indication of the device.

AND Gate

The AND gate accepts as input two signals A and B and outputs a signal which is given by $C = A \cdot B$. The symbol for this element is given in Figure A.2.1.

OR Gate

The OR gate takes as input two signals A and B and outputs a signal which is given by $C = A + B$. The symbol for this device is given in Figure A.2.1.

Inverter

(NOT) The inverter, which is sometimes referred to as a NOT operation, accepts as an input a signal A and outputs an inverted version of this signal $\overline{A}$. The symbol for this element is given in Figure A.2.1. This is equivalent to providing the complement of a variable.

There are two additional building blocks which are frequently employed in electronic digital testing. These elements are used because of the ease with which they may be implemented using the popular circuit families.

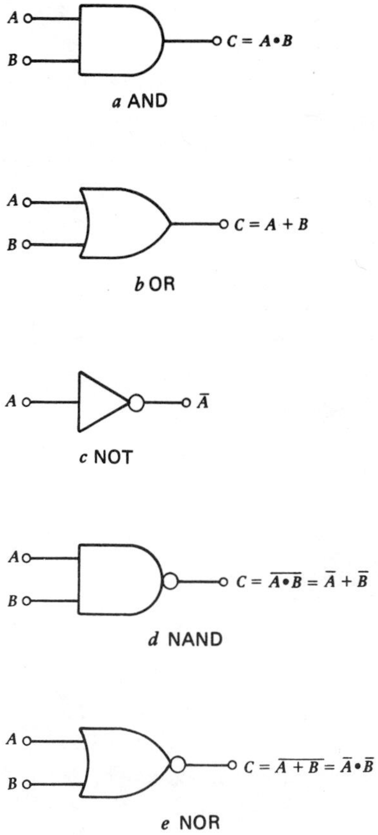

$C = A \cdot B$

a AND

$C = A + B$

b OR

$\bar{A}$

c NOT

$C = \overline{A \cdot B} = \bar{A} + \bar{B}$

d NAND

$C = \overline{A + B} = \bar{A} \cdot \bar{B}$

e NOR

FIGURE A.2.1   Digital gate symbols.

NAND Gate

The NAND gate is essentially a device that performs the AND of two inputs and then inverts the output of that operation. Thus, if the inputs are A and B, the output will be $C = \overline{A \cdot B}$. We note from De Morgan's theorem that this operation is identical to $\bar{A} + \bar{B}$. The symbol for this device is shown in Figure A.2.1.

NOR Gate

As in the case above, the NOR gate performs an OR operation between two inputs and then inverts the results. Thus, if the inputs are A and B, the output will be $C = \overline{A + B}$, which we know again from De Morgan's theorem can be expressed as $C = \bar{A} \cdot \bar{B}$. The symbol for this device is shown in Figure A.2.1. Using the five logic gates described above, most of the basic logic equations developed from problems and using Boolean algebra can be implemented.

APPENDIX

3

THERMOCOUPLE
TABLES

The following tables give the output voltage of several thermocouple (TC) types over a range of temperature. In each case, the TC reference temperature is 0°C. The first-named material will be the positive terminal, as *iron-constantan;* the iron will be the positive lead when the reference temperature is lower than the measurement. The temperature is in °C and the output is mV.

TYPE J: IRON-CONSTANTAN

	0	5	10	15	20	25	30	35	40	45
-150	-6.50	-6.66	-6.82	-6.97	-7.12	-7.27	-7.40	-7.54	-7.66	-7.78
-100	-4.63	-4.83	-5.03	-5.23	-5.42	-5.61	-5.80	-5.98	-6.16	-6.33
- 50	-2.43	-2.66	-2.89	-3.12	-3.34	-3.56	-3.78	-4.00	-4.21	-4.42
- 0	0.00	-0.25	-0.50	-0.75	-1.00	-1.24	-1.48	-1.72	-1.96	-2.20
+ 0	0.00	0.25	0.50	0.76	1.02	1.28	1.54	1.80	2.06	2.32
50	2.58	2.85	3.11	3.38	3.65	3.92	4.19	4.46	4.73	5.00
100	5.27	5.54	5.81	6.08	6.36	6.63	6.90	7.18	7.45	7.73
150	8.00	8.28	8.56	8.84	9.11	9.39	9.67	9.95	10.22	10.50
200	10.78	11.06	11.34	11.62	11.89	12.17	12.45	12.73	13.01	13.28
250	13.56	13.84	14.12	14.39	14.67	14.94	15.22	15.50	15.77	16.05
300	16.33	16.60	16.88	17.15	17.43	17.71	17.98	18.26	18.54	18.81
350	19.09	19.37	19.64	19.92	20.20	20.47	20.75	21.02	21.30	21.57
400	21.85	22.13	22.40	22.68	22.95	23.23	23.50	23.78	24.06	24.33
450	24.61	24.88	25.16	25.44	25.72	25.99	26.27	26.55	26.83	27.11
500	27.39	27.67	27.95	28.23	28.52	28.80	29.08	29.37	29.65	29.94
550	30.22	30.51	30.80	31.08	31.37	31.66	31.95	32.24	32.53	32.82
600	33.11	33.41	33.70	33.99	34.29	34.58	34.88	35.18	35.48	35.78
650	36.08	36.38	36.69	36.99	37.30	37.60	37.91	38.22	38.53	38.84
700	39.15	39.47	39.78	40.10	40.41	40.73	41.05	41.36	41.68	42.00

TYPE K: CHROMEL-ALUMEL

	0	5	10	15	20	25	30	35	40	45
−150	−4.81	−4.92	−5.03	−5.14	−5.24	−5.34	−5.43	−5.52	−5.60	−5.68
−100	−3.49	−3.64	−3.78	−3.92	−4.06	−4.19	−4.32	−4.45	−4.58	−4.70
− 50	−1.86	−2.03	−2.20	−2.37	−2.54	−2.71	−2.87	−3.03	−3.19	−3.34
− 0	0.00	−0.19	−0.39	−0.58	−0.77	−0.95	−1.14	−1.32	−1.50	−1.68
+ 0	0.00	0.20	0.40	0.60	0.80	1.00	1.20	1.40	1.61	1.81
50	2.02	2.23	2.43	2.64	2.85	3.05	3.26	3.47	3.68	3.89
100	4.10	4.31	4.51	4.72	4.92	5.13	5.33	5.53	5.73	5.93
150	6.13	6.33	6.53	6.73	6.93	7.13	7.33	7.53	7.73	7.93
200	8.13	8.33	8.54	8.74	8.94	9.14	9.34	9.54	9.75	9.95
250	10.16	10.36	10.57	10.77	10.98	11.18	11.39	11.59	11.80	12.01
300	12.21	12.42	12.63	12.83	13.04	13.25	13.46	13.67	13.88	14.09
350	14.29	14.50	14.71	14.92	15.13	15.34	15.55	15.76	15.98	16.19
400	16.40	16.61	16.82	17.03	17.24	17.46	17.67	17.88	18.09	18.30
450	18.51	18.73	18.94	19.15	19.36	19.58	19.79	20.01	20.22	20.43
500	20.65	20.86	21.07	21.28	21.50	21.71	21.92	22.14	22.35	22.56
550	22.78	22.99	23.20	23.42	23.63	23.84	24.06	24.27	24.49	24.70
600	24.91	25.12	25.34	25.55	25.76	25.98	26.19	26.40	26.61	26.82
650	27.03	27.24	27.45	27.66	27.87	28.08	28.29	28.50	28.72	28.93
700	29.14	29.35	29.56	29.77	29.97	30.18	30.39	30.60	30.81	31.02
750	31.23	31.44	31.65	31.85	32.06	32.27	32.48	32.68	32.89	33.09
800	33.30	33.50	33.71	33.91	34.12	34.32	34.53	34.73	34.93	35.14
850	35.34	35.54	35.75	35.95	36.15	36.35	36.55	36.76	36.96	37.16
900	37.36	37.56	37.76	37.96	38.16	38.36	38.56	38.76	38.95	39.15
950	39.35	39.55	39.75	39.94	40.14	40.34	40.53	40.73	40.92	41.12
1000	41.31	41.51	41.70	41.90	42.09	42.29	42.48	42.67	42.87	43.06
1050	43.25	43.44	43.63	43.83	44.02	44.21	44.40	44.59	44.78	44.97
1100	45.16	45.35	45.54	45.73	45.92	46.11	46.29	46.48	46.67	46.85

	0	5	10	15	20	25	30	35	40	45
1150	47.04	47.23	47.41	47.60	47.78	47.97	48.15	48.34	48.52	48.70
1200	48.89	49.07	49.25	49.43	49.62	49.80	49.98	50.16	50.34	50.52
1250	50.69	50.87	51.05	51.23	51.41	51.58	51.76	51.94	52.11	52.29
1300	52.46	52.64	52.81	52.99	53.16	53.34	53.51	53.68	53.85	54.03
1350	54.20	54.37	54.54	54.71	54.88					

TYPE T: COPPER-CONSTANTAN

	0	5	10	15	20	25	30	35	40	45
−150	−4.603	−4.712	−4.817	−4.919	−5.018	−5.113	−5.205	−5.294	−5.379	
−100	−3.349	−3.488	−3.624	−3.757	−3.887	−4.014	−4.138	−4.259	−4.377	−4.492
− 50	−1.804	−1.971	−2.135	−2.296	−2.455	−2.611	−2.764	−2.914	−3.062	−3.207
− 0	0.000	−0.191	−0.380	−0.567	−0.751	−0.933	−1.112	−1.289	−1.463	−1.635
+ 0	0.000	0.193	0.389	0.587	0.787	0.990	1.194	1.401	1.610	1.821
50	2.035	2.250	2.467	2.687	2.908	3.132	3.357	3.584	3.813	4.044
100	4.277	4.512	4.749	4.987	5.227	5.469	5.712	5.957	6.204	6.453
150	6.703	6.954	7.208	7.462	7.719	7.987	8.236	8.497	8.759	9.023
200	9.288	9.555	9.823	10.093	10.363	10.635	10.909	11.183	11.459	11.735
250	12.015	12.294	12.575	12.857	13.140	13.425	13.710	13.997	14.285	14.573
300	14.864	15.155	15.447	15.740	16.035	16.330	16.626	16.924	17.222	17.521
350	17.821	18.123	18.425	18.727	19.032	19.337	19.642	19.949	20.257	20.565
400										
450										
500										
550										
600										
650										
700										

TYPE S: PLATINUM-PLATINUM/10% RHODIUM

	0	5	10	15	20	25	30	35	40	45
−150										
−100										
−50										
−0										
+0	0.000	0.028	0.056	0.084	0.113	0.143	0.173	0.204	0.235	0.266
50	0.299	0.331	0.364	0.397	0.431	0.466	0.500	0.535	0.571	0.607
100	0.643	0.680	0.717	0.754	0.792	0.830	0.869	0.907	0.946	0.986
150	1.025	1.065	1.166	1.146	1.187	1.228	1.269	1.311	1.352	1.394
200	1.436	1.479	1.521	1.564	1.607	1.650	1.693	1.736	1.780	1.824
250	1.868	1.912	1.956	2.001	2.045	2.090	2.135	2.180	2.225	2.271
300	2.316	2.362	2.408	2.453	2.499	2.546	2.592	2.638	2.685	2.731
350	2.778	2.825	2.872	2.919	2.966	3.014	3.061	3.108	3.156	3.203
400	3.251	3.299	3.347	3.394	3.442	3.490	3.539	3.587	3.635	3.683
450	3.732	3.780	3.829	3.878	3.926	3.975	4.024	4.073	4.122	4.171
500	4.221	4.270	4.319	4.369	4.419	4.468	4.518	4.568	4.618	4.668
550	4.718	4.768	4.818	4.869	4.919	4.970	5.020	5.071	5.122	5.173
600	5.224	5.275	5.326	5.377	5.429	5.480	5.532	5.583	5.635	5.686
650	5.738	5.790	5.842	5.894	5.946	5.998	6.050	6.102	6.155	6.207
700	6.260	6.312	6.365	6.418	6.471	6.524	6.577	6.630	6.683	6.737

	0	5	10	15	20	25	30	35	40	45
750	6.790	6.844	6.897	6.951	7.005	7.058	7.112	7.166	7.220	7.275
800	7.329	7.383	7.438	7.492	7.547	7.602	7.656	7.711	7.766	7.821
850	7.876	7.932	7.987	8.042	8.098	8.153	8.209	8.265	8.320	8.376
900	8.432	8.488	8.545	8.601	8.657	8.714	8.770	8.827	8.883	8.940
950	8.997	9.054	9.111	9.168	9.225	9.282	9.340	9.397	9.455	9.512
1000	9.570	9.628	9.686	9.744	9.802	9.860	9.918	9.976	10.035	10.093
1050	10.152	10.210	10.269	10.328	10.387	10.446	10.505	10.564	10.623	10.682
1100	10.741	10.801	10.860	10.919	10.979	11.038	11.098	11.157	11.217	11.277
1150	11.336	11.396	11.456	11.516	11.575	11.635	11.695	11.755	11.815	11.875
1200	11.935	11.995	12.055	12.115	12.175	12.236	12.296	12.356	12.416	12.476
1250	12.536	12.597	12.657	12.717	12.777	12.837	12.897	12.957	13.018	13.078
1300	13.138	13.198	13.258	13.318	13.378	13.438	13.498	13.558	13.618	13.678
1350	13.738	13.798	13.858	13.918	13.978	14.038	14.098	14.157	14.217	14.277
1400	14.337	14.397	14.457	14.516	14.576	14.636	14.696	14.755	14.815	14.875
1450	14.935	14.994	15.054	15.113	15.173	15.233	15.292	15.352	15.411	15.471
1500	15.530	15.590	15.649	15.709	15.768	15.827	15.887	15.946	16.006	16.065
1550	16.124	16.183	16.243	16.302	16.361	16.420	16.479	16.538	16.597	16.657
1600	16.716	16.775	16.834	16.893	16.952	17.010	17.069	17.128	17.187	17.246
1650	17.305	17.363	17.422	17.481	17.539	17.598	17.657	17.715	17.774	17.832
1700	17.891	17.949	18.008	18.066	18.124	18.183	18.241	18.299	18.358	18.416
1750	18.474	18.532	18.590	18.648						

4

MECHANICAL REVIEW

A.4.1 MOTION

The mechanical description of motion is based on a specification of the position, velocity, and acceleration of an object relative to some reference section. If the object moves in an arbitrary manner with respect to the reference, we speak of linear or rectilinear motion. If the object is constrained to move *about* the reference, as in rotation, we speak of angular motion. The relationships between position, velocity, and acceleration are given in Table A.4.1. Note that linear position is measured in meters (m) with velocity in m/s and acceleration m/s². Angular motion is measured in radians (rad), angular velocity in rad/s, and angular acceleration in rad/s².

In some cases, a special unit of measure is used for linear acceleration. This unit is the **g** which is defined as the acceleration of a freely falling body at the earth's

TABLE A.4.1   POSITION, VELOCITY,
AND ACCELERATION

Motion	Position	Velocity	Acceleration
Linear	x	$v = \dfrac{dx}{dt}$	$a = \dfrac{d^2x}{dt^2}$
Angular	θ	$\omega = \dfrac{d\theta}{dt}$	$\alpha = \dfrac{d^2\theta}{dt^2}$

surface. The value of **g** is given by **g** = 9.8 m/s². In this context an acceleration of 50 **g** in any direction would be

$$a = (50 \text{ g}) (9.8 \text{ m/s}^2)$$
$$a = 490 \text{ m/s}^2$$

Special cases of motion frequently occur under constant speed or constant acceleration. In these cases, the motion can be described in terms of the speed or acceleration as shown in Table A.4.2.

A.4.2 FORCE

If an object is in uniform motion, it is found that a change in this motion (acceleration) can be effected through the action of some external agent with appropriate exchange of energy. The action which is required is defined as a *force*, and the resultant energy exchange is defined as *work*. The unit of *force* is the newton (N), a derived quantity, equal to 1 kg-m/s². The measurement of force is a significant part of the process-control technology. The relationship of force to other mechanical variables can be formulated only after definition of a property of objects called *mass*.

Mass

Any object at rest or in uniform motion resists attempts to cause a change of its state. This resistance is indeed a reflection of the need for work to be done before a state change can occur. Even in a zero gravity condition (0 **g**) in deep space, we would find it much more difficult to move a locomotive than a ball bearing. This resistance or *inertia* is reflected in a characteristic of an object called *mass*. Mass is a pure property, independent of the location or state of motion (neglecting relativity) of the object. Because a locomotive has much more mass than a ball bearing, it is more difficult to move. Mass is measured in the SI unit of kilogram (kg).

TABLE A.4.2 SPECIAL MOTION CASES

Motion	Linear	Angular
Constant Speed	v	ω
Position	$x = vt + x_0$	$\theta = \theta_0 + \omega t$
Constant Acceleration	a	α
Speed	$v = v_0 + at$	$\omega = \omega_0 + at$
Position	$x = x_0 + v_0 t + \dfrac{at^2}{2}$	$\theta = \theta_0 + \omega_0 t + \dfrac{\alpha t^2}{2}$

$x_0, \theta_0, v_0, \omega_0$ are initial positions and speed

Newton's Law

The relationship between the force applied to an object, its mass, and the resultant change in motion expressed in terms of acceleration is formulated as *Newton's law*.

$$F = ma \qquad (A.4\text{-}1)$$

Where

F = net force in N
m = total mass in kg
a = acceleration in m/s²

EXAMPLE A.4.1

An accelerometer indicates that a 2-kg object is accelerating at 2.2 m/s². Find the net force which acts on the object in newtons and pounds.

SOLUTION

We can find this force by using

$$F = ma$$

$$F = (2 \text{ kg}) (2.2 \text{ m/s}^2) \times \frac{1\text{N}}{\text{kg-m/s}^2} \qquad (A.4\text{-}1)$$

$$F = 4.4 \text{ N}$$

Note we have used the fact that 1 kg-m/s² = 1 N and Appendix A-1 shows that 1 lb = 4.448 N so that

$$F = (4.4 \text{ N}) \left(\frac{1 \text{ lb}}{4.448 \text{ N}}\right)$$

$$F = 0.989 \text{ lb}$$

Gravity and Weight

A fundamental law of nature states that all objects with finite mass exert a force of attraction on each other. In the case of the massive earth and objects close to the surface, this force is (approximately)

$$F_g = mg \qquad (A.4\text{-}2)$$

Where

F_g = force of gravity in N
m = mass in kg
g = acceleration due to gravity defined earlier (9.8 m/s²)

Equation (A.4-2) expresses the *weight* an object possesses, measured as the *force* with which the earth attracts an object close to its surface. The proper measure of weight is also the newton because it is a force. Scales used for *weighing* read weight. These are, of course, calibrated to read the force of the earth on a mass and thus *do not* measure mass. A metric scale may indicate the weighed object mass in kg in terms of F_g/g. This is a true measure of mass because the scale has been calibrated for the earth's surface. An English system scale may indicate a weight in pounds, and this is actually, again, the force of the earth's attraction of the object.

Hooke's Law and Springs

A very important relationship exists between force and displacement for the special case of objects defined as *springs*. A traditional spring of coiled steel wire serves as a good descriptive example. Such a spring is shown in Figure A.4.1a in the equilibrium or relaxed state. Here no forces act on the spring and its length is x_0 as shown. In Figure A.4.1b a force F, is applied to the spring. Application of this force caused the spring to extend from x_0 to a new length x, where the system has again become stationary. The net force must be zero and hence the spring must be pulling back with an equal force F. *Hooke's law* states that the equilibrium force of a spring under compression or extension is given by

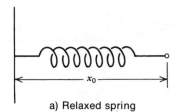

a) Relaxed spring

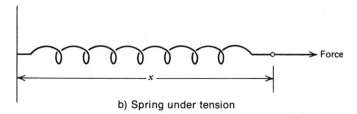

b) Spring under tension

FIGURE A.4.1   A force applied to a spring extends the length from the relaxed x_0 to a new x.

$$F = -k\Delta x \qquad \text{(A.4-3)}$$

Where

F = force in N
k = spring constant in N/m
$\Delta x = x - x_0$ = change in length

The significance of the negative sign is to indicate that the spring force is *opposite* the applied force so that they balance. The spring constant k is a specification of the spring itself and is found from calibration or from a manufacturer specification sheet. Note that the term *spring* is used loosely here. The devices described under Hooke's law could be a bellows, diaphragm, coil, spring, leaf spring, or indeed any number of configurations that relate displacement to force.

EXAMPLE A.4.2

Find the extension which a spring with a spring constant of 5 N/m experiences when a force of 14 N is applied.

SOLUTION

The equation for Hooke's law is

$$F = -k\Delta x$$

$$\Delta x = \frac{14 \text{ N}}{5 \text{ N/m}} \qquad \text{(A.4-3)}$$

$$\Delta x = 2.8 \text{ m}$$

5

P & ID SYMBOLS

A.5.1 INTRODUCTION

Just as electronics has standard symbols to represent components in circuit schematics, process control has symbols to represent the elements of a process-control system. Instead of a schematic, we call the process-control diagram a Piping and Instrumentation Drawing or simply P&ID. The symbols used are standards which have been developed through the years and are accepted by the Instrument Society of America (ISA). In this appendix the principal symbols used in P & ID are presented.

A.5.2 INTERCONNECTIONS

Interconnections in a P&ID can involve many different types of signals and the flow of the process itself. For this reason, a simple drawn line, such as is used in an electronic schematic, is insufficient. Instead, we use the symbol of a line to denote the nature of the signal as shown in Figure A.5.1. Further information may be necessary on the P&ID document itself to indicate the specific nature of the signal, such as a 4 to 20 mA signal as opposed to a 10 to 50 mA signal. Both would have the dashed line on the drawing.

A.5.3 BALLOON SYMBOLS

To denote the actual instruments used in process control, we use a circle called a balloon. Inside the balloon will be a code that tells the function and nature of the device represented. The code is given by letters, such as FC, and numbers, such as 117. The letters tell what the device is—for example, FC would mean flow control-

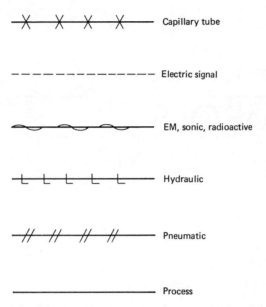

Capillary tube

Electric signal

EM, sonic, radioactive

Hydraulic

Pneumatic

Process

FIGURE A.5.1   P&ID signal and process lines.

ler and the number identifies the process control loop which that element serves. The balloon also tells where the instrument is located, that is, local or control room. In Figure A.5.2, the balloon symbol is shown for the three cases of local (meaning in the plant), control room board, and control room, but behind the board.

Locally mounted

Control room board mounted

Control room mounted behind board

FIGURE A.5.2   Letter codes and ballon symbols.

Figure A.5.3 summarizes the meaning of the various letters which may appear in the balloon.

A.5.4 INSTRUMENT SYMBOLS

The transducer employed in the process-control loop is often broken down into a primary element (second letter E in balloon) and a transmitter (second letter T in balloon) to account for the actual transducer and the signal conditioning. The final control element may also be broken down into several elements, but the actual final control element itself will be in the process line. Figure A.5.4 shows some of the special symbols used for the transducers and final control elements.

REFERENCES

Andrew, *Applied Instrumentation in the Process Industries*, Gulf Publishing Co., 1974 (three volumes).

Bateson, *Introduction to Control System Technology*, Merrill, 1973.

First Letter		Second Letter
A	Analysis	Alarm
B	Burner	
C	Conductivity	Control
D	Density	
E	Voltage	Primary element
F	Flow	
G	Gaging	Glass (sight tube)
H	Hand	
I	Current (electric)	Indicate
J	Power	
K	Time	Control station
L	Level	Light
M	Moisture	
O		Orifice
P	Pressure	Point
Q	Quantity	
R	Radioactivity	Record
S	Speed	Switch
T	Temperature	Transmit
U	Multivariable	Multifunction
V	Viscosity	Valve
W	Weight	Well
Y		Relay (transformation)
Z	Position	Drive

FIGURE A.5.3   Standard meanings of letter codes.

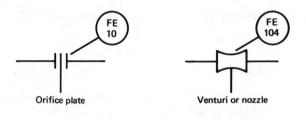

Orifice plate   Venturi or nozzle

Magnetic   Rotameter

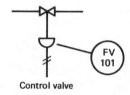

Control valve

FIGURE A.5.4   P&ID symbols for transducers and elements.

Byran, *Control Systems for Technicians,* University of London Press, 1967.

Cerni and **Foster,** *Instrumentation for Engineering Measurement,* Wiley, 1962.

Considine, *Process Instrumentation and Control Handbook,* McGraw-Hill, 1957.

Considine and **Ross,** *Handbook of Applied Instrumentation,* McGraw-Hill, 1964.

Eckman, *Automatic Process Control,* Wiley, 1958.

Grabbe, Ramo, and **Wooldridge,** *Handbook of Automatic Computation and Control,* Wiley, 1958 (three volumes).

Lee, Adams, and **Gaines,** *Computer Process Control,* Wiley, 1968.

Liptak, *Instrument Engineers Handbook,* Volumes I and II, Chilton Book Company, 1970.

Norton, *Handbook of Transducers for Electronic Measuring Systems,* Prentice-Hall, 1969.

Oliver, *Practical Instrumentation Transducers,* Hayden Book Co., 1971.

GLOSSARY

Accelerometer.   A transducer that measures the acceleration of the object to which it is attached.

Actuator.   A part of the final control element that translates the control signal into action of the final control device in the process.

ADC.   An analog to digital converter that converts an analog input of electric voltage or current into a proportional digital signal.

Alarm.   In process control, an indicator that some process variable has exceeded preset limits.

Bellows.   A pressure transducer that converts pressure into a nearly linear displacement.

Binary.   A number representation system of base 2. This is the working system of digital computers.

Bode plot.   A graph of transfer function versus frequency wherein the gain (often in decibels) and phase (in degrees) are plotted against the frequency on a log scale.

Bourdon tube.   A pressure transducer that converts pressure to a displacement. The device is essentially a coiled, flattened tube that tends to straighten when pressure is applied.

Cascade control.   A control system composed of two loops where the setpoint of one loop (the inner loop) is the output of the controller of the other loop (the outer loop).

Controlled variable.   The process variable regulated by the process control loop.

Controller.   The element in a process control loop that evaluates error of the controlled variable and initiates corrective action by a signal to the controlling variable.

Controlling variable.   The process variable changed by the final control element under command of the controller to effect regulation of the controlled variable.

Cyclic.   A condition of either steady-state or transient oscillation of a signal about the nominal value.

DAC.   A digital to analog converter that converts a digital signal, often from a computer, into a proportional analog voltage or current.

DAS.   A data acquisition system that interfaces many analog signals, called channels, to a computer. All switches, controls, and the ADC is included in the system.

DDC.   Stands for direct digital control wherein a computer performs all the functions of error detection and controller action.

Derivative control mode.   A controller mode in which controller output is directly proportional to the rate of change of controlled variable error.

DP cell.   A pressure transducer that responds to the difference in pressure between two sources. Most often used to measure flow by the pressure difference across a restriction in the flow line.

Dynamic variable.   Process variables that can change from moment to moment due to unspecified or unknown sources.

Error.   The algebraic difference between the measured value of a variable and the ideal value. A positive error indicates the measured value is greater than the ideal value.

Floating control mode.   A controller mode in which an error in the controlled variable causes the output of the controller to change at a constant rate. The error must exceed preset limits before controller change starts.

Foreground/background.   A control system that uses two computers, one performing the control functions and the other used for data logging, offline evaluation of performance, financial operations, and so on. Either computer is able to perform the control functions.

Frequency response method.   A method of tuning a process control loop for optimum operation by proper selection of controller settings. This method is based on a study of the frequency response of the open process control loop.

Gas thermometer.   A temperature transducer that converts temperature to pressure of gas in a closed system. The relation between temperature and pressure is based on the gas laws at constant volume.

Hardware.   When used in the context of computers hardware refers to the physical equipment associated with the computer (for example, ICs, printed circuit boards, cables, and so on).

Hex.   A number of representation system of base 16. The hex number system is very useful in cases where computer words are composed of multiples of four bits (that is, 4-bit words, 8-bit words, 16-bit words, and so on).

Hysteresis.   The tendency of an instrument to give a different output for a given input, depending on whether the input resulted from an increase or decrease from the previous value.

Integral control mode.   A controller mode in which the controller output increases at a rate proportional to the controlled variable error. Thus, the controller output is the integral of the error over time with a gain factor called the integral gain.

Ionization gauge.   A pressure transducer based on conduction of electric current through ionized gas of the system whose pressure is to be measured. Useful only for very low pressures (for example, below 10^{-3} atm).

I/P converter.   A device that linearly converts electric current into gas pressure (for example, 4–20 mA into 3–15 psi).

LASER.   Stands for light amplification by stimulated emission of radiation. It is a source of EM radiation generally in the IR, visible, or UV bands and is characterized by small divergence, coherence, monochromaticity, and high colimation.

Linearity.   The closeness to which the curve relating two variables approximates a straight line. It is usually expressed as the maximum deviation between the actual curve and the best fit straight line.

Load.   The process load is a term to denote the nominal values of all variables in a process that affect the controlled variable.

Load cell.   A transducer for the measurement of force or weight. Action is based on strain gages mounted within the cell on a force beam.

LVDT.   A linear variable differential transformer that measures displacement by conversion to a linearly proportional voltage.

Microcomputer.   A computer based on the use of a microprocessor integrated circuit. The entire computer often fits on a small printed circuit board and works with a data word of 4, 8, or 16 bits.

Microprocessor.   A large-scale integrated circuit that has all the functions of a computer, except memory and input/output systems. The IC thus includes the instruction set, ALU, registers and control functions.

Multiplexer.   This device allows selection of one of many input channels of analog data under computer control. The device is often an integral part of a DAS.

Nozzle.   A particular type of restriction used in flow system to facilitate flow measurement by pressure drop across a restriction.

Nozzle/flapper.   A fundamental part of pneumatic signal processing and pneumatic control operations. Basically, the device converts a displacement of the flapper to a pressure signal.

Octal.   A number representation system of base 8. The octal system is most useful for computer systems with digital words of multiples of three bits (that is, 3-bits, 6-bits, 9-bits, 12-bits, and so on).

Orifice plate.   A type of restriction used in flow systems to facilitate measurement of flow by the pressure drop across a restriction.

PD.   The designation of a controller operating in the proportional-derivative mode combination.

Photoconductive cell.   A transducer that converts the intensity of EM radiation, usually in the IR or visible bands, into a change of cell resistance.

Photovoltaic cell. A transducer that converts the intensity of EM radiation, usually in the IR or visible bands into a voltage.

PI. The designation of a controller operating in the proportional-integral mode combination.

P/I. A pressure to current converter linearly converts a signal pressure range into a signal current range (for example, 3-15 psi into 4-20 mA).

PID. The designation of a controller operating in the proportional-integral-derivative mode combination. The PID is also called a three-mode controller.

Pirani gauge. A pressure transducer based on measurement of the resistance of a heated wire as a function of wire temperature that is a function of the gas pressure. Useful primarily for pressures less than one atmosphere.

P&ID. Stands for piping and instrumentation drawing, which is the primary schematic drawing used for laying out a process control installation.

Pneumatic. Systems that employ gas, usually air, as the carrier of information and the medium to process and evaluate information.

Process. Any system comprised of dynamic variables, usually involved in manufacturing and production operations.

Process reaction method. A method of determination of optimum controller settings when tuning a process control loop. The method is based on the reaction of the open loop to an imposed disturbance.

Proportional band. The change in input of a proportional controller mode required to produce a full-scale change in output. Thus, if 10% change in error causes a 100% change in controller output, the PB is 10.

Proportional control mode. A controller mode in which the controller output is directly proportional to the controlled variable error.

Pyrometer. A temperature transducer that measures temperature by the EM radiation emitted by an object, which is a function of the temperature.

Quarter amplitude. A process-control tuning criteria where the amplitude of the deviation (error) of the controlled variable, following a disturbance, is cyclic so that the amplitude of each peak is one quarter of the previous peak.

Range. The region between the limits within which a variable is to be measured. Thus, a temperature is to be measured in the range of 20°C to 250°C defines the range. (See also Span.)

Rate action. Another name for the derivative control mode.

Reset action. Another name for the integral control mode.

Resolution. The minimum detectable change of some variable in a measurement system.

Rotameter. A flow measurement system that is based on the proportionality of the rise of a float in a tapered tube, arranged vertically, placed in the flow system.

RTD. A temperature transducer that provides temperature information as the change in resistance of a metal wire element, often platinum, as a function of temperature.

Self-regulation. The property of some variables in a process to adopt a stable value under given load conditions without regulation via a process control loop.

Sensitivity. The ratio of the change in output magnitude to the change of input magnitude, under steady-state conditions, for a measurement device.

Setpoint. The desired value of a controlled variable in a process-control loop.

Software. When used in the context of computers software refers to the programs that provide the instructions to the computer on operations and calculations to be performed.

Span. The algebraic difference between the upper range value and the lower range value. Thus, a temperature in the range of 20°C to 250°C has a span of 230°C. (See also Range.)

Strain gage. A transducer that converts information about the deformation of solid objects, called the strain, into a change of resistance.

Supervisory control. A process-control installation for which a computer oversees the operation of the control systems and provides the setpoint for the process control loops, which are themselves still analog.

Thermistor. A temperature transducer constructed from semiconductor material and for which the temperature is converted into a resistance, usually with negative slope and highly nonlinear.

Thermocouple. A temperature transducer in which a voltage is produced nearly linearly with the difference in temperature to be measured and a known reference.

Three-mode controller. Another name for a PID controller.

Time constant. A number characterizing the time required for the output of a device to reach approximately 63% of the final value following a step change of its input.

Transducer. A device that converts variation of one variable into variation of another variable (for example, temperature change into resistance change).

Transmitter. In process control, a device that converts a variable into a form suitable for transmission of information to another location (for example, resistance changed to current that is propagated on wires to a control installation).

Transfer function. The response of an element of a process-control loop that specifies how the output of the device is determined by the input.

Ultimate cycle method. See Ziegler-Nichols method.

Vapor pressure thermometer. A temperature transducer for which the pressure of vapor in a closed system of gas and liquid is a function of temperature.

Venturi. A type of restriction used in flow systems to facilitate measurement of flow by the pressure drop across a restriction.

Ziegler-Nichols method. A method of determination of optimum controller settings when tuning a process-control loop (also called the ultimate cycle method). It is based on finding the proportional gain which causes instability in a closed loop.

SOLUTIONS TO ODD PROBLEMS

Chapter 1

1-3   Maximum error = 22 °C, settling time = 8.5 s, residual error = 1 °C.

1-5   37.5 mA.

1-7   0.0390625 V.

1-9   ± 7.5 Ω.

1-11   (a) 50.9 ft/s², (b) 15.5 m/s², (c) 111.8 m/s, (d) 5.67×10^6 J.

1-13   12.2 s.

1-15   10.3 gal/min, 1.29 gal/min.

1-17   1.88 m, 925 N, 94 kg.

1-19 $p = 1.97L - 5.87$ psi, 12.3 ft.

Chapter 2

2-1   2790 Ω.

2-3   0.21μF.

2-5   In Figure 2-10: $R_2 = 12.5002$ kΩ, R > 77 kΩ pot.

2-7   In Figure 2-16 any R_2 and R_1 such that $R_2/R_1 = 22$.

2-11   600 Ω, 600.64 Ω.

2-14   See Figure S-1.

2-15   1392 Ω

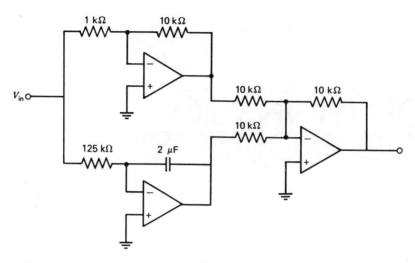

FIGURE S-1

2-17 See Figure S-2, not linear.

2-19 See Figure S-3.

Chapter 3

3-1 10_{10}, 59_{10}, 22_{10}.

3-3 (a) 10101_2, 25_8, $16h$, (b) 1001110110_2, 1166_8, $276h$, (c) 110101011_2, 653_8, $1ABh$.

3-5 10101_2, 101010100_2.

3-9 See Figure S-4.

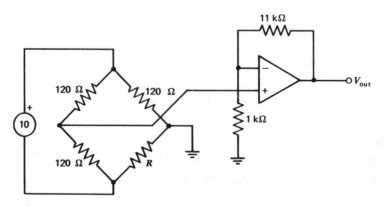

FIGURE S-2

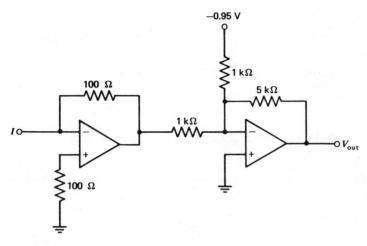

FIGURE S-3

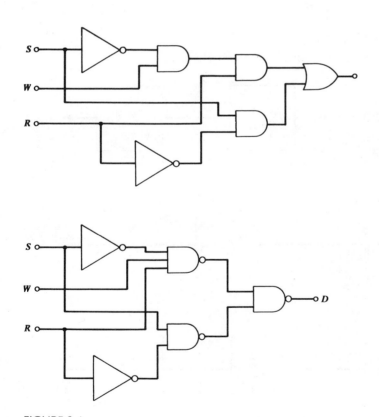

FIGURE S-4

3-11 01100001_2.

3-15   See Figure S-5, resolutions: 0.31 °C/LSB, 0.39 psi/LSB, 0.23 gal/min/LSB.

3-17   −3.203125 V, +3.90625 V, 1.992187 V, 0.0390625 V.

Chapter 4

4-1   251.7 K, −6.5 °F, −21.46°C.

4-3   18.6 °C.

4-5   108.12 Ω.

4-7   0.263 V.

4-9   411.78 °C

4-11   27.89 mV.

4-13   See Figure S-6.

4-15   238 mV.

4-17   196.7 °C.

4-19   64 psi to 420 psi.

4-21   See Figure S-7, resolution = 0.586 °C/LSB.

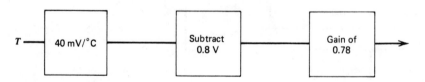

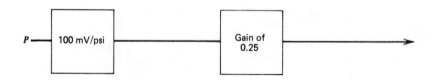

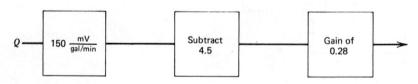

FIGURE S-5

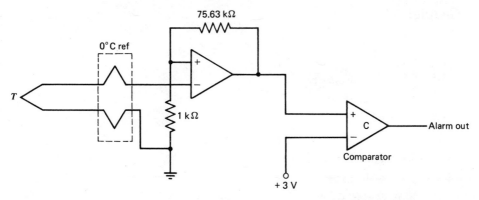

FIGURE S-6

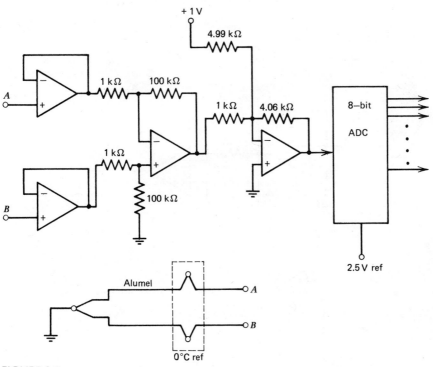

FIGURE S-7

Chapter 5

5-1   70 pF.

5-3   0.037 Ω.

5-5   0.5 $\mu\Omega$/lb, decrease.

5-7   49 m/s, 122.5 m.

5-9   2.2 m/s².

5-11   13.3 Hz.

5-13   See Figure S-8.

5-15 2.39×10^{-3} Ω/g.

5-17   (a) 0.32 atm, (b) 4.2 atm, (c) 32.3 kPa, 426 kPa.

5-19   19.4 lb.

5-21   198 kg/hr, 0.027 m/s.

Chapter 6

6-1 10^{10} Hz, microwave.

6-3   0.57 μs with liquid, 0.33 μs without.

6-5   212.5 lux, 12.1×10^{-3} lux.

6-7   136.6 kΩ.

6-9   See Figure S-9, maximum = 0.504 V.

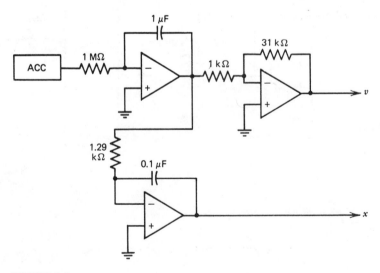

FIGURE S-8

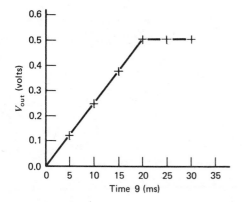

FIGURE S-9

6-11   3.35 W/m².
6-13   See Figure S-10.

Chapter 7

7-1   V = 1250I + 15
7-3   min = 01010_2, max = 10101_2, 93.75 rpm.

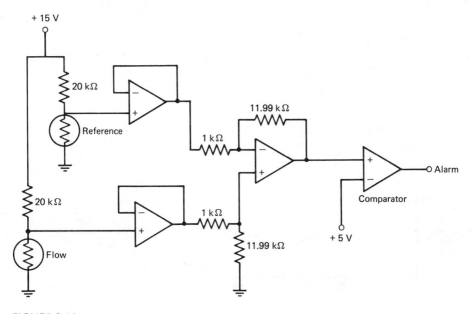

FIGURE S-10

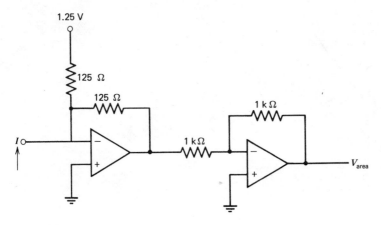

FIGURE S-11

7-5   270 N.

7-7 3.9×10^4 Pa.

7-9   See Figure S-11.

7-11   31.5 m³/hr, 50 m³/hr.

7-13   min = 11.9 m³/hr, max = 356 m³/hr, s = 3.1 cm.

7-15   3.6 s, 17.6 s.

Chapter 8

8-1   −13.75%

8-3   See Figure S-12, 39.2 min.

8-5   See Figure S-13,   min.

8-7   18.75%, 100%, 71.25 + 12.5t%.

8-9 $3.87\cos(0.04t)\%$

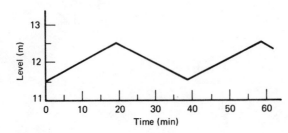

FIGURE S-12

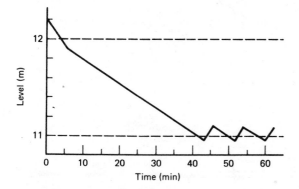

FIGURE S-13

8-11 See Figure S-14.
8-13 See Figure S-15.
8-15 0.72 min.

Chapter 9

9-3 See Figure S-16.
9-5 See Figure S-17.
9-7 0.862 V, 4 V.

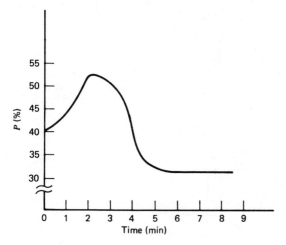

FIGURE S-14

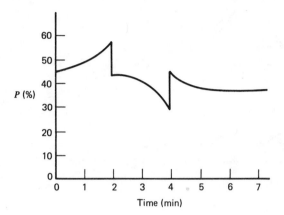

FIGURE S-15

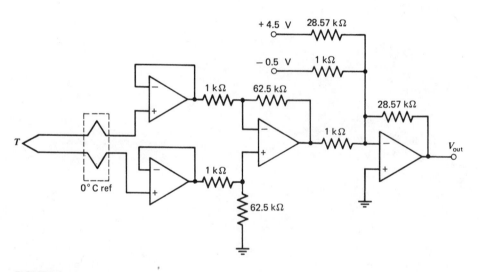

FIGURE S-16

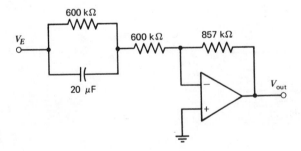

FIGURE S-17

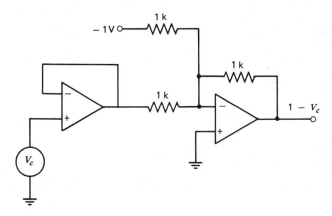

FIGURE S-18

9-9 See Figure S-18 and use with the two position, Figure 9-4, with $R_3 =$ 10 kΩ, $R_1 = 1.56$ kΩ, $R_2 = 13.9$ kΩ, and any R.

9-11 4.35.

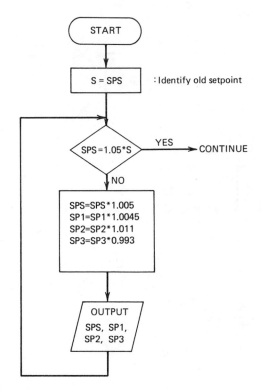

FIGURE S-19

Chapter 10

10-1 Use a comparator set at 2.31 volts.

10-3 $A = \bar{S} \cdot L \cdot R + S \cdot \bar{L}$.

10-5 $V_{in} = \bar{T} \cdot \bar{L}$, $V_{out} = T \cdot L + P \cdot \bar{T} \cdot \bar{L}$.

10-7 90.5 samples/s.

10-9 See Figure S-19.

10-11 See Figure S-20.

10-13 (a) -126.5%-s, (b) -96%-s, (c) -125.5%-s.

10-15 See Figure S-21.

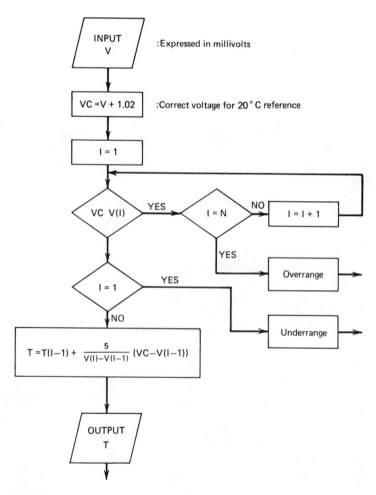

FIGURE S-20

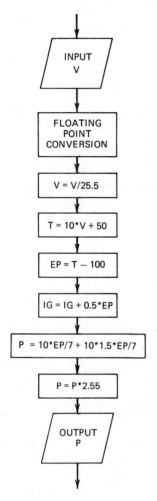

FIGURE S-21

Chapter 11

11-1   See Figure S-22.

11-3   See Figure S-23.

11-5   See Figure S-24.

11-7   (a) $K_p = 3.97$, $T_I = 5.83$ min, (b) $K_p = 6.42$, $T_I = 3.6$ min, $T_D = 0.58$ min.

11-9   Unstable: phase $= -190°$ at unity gain and gain greater than unity when the phase is $-180°$.

11-11   See Figure S-25.

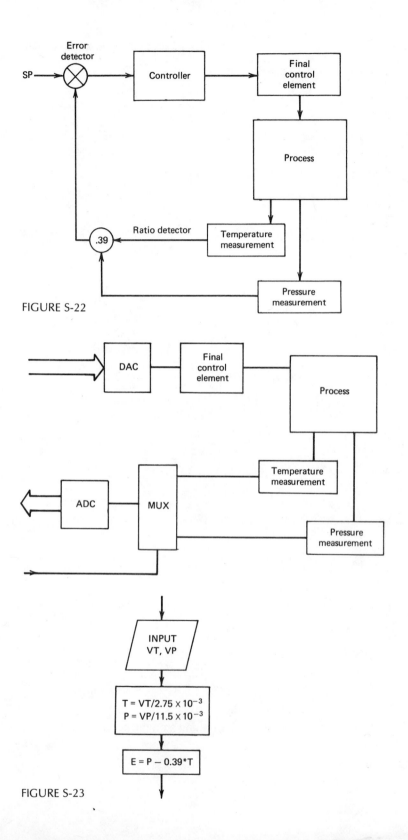

FIGURE S-22

INPUT
VT, VP

$T = VT/2.75 \times 10^{-3}$
$P = VP/11.5 \times 10^{-3}$

$E = P - 0.39*T$

FIGURE S-23

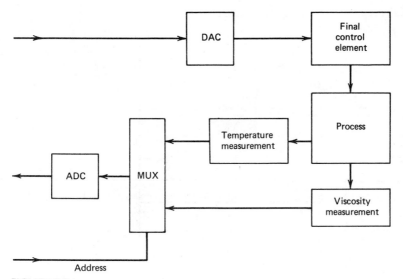

FIGURE S-24

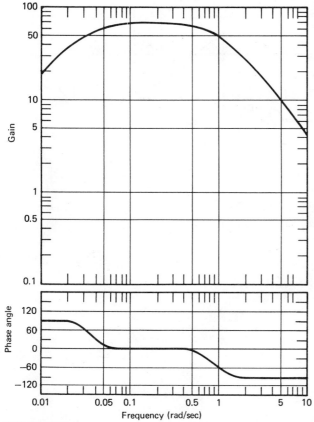

FIGURE S-25 .

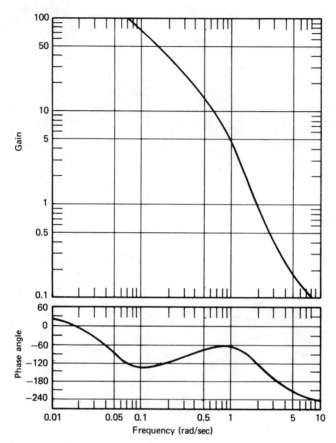

FIGURE S-26

11-13   See Figure S-26.

11-15   Instability for $K_p = 5$. Ultimate cycle gives $K_p = 2.5$, gain margin of 0.5 or about 6 dB and a phase margin of 130°.

INDEX